Reading Ruse

Reading Ruse

Michael Ruse on Darwinism,
Science, and Faith

Michael Ruse

EDITED BY
Bradford McCall

 CASCADE *Books* • Eugene, Oregon

READING RUSE
Michael Ruse on Darwinism, Science, and Faith

Copyright © 2024 Michael Ruse and Bradford McCall. All rights reserved. Except for brief quotations in critical publications or reviews, no part of this book may be reproduced in any manner without prior written permission from the publisher. Write: Permissions, Wipf and Stock Publishers, 199 W. 8th Ave., Suite 3, Eugene, OR 97401.

Cascade Books
An Imprint of Wipf and Stock Publishers
199 W. 8th Ave., Suite 3
Eugene, OR 97401

www.wipfandstock.com

PAPERBACK ISBN: 978-1-6667-3344-0
HARDCOVER ISBN: 978-1-6667-2904-7
EBOOK ISBN: 978-1-6667-2905-4

Cataloguing-in-Publication data:

Names: Ruse, Michael [author]. | McCall, Bradford [editor].

Title: Reading Ruse : Michael Ruse on Darwinism, science, and faith / by Michael Ruse ; edited by Bradford McCall.

Description: Eugene, OR: Cascade Books, 2024 | Includes bibliographical references.

Identifiers: ISBN 978-1-6667-3344-0 (paperback) | ISBN 978-1-6667-2904-7 (hardcover) | ISBN 978-1-6667-2905-4 (ebook)

Subjects: LCSH: Human evolution—Philosophy. | Darwin, Charles, 1809–1882—Influence. | Religion and science. | Evolution—Religious aspects. | Biological evolution. | Atheism. |

Classification: BL240.3 R87 2024 (paperback) | BL240.3 (ebook)

VERSION NUMBER 09/12/24

Scripture quotations marked KJV are from the King James or Authorized Version.

Scripture quotations marked NKJV are from the New King James Version, copyright © 1982 by Thomas Nelson, Inc. Used by permission. All rights reserved.

Scripture quotations marked NASB are from the (NASB®) New American Standard Bible®, copyright © 1995 by The Lockman Foundation. Used by permission. All rights reserved. lockman.org

Scripture quotations marked NRSV are from the New Revised Standard Version, copyright © 1989, Division of Christian Education of the National Council of the Churches of Christ in the United States in the United States of America. Used by permission. All rights reserved.

Contents

PART I: **Introducing Michael Ruse**

Chapter 1: Editorial Introduction | 3
Chapter 2: Mummy's Boy | 17

PART II: **The Readings**

Chapter 3: Atheism, Belief, and Faith
    Reading 3.1 The Arkansas Creationism Trial Forty Years On | 48
    Reading 3.2 God and Humans | 59
    Reading 3.3 Belief | 71
    Reading 3.4 The Unraveling of Belief | 80
    Reading 3.5 Why Atheism? | 94

Chapter 4: Darwinism, Belief, and Religion
    Reading 4.1 The Origins of Religion | 106
    Reading 4.2 God | 118
    Reading 4.3 Darwinism and Belief | 131
    Reading 4.4 Darwinism as Religion | 140
    Reading 4.5 Darwinism Explains Religion (?) | 149

Chapter 5: Darwin, Darwinism, and Darwinian Thought
    Reading 5.1 The History of Evolutionary Thought | 161
    Reading 5.2 The Origin of the *Origin* | 177

CONTENTS

  Reading 5.3 Charles Darwin and the *Origin of Species* | 187
  Reading 5.4 Darwinian Evolution | 201
  Reading 5.5 Darwinism | 210

Chapter 6: Progress and Directionality in Evolution
  Reading 6.1 Progress | 223
  Reading 6.2 The Problem of Progress | 240
  Reading 6.3 Evolution and Progress | 252
  Reading 6.4 Charles Darwin and Progress | 261
  Reading 6.5 Evolutionary Directionality: No Direction to Evolution | 278

Chapter 7: Design, Telos, and Purpose in the Natural World
  Reading 7.1 The Argument from Design: A Brief History | 288
  Reading 7.2 Two Thousand Years of Design | 299
  Reading 7.3 Design | 310
  Reading 7.4 Darwin and Design: Darwin Destroys Design | 319
  Reading 7.5 Design as a Metaphor | 329

Chapter 8: Naturalism, Sociobiology, and Their Entailments
  Reading 8.1 Naturalism | 342
  Reading 8.2 Naturalism and the Scientific Method | 353
  Reading 8.3 Naturalistic Explanations | 363
  Reading 8.4 Sociobiology | 375
  Reading 8.5 Social Darwinism | 386

Chapter 9: Darwinian Ethics and Morality
  Reading 9.1 Darwinian Ethics | 397
  Reading 9.2 Evolutionary Ethics: The Debate Continues | 408
  Reading 9.3 Darwinian Evolutionary Ethics | 417
  Reading 9.4 Morality for the Mechanist | 425
  Reading 9.5 Morality | 435

Bibliography | 447

PART I
Introducing Michael Ruse

Chapter 1

Editorial Introduction

BRADFORD MCCALL

MICHAEL RUSE, A PHILOSOPHER and historian of science, is author or editor of nearly four dozen books, as well as author of some two hundred peer-reviewed articles. While I do contend that Ruse needs no real introduction, per se, Ruse is a nonbelligerent nonbeliever. His writings come across as erudite, learned, and yet somewhat accessible. Further, he is one of the most pious men that I know. I point this last aspect out because he is not in the mode of the "new atheists," who consistently tear things down, but then build nothing to replace it.

Ruse was born in 1940 in Birmingham, England. His father, William, was a civil servant and school bursar, and his mother, Margaret, a schoolteacher who died when Michael was thirteen years old. William was a conscientious objector to the war, and after World War II, the family became involved in the Society of Friends (Quakers). Michael Ruse attended a Quaker boarding school in York and acknowledges his association with the Junior Friends as a significant influence on his early years. I make mention of this "background data" only because I think it complexifies the "location" of Ruse in contemporary philosophy of science, as this reader itself will attempt to demonstrate.

In 1962, after completing a BA in philosophy and mathematics at Bristol University, Ruse emigrated to Canada and was awarded an MA in philosophy from McMaster University (1964). Ruse returned to Bristol for his doctoral work and was awarded a PhD in 1970 for his dissertation, entitled "The Nature of Biology." Notably, he taught at the University of Guelph, Ontario, Canada, from 1965 to 2000. Since 2000, Ruse has served as both the Lucyle T. Werkmeister Professor of Philosophy and Director of the History and Philosophy of Science program at Florida State University. He retired after his fifty-fifth year in the academy in 2020. Notably, while Ruse himself is an "nonbeliever," he is highly interested in the science-religion relationship, and in fact contends that one can be perfectly "religious" even in this age of science. So then, he argues for a sort of "middle" position with regard to matters of faith and science, particularly with respect to evolutionary biology.

In my opinion, there are four main time periods in Ruse's scholarly output: (1) the *early* years, with him grappling with evolutionary theory and its meaning; (2) the *middle* years, with him concerning himself with the nature and structure of evolution (laws in biology, species questions, and then teleology and whether these make biology different from physics); (3) the *human* years, dealing first with sociobiology and then on to evolutionary epistemology and ethics; and then (4) the *science and religion* years, dealing with faith and science issues. This collection of writings will bring into one volume representative samples of the broad range of Ruse's oeuvre, as represented in his academic books, mainly from post-2000, which would focus upon the fourth area delineated above: science and religion issues. The decision to re-present his writings from this period forward and from this publishing venue is due to manifold concerns of my own. For example, one reason is to collect his voluminous writings from around the turn, and in the new, millennium, in one place. Yet another reason is because his "flare" in writing increased around the turn of the millennium, and thus this volume would make for an extremely exciting and interesting "read." Understanding those preceding points, this book will present Ruse's writings in this period under seven headings: (1) Atheism, Belief, and Faith; (2) Darwinism, Belief, and Religion; (3) Darwin, Darwinism, and Darwinian Thought; (4) Progress and Directionality in Evolution; (5) Design, Telos, and Purpose in the Natural World; (6) Naturalism, Sociobiology, and Their Entailments; and (7) Darwinian Ethics and Morality—each with five readings, for a total of thirty-five readings in this edited volume.

Throughout his voluminous writings, Ruse develops a constructive Darwinian worldview that is in dialogue with metaphysics, modern philosophy, postmodernity, and science writ large. (This chapter will use portions of Michael Ruse's writings rather liberally, including *Taking Darwin Seriously*; *Monad to Man*; *Mystery of Mysteries*; *Darwinian Revolution*; *Can Darwinian Be Christian?*; *Darwin and Design*; *Evolution-Creation Struggle*; *Darwinism and Its Discontents*; *Charles Darwin*; *Philosophy after Darwin*; *Philosophy of Human Evolution*.) Ruse, I contend, is a highly important figure in numerous areas of concern for philosophy of evolutionary biology, science more broadly, and matters of "faith in dialogue with biology." He has distinguished himself in these areas over the last half a century: he has, simply, published at an amazing rate since the mid-1990s. For example, Ruse has distinguished himself in at least the following five areas: by 1998, he had coedited the singular best anthology in philosophy of biology for at least the next fifteen years, if not more (Hull and Ruse 1998); this title, more than any other single title, raised him to be the *most* significant theorist in philosophy of biology as a whole, to be sure. Additionally, he is perhaps the single best representative of Darwinian thought and historical Darwinism within the twentieth century (and his wake is still being felt today).

Third, he became a leading voice in the "science and religion" debate, which almost goes without saying, even though he may not be comfortable with that terminology; in fact, in response to what I term the idiocy of the intelligent design debates in and around the turn of the millennium, Ruse sought a more peaceful or irenic resolution that does not disrespect biology, but that accommodates it instead. Fourth, he has become a highly productive theorist in a positive role regarding spirituality and values. Fifth, his concern with evolutionary ethics is unsurpassed, even to this day. Indeed, many of the areas he has written in are core areas for the modern debates about Darwin, Darwinism, and naturalistic explanations and ethics, and even sociobiology. One may say that the earlier part of his career established his bona fides in evolutionary biology, and the last twenty or so years, he has been working on making connections between the first part of his career into an overarching narrative of sorts.

Ruse does not think that a spirituality should be founded on a requisite number of beliefs, as if they were the foundations and everything should be based around them; rather, he thinks there is a sort of immediate, intuitive apprehension of something beyond us, which is but one reason why he is so very wise and effective in the "religion and science" (though I have some apprehension with reference to the usage of "religion" and Ruse in the same sentence!) discussion. But he is not wedded to individual beliefs, per se. I contend, however, that faith is crucial to his position, in a sense, which is demonstrable by his own works (cf. Ruse 2010b). As a thoroughgoing Darwinian, Ruse defends a Darwinian worldview against all comers, from both inside and outside the movement.

In his writings, Ruse highlights three main stages of evolutionary thought through the twenty-first century: (1) the pre-Darwinian era, in which evolution was seen to be a pseudoscience; (2) the Darwinian revolution period from the *Origin* to the full incorporation of Mendelism into evolutionary thinking; and (3) an era following the neo-Darwinian synthesis in or around 1930 (which persists to this day). Ruse points out that the concept of evolution gained true traction only *after* the Enlightenment, and is in fact a child of the Enlightenment (Ruse 2013a, 39). Ruse especially highlights the advancement of the concept of evolution that Charles Darwin made in identifying the mechanism of evolution—natural selection—which explains the design-like aspect of living beings (Ruse and Travis 2009, 25). However, Ruse is keen to point out that the idea of progress in biological advancement was not quick to disappear, persisting at least until the modern synthesis in the middle of the twentieth century (some would contend that the notion still persists unto this day).

Ruse notes that the ancient Greeks were not evolutionists. It was not that they had an a priori prejudice against a gradual developmental origin for organisms (including humans), but that they saw no real evidence for such. More importantly, they could not see how blind law—that is to say, natural law without a guiding intelligence—could lead to the intricate complexity of the world, in particular complexity serving the ends of things and organisms. This need to think in terms of consequences or purposes is what Aristotle called "final causes," which was taken to speak definitively against solely natural origins. It was not until the seventeenth century—what is known as the Age of the Enlightenment—that we get the beginnings of evolutionary thinking (Ruse 2013a, 1). This could have happened only if there were something—an ideology—sufficiently strong to overcome the worry about ends. Such an ideology did appear, and it was one of progress: the belief that through unaided effort humans could themselves improve both society and culture. It was natural for many in this period to move straight from progress in the social world to progress in the biological world, and so people argued for a full-scale climb upward from primitive forms, all the way up to the finest and fullest form of being, Homo sapiens: from "monad to man," one might say. It was not a genuinely atheistic doctrine, but more one in line with "deism," that is, the belief that God works through unbroken law. That said, such thoughts did increasingly challenge biblical reading of the past and it went against evangelical claims about providence, the belief that we humans unaided can do nothing except for the sacrifice of Jesus on the cross.

Critics of evolution, notably—for example—the German philosopher Immanuel Kant and his French champion, the comparative anatomist Georges Cuvier, averred that final causes stand in the way of all such speculations. Moreover, particularly after the French Revolution, many thought the idea of progress to be both false and dangerous. For this reason, evolution was hardly a respectable notion; rather, it had all of the markings of a "pseudoscience" (Ruse 2013a, 40). *On the Origin of Species* (1859) changed all of this. It is important therefore, from the beginning, to get Charles Darwin right. Darwin had an inventive and discerning eye for a good hypothesis. Added to this is the fact that he was somewhat ruthless in his pursuit of an idea and its supporting facts, using others to gather information for his speculations (Browne 1995, 2002). For Darwin, becoming an evolutionist was a bit gradual. There is no question but that a major influence, along with the geology that was making him think about the operation of laws in nature and implications for such things as time and place, were the fossils that he was collecting on the *Beagle* trip in the 1830s. His finds were almost forcing him to think about origins and changes and causes. Along with the fossils he found on the voyage, Darwin was certainly set on the path to evolution by the distributions of the organisms—birds and reptiles particularly—that he saw when the *Beagle* visited the Galápagos Archipelago in 1835 in the mid-Pacific. Even more certainly, his thinking solidified early in 1837 when the taxonomist studying his bird collection confirmed that from island to island there are genuinely different species (Ruse 2013a, 5). It was at this point that Darwin opened a series of notebooks (the key species notebooks are B through E, and the human notebooks are M and N) and wrote down thoughts on evolution.

The key insight leading to the discovery of the mechanism of natural selection—the systematic differential reproduction of organisms brought on by the limited supplies of food and space—came late in September 1838 when Darwin read the *Essay on a Principle of Population* (1826) by the Reverend Thomas Robert Malthus. Malthus had argued that the population pressures in humans lead to inevitable struggles for existence. Darwin generalized this concept to all species and then argued that success in the struggle will (on average) be a function of the different variations of the competitors, and that this will lead to ongoing change—change of a particular kind, namely in the direction of features or characteristics (like the hand and the eye) that aid their possessors. In other words, this process of natural selection (though the term is not used by him for another two or three years) produces contrivances or adaptations—things that seem as if designed for the ends they serve currently in species. That is to say, the process gives a natural (in the sense of blind, unguided law) explanation of Aristotelian final causes (Ruse 2013a, 5). There is no need to suppose outside divine intervention in such instances.

For Darwin, continued change (or even trying to keep things stable) makes for inadvertent differences, so that later forms are different from earlier ones, and groups separated simply tend to move away from each other. "Unconscious selection" thus changes the forms of organisms quite without our knowledge or desire (Ruse 2013a, 96). Obviously, by pointing out this, Darwin intentioned to prepare the way yet more strongly for a natural form of selection. Differential reproduction can have cumulative effects even without intelligent forethought. The way prepared as such, Darwin was now ready to start the argument for the main mechanism of natural selection.

First, Darwin had to convince the reader that there is widespread variation in the natural world. Without this widespread variation, no sustained change would be possible; he therefore pontificated widely over the world of animals and plants showing that whenever organisms are looked at in any detail, they exhibit a great deal of variation. Darwin always believed this, but no doubt his extended eight-year study of barnacles confirmed his conviction that no two forms of an individual are ever exactly identical.

Next came two crucial chapters. First, Darwin argued that there is always an ongoing struggle for existence, which he lifted from Malthus, as mentioned above. Population pressures put everything under a strain.

> A struggle for existence inevitably follows from the high rate at which all organic beings tend to increase. Every being, which during its natural lifetime produces several eggs or seeds, must suffer destruction during some period of its life, and during some season or occasional year, otherwise, on the principle of geometrical increase, its numbers would quickly become so inordinately great that no country could support the product. Hence, as more individuals are produced than can possibly survive, there must in every case be a struggle for existence, either one individual with another of the same species, or with the individuals of distinct species, or with the physical conditions of life. It is the doctrine of Malthus applied with manifold force to the whole animal and vegetable kingdoms; for in this case there can be no artificial increase of food, and no prudential restraint from marriage. Although some species may be increasing, more or less rapidly, in numbers, all cannot do so, for the world would not hold them. (Darwin 1859, 63–64)

And so to natural selection (Ruse 2013a, 97). With the struggle and with variations, a differential survival and reproduction follows automatically.

> How will the struggle for existence . . . act in regard to variation? Can the principle of selection, which we have seen is so potent in the hands of man, apply in nature? I think we shall see that it can act most effectually. Let it be borne in mind in what an endless number of strange peculiarities our domestic productions, and, in a lesser degree, those under nature, vary; and how strong the hereditary tendency is. . . . Can it, then, be thought improbable, seeing that variations useful to man have undoubtedly occurred, that other variations useful in some way to each being in the great and complex battle of life, should sometimes occur in the course of thousands of generations? If such do occur, can we doubt (remembering that many more individuals are born than can possibly survive) that individuals having any advantage, however slight, over others, would have the best chance of surviving and of procreating their kind? On the other hand, we may feel sure that any variation in the least degree injurious would be rigidly destroyed. This preservation of favourable variations and the rejection of injurious variations, I call Natural Selection. (C. Darwin, 1859, 80–81)

Along with natural selection, Darwin introduced a subsidiary mechanism that he termed "sexual selection." He divided this further into two kinds: (1) sexual selection brought about by male combat, and

(2) sexual selection brought about by female choice. The former produces such things as the antlers of deer and the latter such things as the remarkable tail feathers of species of peacocks. It was made clear that the division between natural selection and sexual selection was based on the different intents of human breeders (Ruse 2013a, 97). Some breed for profit, for such things as fleshier cattle and shaggier sheep, and others breed for pleasure, for such things as more vicious fighting cocks and more beautiful birds.

Darwin took advantage of this part of his discussion to give his opinion on a matter which has divided biologists from the time of Aristotle to the present day: form versus function.

> It is generally acknowledged that all organic beings have been formed on two great laws—Unity of Type, and the Conditions of Existence. By unity of type is meant that fundamental agreement in structure, which we see in organic beings of the same class, and which is quite independent of their habits of life. On my theory, unity of type is explained by unity of descent. The expression of conditions of existence, so often insisted on by the illustrious Cuvier, is fully embraced by the principle of natural selection. For natural selection acts by either now adapting the varying parts of each being to its organic and inorganic conditions of life; or by having adapted them during long-past periods of time: the adaptations being aided in some cases by use and disuse, being slightly affected by the direct action of the external conditions of life, and being in all cases subjected to the several laws of growth. Hence, in fact, the law of the Conditions of Existence is the higher law; as it includes, through the inheritance of former adaptations, that of Unity of Type. (C. Darwin, 1859, 206)

There are, to this day, detractors who attempt to claim that Darwin only came along at the end of a long process to inherit all of the glory of the discovery of evolution. There is little need to spend much time on those claims because basically they do not hold much water. In no wise did Darwin steal ideas from Alfred Russel Wallace. Darwin's ideas—the ideas of the *Origin* that is—are all right there in the thirty-five-page sketch of his ideas that Darwin wrote in 1842 (C. Darwin and Wallace 1958). There was some tweaking about the nature of adaptation; perhaps he hit upon the principle of divergence in the early 1850s—although there are certainly hints of that in the species notebooks—but the mechanisms (natural and sexual selection) are there, as is the structure of the argument of the *Origin* (Ruse 2013a, 5). During this time frame, the private Darwin was thinking furiously and by 1842 felt sufficiently confident to put his ideas on paper in a thirty-five-page preliminary essay (usually known as the "sketch"), and then some two years later in 1844 he expanded his ideas to a much longer, 230-page essay (usually known as the "essay.") The activity rather slowed as Darwin—the professional, public Darwin—turned increasingly away from geology and toward the life sciences.

For Darwin, especially for a Darwin whose thinking about evolution was ever influenced by those Galápagos organisms' distribution from island to island and changing as they went and thus making a treelike history to life (unlike Lamarck's parallel upward progressions), it was almost a truism that his developmental thinking was the explanation of the fanlike distributive pattern that epitomized Linnaeus's system. It is more than probable that it was taxonomic thinking that pushed Darwin to what he considered the major conceptual addition to his theory—the "principle of divergence"—that occurred in the years from the "essay" to the *Origin* in 1859. Why should there be the range of different forms that we find—is it just accidental or is there a deeper reason? In the notebooks, things seem to happen almost by necessity. "The enormous number of animals in the world depends on their varied structure and complexity; hence as the forms became complicated, they opened fresh means of adding to their complexity; but yet there is no necessary tendency in the simple animals to become complicated although all perhaps will have done so from the new relations caused by the advancing complexity of others" (C. Darwin 1987, E 95).

Then, Darwin saw how this all comes about by selection, because it is of advantage to organisms to differ from potential competitors and thus occupy different niches reducing conflict. Darwin writes,

> The same spot will support more life if occupied by very diverse forms. . . . Each new variety or species, when formed will generally take the place of and so exterminate its less well-fitted parent. This, I believe, to be the origin of the classification or arrangement of all organic beings at all times. These always *seem* to branch and sub-branch like a tree from a common trunk; the flourishing twigs destroying the less vigorous,—the dead and lost branches rudely representing extinct genera and families. (C. Darwin 1860b, point 6; emphasis original)

In the summer of 1858, Alfred Russel Wallace, a young naturalist and professional collector, formerly in Brazil and now in the Far East—someone with whom Darwin had been corresponding—sent to Darwin a short essay with exactly the same ideas that had been fermenting in Darwin for nearly twenty years. Darwin's friends—Lyell and Hooker—came to his rescue: Wallace had to be acknowledged as a source of research validating natural selection, but there must be no nonsense about Darwin's priority being removed, as the various letters of Darwin outlining selection were published in the *Proceedings of the Linnean Society of London*, and then Darwin sat down to write an overview of his theory. Thus it was in the late fall of 1859 that the *Origin of Species* arrived on the intellectual arena.

Evolution after the *Origin* was nearly a truism. The mechanism of it, however, was another matter. No one denied natural selection, per se. Very few, however, accepted that it could be as powerful as Darwin suggested. That said, people became evolutionists in droves. The number of pure Darwinians, as we might name those who adhered unto selection as primary, was very few, and the most prominent after Darwin himself was Alfred Russel Wallace, who became enamored of spiritualism in the 1860s and thereafter started to deny selection as applied to humans (Wallace 1870a).

The reasons for this halfway acceptance are well known. On one side, there were scientific problems with selection. It was thought that it could never be strong enough to overcome the supposed averaging nature of heredity as it was then understood to be. Even the best new variations would be swamped into equilibrium in a generation or two (Greg 1868). Added to this was the physicists (ignorant as they were of the warming effects of radioactive decay), who denied that there was time enough for such a slow process as natural selection (Burchfield 1975). On the other side, there was the matter of adaptation. Selection does not just bring about change, for it brings about adaptive change. Heavily Christian evolutionists like American botanist Asa Gray thought that selection could not fully explain adaptation and so they wanted (God-) directed variations, with a basis in design (Gray 1876). But this, Darwin said, rather made natural selection redundant. Nevertheless, there was much ink spilled on the notion of design, to which we now turn.

EDITORIAL INTRODUCTION

Design with and after Darwin

Surprisingly, there is still no consensus on Darwin's dealing with teleology. After *On the Origin of Species* was published, Thomas Huxley praised Darwin's defeat of teleology. Asa Gray, on the other hand, applauded his restoration of teleology. In our own time, scholars are similarly divided. Darwin did not reject the idea of a creator at any time throughout the six editions of the *Origin*, as one of his heroes was the natural theologian William Paley. Teleology, says Ruse, was in bad shape when Darwin's *Origin* appeared. Darwin investigated Paley's claims by seeking proximate, or secondary, causes rather than ultimate causes. And in seeking those explanations, he made no qualms about invoking final causes—that is, teleology. The complexity of life, he believed, was there for a purpose.

David Hume struck a presumed fatal blow to Christian teleology by saying it could infer a multitude of designers, not just one; or one—perhaps even a "stupid mechanic"—who simply copied others who had perhaps produced something only after lengthy trial and error. Moreover, the abundance of evil argues for a kind of God that no one would want anyway. Teleology thus became an idea "unneeded in science, riddled with paradox in philosophy, and obstructive of genuine belief in religion ... destined for the slag heap ... along with phlogiston" (Ruse 2003d, 28). Anglican churchmen such as William Paley and other "natural theologians" responded to Hume with a resounding restatement of the *argument from design* (that is, design exists and only a designer can explain it). Just as the complex machinery of a watch working together for the purpose of telling time points to the existence of a watchmaker, so the marvelous contrivances of the eye for the purpose of seeing point to the existence of an eye-maker.

One failing of Paley's argument, says Ruse, was that he had no credible alternative explanation (apart from chance, which may, he argued, give us a wart or a pimple but not an eye) so his proposition became true simply by definition (Ruse 2003d, 44)—the implication is that Darwin did provide a credible alternative so we can now safely conclude that Paley was wrong. "Like Paley, Darwin was looking at the organic world as if it were an object of design" (Ruse 2003d, 121). Darwin's originally orthodox Anglican belief gave way to deism and finally to agnosticism, but he did not go so far as to atheism. He endorsed the argument to complexity (life exhibits evidence of design), but his work planted a "bomb" within the argument to design (that design requires a designer) (Ruse 2003d, 128).

While Darwin's thesis of descent with modification as the explanation of biological diversity was widely accepted after publication of *The Origin* in 1859, his mechanism of natural selection did not fare as well. Oddly, evolution as a philosophy made great gains, but evolution as science did not. In biology, evolutionary thinking was taken over by the German school led by Ernst Haeckel, which occupied itself with working out phylogenies, whereas in England, it was "social Darwinism" that took over. It was Huxley who turned Darwinism against teleology, as he pointed out that teleology was like "a rifle bullet fired straight at a mark; [but] according to Darwin, organisms are like grapeshot of which one hits something and the rest fall wide" (Ruse 2003d, 139). Asa Gray, however, pointed out that even the grapeshot points to design—natural selection may choose among variants but it does not explain the origin of those variants. Darwin was not impressed. He countered Gray by saying:

> The view that each variation has been providentially arranged seems to me to make natural selection entirely superfluous. . . . It seems to me that variations . . . are due to unknown causes and are without purpose . . . they become purposeful only when they are selected. (Ruse 2003d, 147)

Darwin lacked an explanation of biological variation, so it is not surprising that natural selection made little scientific headway immediately after 1859. The rediscovery of Mendel's work in genetics at the turn of the twentieth century caused a further eclipse of natural selection in favor of saltationism—marked by those who believed evolutionary change occurred in big jumps rather than Darwin's gradualism. The rediscovery of Mendel's ideas moved evolution from its past toward its future. How much Mendel himself truly realized what he had done, and how much later thinkers read back into his work what they wanted to find, is still a matter of debate. The point is that now the way was being opened for an adequate theory of heredity, something so lacking and so needed by the theory of the *Origin*. What was necessary was that the genetics be extended from individual organisms to factors of heredity working in populations. Unlike Lamarckism, to take an example, natural selection is something that does not act on the individual but is meaningful only in groups. Thanks to some very mathematically gifted biologists, this work was done, and so by around 1930, the framework of a full theory or paradigm of evolutionary change was starting to emerge.

With the discovery of random mutations and further developments in genetics, a school of thought emerged based on nonadaptive random variation. This further relegated selection as secondary because Darwin had been concerned to explain adaptation—the fit of organisms to their special way of life. By the 1930s, however, genetics and natural selection were brought together in what has since been called the "synthetic" or "neo-Darwinian" theory of evolution. Mutations provided the raw material of evolution, and natural selection adapted it to the needs of organisms. In the USA, the key figure was the Russian-born geneticist Theodosius Dobzhansky. He took a proposal by the American geneticist Sewall Wright—the so-called "shifting balance theory," or at least he took the version that used the pictorial metaphor of an "adaptive landscape"— and used it to pursue studies in the wild and in the laboratory. Following his teacher Morgan in taking the little fruit fly as the model organism, Dobzhansky, his associates and his students followed in detail the physical and chromosomal changes over generations, trying to work out how forces of selection and of drift bring on changes. The little fruit fly Drosophila showed itself a perfect organism for genetical studies—it breeds easily and quickly, it requires minimal maintenance, it has no odd sexual system, it has giant chromosomes easy to study, and it can be found readily in the wild in accessible places. Dobzhansky's work and that of those in his orbit (particularly the taxonomist Ernst Mayr, the paleontologist George Gaylord Simpson, and the botanist G. Ledyard Stebbins) did much to establish Darwinian selection as a major force in the natural world, although there was often a non-Darwinian flavor to the work, especially when their thinking was influenced by deep roots that Wright's thinking had in Herbert Spencer as much as Charles Darwin. Sometimes the thinking of paleontologists was at best neutrally Darwinian and even at times verged on the unfriendly. The well-known theory of "punctuated equilibrium" of Niles Eldredge and Stephen Jay Gould suggested that the course of evolution is not smooth but comes in fits and starts, and went through various incarnations, but in Gould's hands was not particularly selection friendly. The very name of the theory had echoes of a theory, "dynamic equilibrium," from another tradition. In a similar vein, John J. Sepkoski Jr. (student of both S. J. Gould and E. O. Wilson) did sterling work in mapping the major events in life's history, producing "neo-Spencerian" pictures of the repeated upward spurts of complexity, followed by subsequent balance. It was work that could be given Darwinian underpinnings, but not work starting with Darwinism, per se.

In concluding, because he did not want to be accused of dodging the crucial issue, Darwin made brief reference to our own species. "In the distant future I see open fields for far more important researches. Psychology will be based on a new foundation, that of the necessary acquirement of each mental power and capacity by gradation. Light will be thrown on the origin of man and his history" (C. Darwin 1859, 488).

It is fitting to now return to the *Origin* for its most famous closing words. And so to the final, famous paragraph:

> It is interesting to contemplate an entangled bank, clothed with many plants of many kinds, with birds singing on the bushes, with various insects flitting about, and with worms crawling through the damp earth, and to reflect that these elaborately constructed forms, so different from each other, and dependent on each other in so complex a manner, have all been produced by laws acting around us. These laws, taken in the largest sense, being Growth with Reproduction; Inheritance which is almost implied by reproduction; Variability from the indirect and direct action of the external conditions of life, and from use and disuse; a Ratio of Increase so high as to lead to a Struggle for Life, and as a consequence to Natural Selection, entailing Divergence of Character and the Extinction of less-improved forms. Thus, from the war of nature, from famine and death, the most exalted object which we are capable of conceiving, namely, the production of the higher animals, directly follows. There is grandeur in this view of life, with its several powers, having been originally breathed into a few forms or into one; and that, whilst this planet has gone cycling on according to the fixed law of gravity, from so simple a beginning endless forms most beautiful and most wonderful have been, and are being, evolved. (C. Darwin 1859, 489–90)

Chapter 2

Mummy's Boy

Making a Start

I WAS BORN IN England in 1940, the same week as the fall of France. My father was a conscientious objector. After the war, he and my mother joined the Religious Society of Friends, Quakers, and it was within that religion I grew up, even going as a teenager to a Quaker boarding school in York. At about the age of twenty, my belief in the existence of God faded away. I had no Saul on the road to Damascus experience in reverse. Less dramatically, Lewis Carroll knew the score.

> In the midst of the word he was trying to say,
>   In the midst of his laughter and glee,
> He had softly and suddenly vanished away—
>   For the Snark *was* a Boojum, you see.
> (L. Carroll 1950, final stanza)

I used to think that, by seventy, I would again believe in him. At that age you cannot be too careful. But here I am, in my ninth decade and no belief. For a while, I was a fairly militant atheist. When I was thirteen, my mother died suddenly of yellow jaundice. She was thirty-four. She had been orphaned as a child, unable to develop her intellectual abilities to the full, so she put her energies into making sure her bright little boy had a loving childhood, one that guided him strongly to develop his intellectual abilities to the full. I will say that, unsurprisingly, nonbelief opened the way to hitherto-unacknowledged hate of God, for killing my mother. Hate or not, I see now that, for about fifteen years, from ages thirteen to twenty-seven, I lived in a haze of sadness—until I got married and my newfound joy cleared the clouds. I don't think I was particularly perverse or weak in this. I was just about to enter adolescence, I had just gone away to boarding school, so I was in a strange new environment, separated for the first time from the all-dominant, very-much-loved force in my life, my mother, and now she was gone. My headmaster gave me the news at 3:00 p.m. on a Wednesday. That evening, no one said anything to me. I am not sure they were told. The next morning, at 8:45 a.m., I was in class. That was the therapy and support I got. It was an all-boys school. The first woman I spoke to was two months later at Christmas.

I contrast this with my son Edward, who at twenty-four went to the University of Virginia in Charlottesville to read law. Arriving on the Friday—classes started on the Monday—on the Saturday he went downtown (alone) to explore. He got caught up in a neo-Nazi demonstration and was literally within arm's length of a young woman who was killed when someone drove a car into the crowd. On the Monday morning, when he went on campus at 9:00, the university already had counselors on call. Edward saw one the whole first semester and often thereafter, graduating with honors. These days, when everybody is criticizing higher education, it is comforting to know that sometimes some people get everything absolutely right.

I did not at all dislike my father. His attitude toward me was a combination of pride and jealousy at my achievements, although toward the end of his life his feelings mellowed as he started to feel (a totally justified) satisfaction with what he was able to do with his life. (He was a school bursar.) I am not writing to speak ill of him. But, he did absolutely nothing to aid me through life's journey's after my mother died. Fifteen months later, he had found and married a new partner—an au pair from Germany—and, to be honest, from then on I was an irritation who, fortunately for the happiness of all, was around only for limited periods, namely through school holidays. I always got a job, like working for the painters at my father's school, and was expected to contribute to my keep. If it was hard on me, it was sheer misery for my sister (two years younger than I) who was packed off as a boarder at my father's school, literally just across the road. Talk about something to instill a sense of self-worth.

I will say, talking of senses of self-worth, part of my mother's influence on me was to instill a sense that I was exceptionally talented, one who was going to make a mark in life. My mother's belief was not entirely unjustified. In the state school system at that time, the all-important hurdle was the "eleven-plus," a kind of IQ test one took when one was ten, and if successful (20-percent pass rate) one could go on to "grammar school" with the prospect of secondary and tertiary education on the state's tab. I aced the exam and went on to the top grammar school in town, top class, top five out of a class of thirty. That all changed a couple of years later, when I went away to boarding school. (Why on earth did my parents send me there? The English class system, I am afraid. Private school, known misleadingly as "public school," was—by my lower-middle-class parents, striving to climb the greasy pole—considered a priori as better than anything in the public sector.) From the first, at my boarding school, I was labeled as B material—far behind those boys who had been to private school and already had five years of Latin and Greek under their belts, no good at games, father unimportant, socially difficult adolescent—and with no mother to plead my cause, that was the end of that. Given my depressed state, I was not about to change things. And yet, the influence was still there. Deep down I knew I was special, if not showing it then. Not deep enough, I am afraid. By the time, six years later, I went to university, I came to accept what the world was telling me. My feeling about unacknowledged talent was quite gone.

I am sure that, had my mother lived, I would have been directed toward a life in medicine. If it sounds a bit arrogant to assume that I would have been good enough to get into medical school, I should explain that, in the fifties, this was not the high hurdle that it became by the seventies. (My suspicion is that television was much responsible for this. Medical shows like *Ben Casey* had a powerful effect. Students in veterinary schools went from all male to all female. After *All Creatures Great and Small*, every little girl wanted to be a vet.) In the fifties, indeed, medicine was generally marked down as something for those "not good with concepts." At my school, it was generally the career choice for those boys whose fathers were GPs, and the understanding was that they would inherit the practice. To be honest, for quite a while I wondered if I had taken the wrong path, or at least not taken the right path. Apart from anything else, I had friends in medical school and I envied the camaraderie of that way of living. Very different from the rather lonely ways of a philosopher.

It was also partly that the English school system narrows one into fixed channels at an early age—sixteen in my case. I followed the herd in doing science rather than humanities—after the successes of WWII, with prospects of intergalactic travel, not to mention the power of nuclear understanding, physics was high profile. Because of my choice, I was locked into mathematics and physics and nothing else. By the middle of the first week at university, I realized that I may have been a good high school mathematician, but I was no university mathematician. I managed, at the end of my first year, to divert my path to a joint degree in mathematics and philosophy. The start to my whole life as a philosopher puts one in mind of those desperate people at the US-Mexican border, willing to do anything to escape their present position. Unsurprisingly, my mental attitude looked upon writing as a chore to be completed as quickly as possible and then forgotten. That said: "God works in mysterious ways." Had my mother lived and I had been steered into a life of medicine, I might now be looking back on a lifetime of urology or something similar. What might have been. (More details of my life as a professional philosopher are given in Ruse 2017c, 2017d, 2017e. For those not looking desperately for material to pad out their dissertations, a cadet version is Ruse 2016.)

Canada

Despite the burdens that weighed down on me after my mother died, there were people to whom I have owed an everlasting debt. People who said: "Mike, you are better than you let yourself be." One was Stephan Körner, my philosophy professor in England, where I did my undergraduate degree. It was he who, on getting a letter from an unknown university in an unknown city in Canada—McMaster University in Hamilton, Ontario—offering support for graduate work, put forward my name. Hence, at the age of twenty-two, in 1962, I emigrated to Canada for my MA. That completed, I moved on to a PhD program at the University of Rochester, in upstate New York. I was not a success. My personal stresses were still a major factor, to which one can add the loneliness of life in a new and rather alien place. After two years, following a summer spent working at "Manpower"—summer jobs as a teenager prepared me for a firm that offered daily working gigs with immediate pay in cash—I left to return to Canada, tail between my legs, with one B and seven "incompletes." That I got a Guggenheim twenty years later suggests that my problems were less about ability and more about psychology. I should say that at Rochester was the second of those people who thought I was better than I let myself be. The epistemologist Keith Lehrer was ever ready to write me a letter of support. I still get pleasure from remembering that the only other philosopher to get a Guggenheim in my year was none other than Keith Lehrer.

In Canada, as elsewhere in the sixties, there was a massive increase in higher education, with the founding of many new universities. The new University of Guelph, in Southern Ontario, was atypical in being founded on the long-existing Ontario College of Agriculture, the Home Economics (Domestic Science) College, and the College of Veterinary Science. There were many students whose grandparents had met at Guelph. The dean of Agriculture, recognizing that things were about to change dramatically, and the cozy little world of hitherto was gone forever, decided to seize the opportunity. He decreed the week before term started that philosophy would be a compulsory subject for all his students. The Philosophy Department was desperate for anyone willing to pitch in and teach first-year courses.

At this point, the third of my benefactors, John Thomas (father of the comedian Dave Thomas), on the faculty at McMaster, stepped up to the plate. He virtually hustled me onto the bus from Hamilton to Guelph. I was the right person, in the right place, at the right time. The job was mine. Having to get up in the morning, wash, and go to work—not to mention marrying the first of the student wives—started to turn me around. It needed to. Although let me not be entirely negative about my years of trial. One very good consequence is that my life as a teacher—something I loved as much as research—was fused with the strong conviction that the mental health of my students—especially my grad students—was as important as their academic prowess. Without wanting to sound like Mother Teresa, the Ruse table had always space for another mouth. For a bed, if needed. As we shall see Darwinism telling us, the truly happy person is the person giving to others. This is infectious. Not one of my five children had any religion as they grew up. All five of them are in jobs serving others.

Back to my new job. I should stress that, while the students in the Agricultural College—or "Cow College" as it was called elsewhere in Ontario—tended not to be overburdened with culture, generally they were very far from stupid. Lively and inquiring, they were great fun to teach. I knew in the first five minutes of the first class that that was what

I was going to do with my life. So it proved. I retired at eighty after fifty-five happy years of teaching. Within a week, my "aggies" were tearing each other's throats out as they argued about Thrasymachus and "might is right." It rather boggles the mind to think of them, back home for Christmas, discussing philosopher kings while milking the cows. The dean of Agriculture was a wise man. He knew that education is more than teaching about fertilizers and the like. It is about developing the whole person. I suspect he had read John Dewey.

Life Becomes Positive

If my job at Guelph were to become permanent, there was still a PhD to be written. Any good mentor will tell one's student to pick a topic on which there is not a lot written, and what there is tends to be rather bad. Don't touch Hume on causation. My background in mathematics and physics still inclined me toward the philosophy of science. More than one person suggested that biology, as opposed to the traditional physical sciences, looked promising. Looking back, I can see that an unacknowledged (and unappreciated) factor was that, in the 1960s, biology was really coming into its own, with the double helix in molecular biology and more formal discussions about groups and how natural selection molds their biological nature. A paradigm was the work in the early 1960s by William Hamilton on the structure of hymenopteran (ants, bees, and wasps) sociality, showing how what came to be known as "kin selection" promoted the sterility of workers, who hereby increased their genetic fitness (ability to pass on one's heritable factors, that is, "genes").

Encouraged by the already-mentioned Stephan Körner, I went back for doctoral work to my undergraduate university (Bristol). The English PhD, unlike the American, has no coursework. It is all dissertation (what is called a "thesis" in England and Canada). There was no escape from it. I was faced with a major writing project. It was at this point, moving on from the frozen state of my thinking and emotions of the previous fifteen years, I discovered (totally unexpectedly) that I have this incredibly strong talent for writing and loved doing it. It was not like being moderately good at games and knowing that, if you worked at it, you could be really good. It was like having the sporting ability I do have and then, overnight, when I kicked the ball, it would go in the direction I intended. I do not stress this newly discovered ability because I am totally immodest—although that helps!—but because it was hitherto totally unknown to me and so very strong. A complete surprise. That it took so long to discover this was, surely, a consequence of the aforementioned haze of sadness within which I still lived. I could function in everyday life (if getting one B and seven incompletes can be called functioning), but creativity and passion were simply unknown. Released from the sadness, I could flourish. Not just the dissertation but, from that day to this, publications of the order of Erle Stanley Gardner, the author of the Perry Mason series.

Came the publication (in 1972) of my first book, *The Philosophy of Biology*. Philosophically, as at the time were most Anglophone philosophers of science, I was a "logical empiricist," meaning that Newtonian mechanics was regarded as the model to be emulated—laws, axiomatic, testable, and so forth. Ernest Nagel (1961) and Carl Hempel (1965) were the standard-bearers. With enthusiasm, from this perspective, I embraced philosophical work in the biological sciences. More accurately, philosophical work in the *evolutionary* side of the biological sciences. I, like others on the same path, took seriously Dobzhansky's aphorism that nothing in biology makes sense except in the light of evolution. Truly though, my focus was more the result of a nonbiologist not knowing there were other parts of biology! I was hugely pleased and relieved when its first review by Anthony Flew was favorable. I should say that, generally, negative reviews of my papers never discouraged me. It was through them that I learned the trade of writing at a professional philosophical level. My colleagues always used to say: "It's okay for you, Mike. They accept anything you send them." To which I would truthfully reply: "I have more rejection letters than the rest of the department put together." "You have a thick skin." "We all have thin skins. It is just that some of us learn to live with them." I never had any choice. Earning the approval of my mother is still my dominant motivating factor. If she were to return from the dead, she would be shocked to see how her memory has motivated my life. Over seventy books written or edited. More articles than I have time to count or to bother to put on my CV. Shocked, but (despite the regret of my not becoming what my father always referred to as a "real doctor") rather proud.

The Influence of Thomas Kuhn

I was launched into the philosophy of biology. But, we are talking now of the late sixties. Like many of my generation, a book written by one who was not a philosopher changed the course of my thinking. I refer to the *Structure of Scientific Revolutions* (1962) by the historian of science Thomas Kuhn. Using the concept of a "paradigm," a kind of scientific theory that included the sociological factors—like training—Kuhn argued for an extremely idealistic view of science, where moving from one incommensurable paradigm to another demanded something akin to a leap of faith. Like all philosophers, I rejected this. What I did absorb and endorse—like many other philosophers—was Kuhn's directive, that in order to do good philosophy of science one had to be able to do good history of science (Callebaut 1993).

In my case, this sparked an interesting paradox. Down the road from Guelph, on the other side of the border, was the long-established, highly regarded University of Michigan, with a Philosophy Department having a deservedly strong reputation. On the door of his office, the chair had pinned a list of the top ten philosophy journals, in order. Imagine being a tenure-track assistant professor in that department. Talk about heavy-handed suggestions! Guelph, being a new university in Canada, had other priorities. A major function was to supply higher education to immigrants or the children of immigrants, a huge population since WWII. Often, over 50 percent of the class would fall into this category, starting with the professor! And immigrants did not come to Canada just to watch hockey. They wanted to seize the opportunities. To better themselves, and—even more—their children. At convocation, there would be father in the suit he wore only at weddings and funerals, mother in corsets so tight she could not sit down, and kid with a T-shirt under the gown. Suddenly, leaving war-torn Europe for twenty years of backbreaking labor in the New World was all worthwhile.

As I discovered, this was certainly no block to being a prolific publisher. The contrary. If anything, living in an atmosphere of commitment to better oneself, it was a spur to action. We had the opportunities. We were going to grab them. I went from untenured lecturer in 1968 to tenured full professor in 1974. Less than twenty-five years after I had arrived penniless in Quebec, I was made a Fellow of the Royal Society of Canada. I felt I had done something to mark the trust the country had shown when it invited me in. Our Lord had given us five talents (Matt

25:14–30). We were able to turn to him and say: "Here are your five talents. And here are five more." No one needed to pin anything on doors telling us what we should do. On my first sabbatical in 1972, I scarpered off for the year to England where I spent the whole twelve months in the manuscripts room of the library of the University of Cambridge, huddled over the unpublished notebooks of Charles Darwin.

The end result was, in 1979, I published my second book—the one that I say truthfully was the book I wish I had had ten years previously, when I had started to take an interest in Darwin. *The Darwinian Revolution: Science Red in Tooth and Claw* sold thirty thousand copies in English, and at least that number in Spanish and Portuguese. One point worth noting is that basically all the books I have written have some connecting theme. My first book, *The Philosophy of Biology*, tried to show that evolutionary biology is good science for the same reason we judge work in the physical sciences to be good—predictive fertility and that sort of thing. The Newtonian science of biology. My first Darwin book was, and still is, notable because I take seriously the influences on Darwin. Predictably, I spent some considerable time exploring the influence of the important philosophers of science of Darwin's day—the empiricist John F. W. Herschel and the rationalist William Whewell. Up to this point, the general assumption was that Darwin's theory was a great step forward and so its methodology must be new (Hull 1973). To the contrary, thanks to those hours spent in the university library, I was able to show that Darwin was immersed in the methods of science as identified by the philosophers, and his success came from carefully following their dictates (spelled out in Ruse 1975.) This was at once accepted as obvious, so much so that most think I deserve little or no credit for the discovery. I can live with that, especially since I had already moved on to the next project!

Moving On

Unsurprisingly, by the end of the seventies, I was starting to wonder if I should make the move from philosophy to history. I was already teaching the history of science, to the biggest class of the History Department. But, to adapt a phrase for the nineteenth-century novel *Eric, or Little by Little* on the dangers of masturbation, the "Silver Cord" could not be broken. "The grass withereth," he murmured, "the flower fadeth, and the glory of his beauty perisheth; but—but the word of the Lord endureth for ever" (Farrar 2007, 1:14, from Isa 40:8). My grass did not wither, my flower did not fade. But the word of the Lord did endure for ever. Less metaphorically, I found I could not leave philosophy behind. Those Quaker mentors of my childhood could not be denied or ignored. I am also a bit of a snob. Philosophy of science was always recognized as legitimate philosophy—Aristotle was a biologist and wrote philosophically on the topic. To this day, history of science, professionalized only in the fifties, is not respected by traditional historians. I was not about to change from a career in high-grade philosophy of science to one of low-grade history of science.

You will indeed learn that I have switched entirely from history of science to history of ideas which, despite its name, is firmly located in philosophy—you are using history to tackle philosophical problems of today—I love history but, like all philosophers—the Philosophy Department at Berkeley would not promote Kuhn to full professor—I am not sure it is a full-time occupation for a grown-up. Less condescendingly, I should say that although philosophically I have been a lifelong logical empiricist, my days in Cambridge, where the dominant historian of science was the American Robert Young (1985), have led me in history to being an equally lifelong "social constructivist," where science is considered an epiphenomenon of the social culture of the day. Not entirely consistent—logical empiricism thinking science the epitome of objectivity and social constructivism thinking science the epitome of subjectivity—but intellectually very fertile.

As it happens, pressed upon me rather than by my uninfluenced choice, two philosophical issues came to the fore around 1980. First, there was the sociobiology controversy, sparked by the publication in 1975 of a book by the Harvard ant specialist, Edward O. Wilson. In *Sociobiology: The New Synthesis*, Wilson argued that all animal behavior, from the ants up to and including humankind, is a function of biology as fashioned by natural selection. When I read the book, shortly after it was published, I thought it a major work about evolutionary biology, one that took behavior seriously and well beyond the now rather crude attempts of the ethologists like Konrad Lorenz. Coming from England, I literally had no idea that looking at humans from an evolutionary perspective might be an ideological issue that many would find exceedingly distasteful. In the 1950s, I had watched television shows like the *Brains Trust*, where on Sunday afternoons, intellectuals like Julian Huxley held forth (and forth and forth and forth) on serious topics. Human evolution by Darwinian forces was taken as a given.

I walked smack bang into a cohort of left-wing critics, notably two of Wilson's colleagues—same department—the geneticist Richard Lewontin and the paleontologist Steven Jay Gould (Gould and Lewontin 1979). I was quite unprepared for this, particularly given that their motivating ideology was Marxism, and, although left wing, my sources were nineteenth-century socialists, John Stuart Mill and then the Fabians—notably Sydney and Beatrice Webb. Naturally, I wrote a book about the topic—by then I was sufficiently confident about my abilities to write books on any topics that interested me—and *Sociology: Sense or Nonsense?* duly appeared. My book duly appeared with two effects. It made me very visible and set me up as a right-wing maverick who endorsed neofascist premises, to the disapproval of many of my fellow philosophers of biology. To bring a long story to a quick conclusion, human sociobiology endured, if under different names like behavioral ecology. A couple of years ago I wrote a book on hatred—*Why We Hate: Understanding the Roots of Human Conflict*—arguing that humans are, by nature, biologically based social animals and that war and prejudice are the result of cultural phenomena, notably the coming of agriculture. No one, including me, thought there was anything ideologically offensive about my strategy.

The other philosophical topic that engaged me around 1980 was the rise of the creationism movement arguing that evolution is false and Genesis, read literally, is true and it is the latter that should be taught in the biology classes of publicly funded schools. I wrote a book on the topic—*Darwinism Defended: A Guide to the Evolution Controversies* (1982)—and, in the fall of 1981, found myself in an Arkansas courtroom, an expert witness, defending evolution as science and criticizing creationism as religion. The judge agreed with me. My testimony was at the heart of his ruling.

> What is science?
>
> It is guided by natural law;
>
> It has to be explanatory by reference to natural law;
>
> It is testable against the empirical world;
>
> Its conclusions are tentative, i.e., are not necessarily the final word; and
>
> It is falsifiable.

In making these points, explicitly the judge referenced me: "Ruse and other science witnesses."

A nice conclusion because, with good reason, convinced that we are the brightest people on campus, we philosophers tend to be rather condescending toward lesser mortals. Just what our lawyers did not need on the witness stand. As it happened, rambunctiousness cowed by the ferocity of our lawyers, all went well. Indeed, judging by the judge's conclusion, very well. I did slip in one good line, avoiding the glaring eyes of our counsel. The state's attorney was trying to trip me up over my religious beliefs, more accurately nonbeliefs, with the intention of showing that I was biased against the law. Finally, I blurted out: "Mr. Williams, surely you can see that I am not an expert witness on my own religious beliefs." Everyone laughed, and when Williams returned to the attack, the judge said: "Mr. Williams, surely you can see, he is not going to give you what you want. Move on."

Expectedly, the world of learning welcomed me (Ruse 2021b). I have always had a suspicion: that I, from a virtually unknown university in Canada, should on my first attempt get that Guggenheim is explicable only because the academic community was saying "thank you." This suspicion is fortified by the fact that I got my honor as a "science writer" and not as a philosopher. This is just fine by me. Among the many philosophers approached, I was the one with the guts—desperate desire for notoriety?—to get up on the stand and speak up for Darwin. Not that I got any thanks from my home discipline. As expectedly, my fellow philosophers went after me. Any kind of Popperian-type attempt to characterize science is anathema to right-thinking intellectuals. Larry Laudan was particularly vociferous in his objections. As one who was just fighting the sociobiological critics at the same time, I was almost indifferent. A major plank of Laudan's criticism was that I should have shown that creationism was bad science, not religion. To which, somewhat condescendingly, I replied: "The First Amendment separates church and state. It forbids the teaching of religion in state schools. It does not forbid the teaching of bad science." On the principle that revenge is a dish best eaten cold, I then published a collection, including my testimony and that of other witnesses (including fellow Arkansas witness Steve Gould with whom, sociobiology or no sociobiology, I always had the closest of relationships) and also Larry's critique and my response—*But Is It Science? The Philosophical Question in the Evolution/Creationism Controversy.* Still in print.

A Darwinian Philosopher

In the mid-eighties, I started the journal *Biology and Philosophy*. In what I suspect is a common practice, I ghosted most of the articles for the first issue. It then picked up rapidly. A number of senior academics hunted around in their bottom drawers and sent in an unsuccessful paper they had given at a conference some years past. I soon realized that a mark of a successful editor was the ability to say no. More accurately, I soon realized that a mark of a satisfyingly successful editor was the ability to say no to self-important human beings. At the personal level, there was one more major enterprise that occupied me in this decade. The Australian, Oxford-residing philosopher J. L. Mackie wrote a very favorable review of my sociobiology book (1980). Coming from such a powerful mind, I was more chuffed by that than any negative reviews could dispel—but, at the end, he was very critical. I had a discussion with Ed Wilson, who had opened his book by saying:

> Camus said that the only serious philosophical question is suicide. That is wrong even in the strict sense intended. The biologist, who is concerned with questions of physiology and evolutionary history, realizes that self knowledge is constrained and shaped by the emotional control centers in the hypothalamus and limbic systems of the brain. These centers flood our consciousness with all the emotions—hate, love, guilt, fear, and others—that are consulted by ethical philosophers who wish to intuit the standards of good and evil. What, we are then compelled to ask, made the hypothalamus and limbic system? They evolved by natural selection. That simple biological statement must be pursued to explain ethics and ethical philosophers, if not epistemology and epistemologists, at all depths. (E. Wilson 2000, 3)

I argued that Wilson was dead wrong. You cannot derive "ought" from "is": the naturalistic fallacy. Mackie argued that ethics is subjective not objective. There is no "is" from which "ought" might or might not be derived. I saw that Mackie was right. You can *explain* ethics using sociobiology. We need ethics to be social. What you cannot do is *justify* it. As I have somewhat controversially said: "Ethics is an illusion put in place by natural selection to make us efficient cooperators" (Ruse and Wilson 1986). In the lingo of philosophers, I am a moral "non-realist." But why should one not say that the fact that something is put in place by natural selection does not mean it is not real? Natural selection, for obvious reasons, leads me to believe that my wife's body is thrillingly different from mine. That does not mean it is not really. How else would you explain our three kids? But that does not prove that selection always hits on the right solution. Here I invoked what has come to be known as the "debunking" argument. I argued that natural selection might lead us to believe different things that might function well. Redolent of the secretary of state under Eisenhower John Foster Dulles's attitude to the Russians in the Cold War, instead of "love your enemy as yourself" you might believe "hate your enemy as much as you can, but remember that your enemy hates you, so you had better get on together." Such a position, active and as highly developed as in humans, would promote exactly the same effect. Darwin had a similar argument in the *Descent*.

> It may be well first to premise that I do not wish to maintain that any strictly social animal, if its intellectual faculties were to become as moral sense as ours. In the same manner as various animals have some sense of beauty, though they admire widely different objects, so they might have a sense of right and wrong, though led by it to follow widely different lines of conduct. If, for instance, to take an extreme case, men were reared under precisely the same conditions as hive-bees, there can hardly be a doubt that our unmarried females would, like the worker-bees, think it a sacred duty to kill their brothers, and mothers would strive to kill their fertile daughters; and no one would think of interfering. Nevertheless the bee, or any other social animal, would in our supposed case gain, as it appears to me, some feeling of right and wrong, or a conscience. For each individual would have an inward sense of possessing certain stronger or more enduring instincts, and other less strong or enduring; so that there would often be a struggle which impulse should be followed; and satisfaction or dissatisfaction would be felt, as past impressions were compared during their incessant passage through the mind. In this case an inward monitor would tell the animal that it would have been better to have followed the one impulse rather than the other. The one course ought to have been followed: the one would have been right and the other wrong. (C. Darwin 1871, 70)

Darwinism implies moral non-realism.

This is only part of the question of ethics—what is generally known as "metaethics," because it talks about foundations, whether these exist or not. There are also questions about what one should do, generally known as "substantive ethics." Darwinian evolution stresses that the success of humans stems from the fact that we are social animals. We are not very fast. We are not very strong. We are not equipped with weapons of attack or defense. We are however, very clever, with an unrivaled power of communication (language). This suggests that our moral implications are toward cooperating, which in turn suggests that being moral will bring us pleasure. If being moral makes us miserable, no one is going to be moral, except under compulsion. You don't need evolution to realize this. John Stuart Mill said: "When people who are fairly fortunate in their material circumstances don't find sufficient enjoyment to make life valuable to them, this is usually because they care for nobody but themselves" (Mill 1863, 9). Jesus did not tell us that we must be miserable in order to get into the kingdom of heaven.

What about epistemology? Is there a Darwin-based argument that parallels ethics? There certainly is. Charles Sanders Peirce. Pragmatism!

> Not man merely, but all animals derive by inheritance (presumably by natural selection) two classes of ideas which adapt them to their environment. In the first place, they all have from birth some notions, however crude and concrete, of force, matter, space, and time; and, in the next place, they have some notion of what sort of objects their fellow-beings are, and of how they will act on given occasions. (Peirce [1883] 1955, 178)

Adding:

> Side by side, then, with the well established proposition that all knowledge is based on experience, and that science is only advanced by the experimental verifications of theories, we have to place this other equally important truth, that all human knowledge, up to the highest flights of science, is but the development of our inborn animal instincts. (Peirce [1883] 1955, 181)

Again, we are simply developing something stated by Darwin. From a notebook. "Plato Erasmus says in Phaedo that our 'necessary ideas' arise from the preexistence of the soul, are not derivable from experience.—read monkeys for preexistence—" (C. Darwin 1987, M 128; the Erasmus referred to here is Darwin's older brother). This is the key to human knowledge. Our brains, to use the modern analogy, are computers, but they are programed by natural selection to think in certain ways rather than in other ways.

This tells us about what one might call the epistemological side to substantive ethics. What about the epistemological side to metaethics? I am not sure that Darwinism is of much help here. Like many, I opt for some form of skepticism. There are basic questions to which we have no philosophical answers. Why is there something rather than nothing? Who can say? I remain mystified. As the biologist J. B. S. Haldane said: "My own suspicion is that the universe is not only queerer than we suppose, but queerer than we can suppose" (Haldane 1927, 286). All is a mystery. We do not know the ultimate meaning of things. There could be nothing. For all this, somewhat inconsistently, I can only say that as one who lives in a culture that produced the *Saint Matthew Passion*, whatever is the conclusion of reason and empirical evidence, emotionally I find such a conclusion hard to accept. When you add in the *Marriage of Figaro*, impossible to accept. Camus suggests that life is "absurd." Too much philosophy clouds his judgment (Ruse 2019).

Riding roughshod over my hesitations, I put everything all together in another book: *Taking Darwin Seriously: A Naturalistic Approach to Philosophy* (1986b). My fellow philosophers greeted it with the same enthusiasm they had shown toward my thoughts on sociobiology. So negative were they that I gathered them up, and printed them all, in a greeting I sent to each one inviting them to come to my fiftieth birthday party, asking them to celebrate fifty years of unbroken success with the prospect of fifty more. I am glad to say that with one exception—who has never forgiven me for a bad review I once wrote of one of his books—they all responded in the friendly spirit intended. I am also glad to say that ten years after the book's first appearance, without my asking, another publisher took it up and produced a second edition. It is unique among my books, since it is the only book where I give, in a unified way, my basic philosophy. Not just philosophy of science, but epistemology and ethics. I try to show how fundamental is a Darwinian approach and how those who do not take this approach are simply wrong. It really matters that we are modified monkeys rather than modified mud.

These days, I am not alone in arguing for this. Richard Rorty, for one, was ahead of me. He was adamant that we humans are at one with the rest of the living world, and that this is a consequence of being a Darwinian. "Darwinism requires that we think of what we do and are as continuous with what amoebas, spiders, and squirrels do and are" (Rorty 1998, 295). And he was openly a pragmatist about all of this, with John Dewey as his hero. He wrote that pragmatists "should see themselves as working at the interface between the common sense of their community, a common sense much influenced by Greek metaphysics and monotheism, and the startlingly counterintuitive self-image sketched by Darwin, and partially filled in by Dewey" (Rorty 1998, 41).

Normal Philosophy

Thus far, I have been stressing the negative reactions to just about everything I published. I may seem a bit paranoid. A more accurate suggestion is that the negativity could have been predicted. Despite the pacifism that infuses Quakerism, I enjoy a good fight. Intellectually, that is. I have an urge to choose topics and endorse positions precisely because they will irritate the staid and successful. I was having so much fun. The epitome of boring philosophy is where we all agree. The response to my invitation to come to my birthday party suggests that even my loudest critics knew what I was up to, and there was respect for this. They thought the same about boring philosophy. Larry Laudan was and continued to be a good friend. This said, I am not about to whine at not getting my Guggenheim for philosophy. If philosophers had anything to do with this, I had brought it on myself. I am not sure I see this as a negative judgment. I am after all the guy who went off to Cambridge to study the Darwin notebooks rather than stay at home and try to publish articles in the top journals.

This all notwithstanding, in a sense, after all the creativity/controversy of the first part of the eighties, my career rather settled down. Cribbing a title from Thomas Kuhn who wrote of scientists settling down into "normal science" when all the shouting is over, I was now ready for normal philosophy. I am sure this in part was because I had met Lizzie, one of my students, over twenty years younger than I. For forty years now, we have been idyllically happy, with three children (now in their thirties). Such a change from my first marriage, which had started well, but soon morphed into something not really enjoyable for either of us. Open marriage does mean you get to sleep with a lot of people other than your spouse, but truly it is not really that satisfying. I don't mean to sound puritanical. It was at times a lot of fun. Romping with the wives of senior administrators has much to commend it.

All said, I think the change in my work patterns and urges to be controversial was because I had done a lot of pioneering work. Now was the time to labor on and from it. This I did for thirty years or more. I don't mean it was just humdrum and dull. One thing that absorbed me—still absorbs me—is the question of values in science. Can a work be—as feminists worry—female hostile (Merchant 1980)? For that matter, can it be female friendly? Or is that an oxymoron like weapons for peace? I was spurred by a presidential address (to the Philosophy of Science Association) by the Catholic philosopher Ernan McMullin (1983). McMullen argued that non-epistemic values like feminism guided early science, but as time goes by, these get expelled by epistemic values like predictive fertility. Phrenology to psychology sort of thing. This sparked my interest, particularly over the issue of progress—non-epistemic—in evolutionary biology. From the beginning of systematic thinking about evolution, in the eighteenth century, progress—acorn to oak, monad to man—had always been the backbone (happy metaphor) of evolutionary thinking. Erasmus Darwin, Charles Darwin's grandfather, an early evolutionist, was explicit.

Biology

> Imperious man, who rules the bestial crowd,
> Of language, reason, and reflection proud,
> With brow erect who scorns this earthy sod,
> And styles himself the image of his God;
> Arose from rudiments of form and sense,
> An *embryon* point, or microscopic *ens*!
> (Erasmus Darwin 1803, 1:11)

Humans

> The idea of evolution "is analogous to the improving excellence observable in every part of the creation; such as the progressive increase of the wisdom and happiness of its inhabitants." (Erasmus Darwin [1794–96] 1801, 2:247–48)

Charles Darwin wrestled, not always successfully, with keeping progress out of his theorizing (Ruse 1996, 2004). Today it is gone. There is no progress in the theorizing in the leading journal *Evolution*. Mendelian genetics, and its successor molecular genetics—epistemic superstars—insist on the randomness of new variations (mutations). Progress is impossible. McMullin was right. Or was he? My counter insight (to McMullin) was that, over time, one did see the expulsion of progress—but this was because we now have the counter non-epistemic value of non-progress guiding evolutionary thought. In an era of nuclear weapons, how could anyone think of progress? I am a naturalist about these sorts of things. Could one like a scientist set up two hypotheses about these claims and then see if the facts—history—decide between them?

A consequence of questions like these was that, in the mid-nineties, I published my biggest, most researched book—on the notion of values in biology. *Monad to Man: The Concept of Progress in Evolutionary Biology* (1996b) involved interviewing a number of leading biologists (thirty or so) and masses of archival work. I found that both initial hypotheses were wrong. I did make the wonderful discovery of how important status is to scientists. The people I dealt with all knew the consequences if they produced science infused with non-epistemic values. Like the progress of life up to the winners—Homo sapiens. "Just pop science." With this judgment would go the hope of major fellowships and prestigious jobs and membership of elite societies. So, although they all believed in progress and that sort of thing, they were meticulously careful about keeping it out of their professional science, reserving it for popular books and the like—museum displays, for example.

I had gone to the headquarters (in Philadelphia) of the confusingly named American Philosophical Society (because it is really about science). I wanted to study the papers/letters of the eminent Russian-born evolutionist Theodosius Dobzhansky. Unfortunately, Dobzhansky late in life moved from NYC to California and threw away most of his papers, and those still there tended to be in Russian. I had two weeks, and after two days I was done. The curator suggested I look at the papers of the paleontologist George Gaylord Simpson. His file was huge—he had kept copies of all the letters sent to and from him. For better or for worse, there was a massive amount of (fascinating) gossip and less science. Naturally, I could not keep my eyes off it, so returned home thinking I had wasted my time. Then in the next week the truth hit me. The gossip was all about establishing evolutionary studies as genuine professional science and how non-epistemic values had to be rejected. Progress had to go. Simpson's referee's reports were hugely revealing. As were the judgments of Ernst Mayr, first editor of *Evolution*. I went back to Philadelphia four months later for another week in the archives. It was all there. I went home and rewrote my (rather dull) manuscript from the beginning, showing that I had the idea all along—the nature of evolutionary biology was driven by the desire for status and not by the nature of the non-epistemic values that shaped it—and I was off to the races. I wrote a book of which I am still very proud.

What Gaia Taught Me about Pseudoscience

My involvement with the creationist movement showed that I had long had an interest in the science-religion relationship—it is hard to work on Darwin without having such an interest. This interest continued and indeed continues to this day. (I have a discussion in my recent book on Darwin, to be mentioned later.) One publication back then that showed how I wrestled with the problem was *Can a Darwinian Be a Christian? The Relationship between Science and Religion* (2000a). My answer was that one can hold both views consistently, but it could be more difficult than people often confidently assume. I think, today, I would make more of the apophatic approach that I mentioned earlier. Indeed, I would be inclined to say that much is clouded in mystery. Faith, not reason, should be one's underlying foundation. I do take very seriously the idea that there is much that we simply cannot understand. As a Darwinian pragmatist I am okay with that. Natural selection cares about us having adaptations to live and reproduce. It really is not that interested in philosophical and theological questions. I should note that I have written extensively on atheism. I am very interested in the topic, but my publications may give the impression that I am more interested than I really am. The publications are, to a great extent, reflections of the wishes of my editors who, understandably, want books that reflect the huge interest in nonbelief sparked by the new atheists like Richard Dawkins. For instance, my *Atheism: What Everyone Needs to Know* (2015) was written at the express wish of Peter Ohlin, my editor at Oxford University Press. I might add incidentally my *Can a Darwinian Be a Christian?* sparked the most polemical anti-Ruse diatribe I have ever encountered (Coyne 2002). My feeling is that the writing was not really about me. It was a plea, by a new atheist wannabe, to be considered a big boy in the playground of nonbelief.

There were other events in the years after *Monad to Man*, notably the Darwin two-hundredth birthday celebrations (2009), for which I wrote books (*Charles Darwin* [2008a]) and edited volumes (*Evolution:*

The First Four Billion Years, with my biological colleague Joe Travis [2009]). And some interesting independent works. I still am very pleased with my book on the idea of the world as an organism: *The Gaia Hypothesis: Science on a Pagan Planet* (2013b). This was the brainchild of my editor at the University of Chicago Press, Karen Darling. People like me are hugely indebted to their top press editors. In my case, particularly the just-mentioned Peter Ohlin at Oxford University Press and Beatrice Rehl at Cambridge University Press (who commissioned the gorgeously illustrated *The Cambridge Encyclopedia of Darwin and Evolutionary Thought* [2013a]).

One thing I saw explicitly from writing the Gaia book is that writing nonfiction is much like writing fiction. You must have a beginning, a middle, and an end. Above all, you must tell a good story. Being invited to write a book on Gaia meant two things. First, I was going to accept the offer. Never turn down a suggestion from an editor of a top university press. (When Peter Ohlin asked me to write on atheism, I went out and bought a bottle of single malt.) Second, I had to find a theme. Karen asked me on the basis of a discussion review I had written, in the *Chronicle of Higher Education*, of two books on Gaia. But that was one thing. A book is different. When the original idea comes from you, you already have a theme you want to be the basis of a book. This I did not have. Not deterred, I plunged into the literature on Gaia, much more or less philosophical. Jim Lovelock had come up with the idea in the late sixties, so by the time of the eighties there was a whole subdiscipline of writing on Gaia, much on whether it was a falsifiable notion in a Popperian sense.

Dutifully, I wrote about 150 pages on this stuff. At which point, I pushed back my chair and said: "This is the most [expletive removed] boring stuff I have ever read." Then I had a brainwave. Scientists hated Gaia, they called it everything from "wrong" to "subversive." The general public reaction was overwhelmingly positive—and not just in California. Gaia massages, Gaia sandals, Gaia organic food. There may even have been Gaia sex—there certainly was if you read Jim Lovelock's autobiographical writings. I spotted that main complaint of the scientists about Gaia was that it is "pseudoscience." But what is pseudoscience? It turns out that it is not a fixed idea, but an epithet that is flung when people are insecure and threatened.

An example. Here in Tallahassee, our doctors and the chiropractors get on just fine. Florida State has a new medical school. Already insecure because people questioned the need of such a fund-draining campus institution. The medics were getting onside by stressing that their main aim was to produce GPs for rural areas. How could a good lefty like me complain? Then a state senator, a chiropractor, got a huge amount of (state) money to fund a college of chiropractic. You can imagine how the medics reacted. There would go their reputation for all time. Charges of pseudoscience were flung about with gay abandon. Then the offer was withdrawn. All returned to normal. Pseudoscience disappeared from the vocabulary, and GPs went back to suggesting chiropractic. It is the same with Gaia. Biologists were feeling very tense, tearing each apart because of things like the sociobiology row. All turned with relief to Gaia and happily joined together in condemning it as pseudoscience, thereby letting the world know how just professional biology is today. The eminent evolutionary biologist John Maynard Smith judged that "Gaia is just an evil religion." Twenty years later, somewhat sheepishly, he admitted:

> Look . . . all the trouble with Gaia is that we've had such agony with vitalism and group selection, and all these other things, and we thought we had it all worked out, and then you came along. You couldn't have chosen a worse moment.

Expectedly, sociobiology exists today in professional circles, under the name of "ethological psychology" or the like. Gaia is "Earth-system science" or some such thing. Scientists have peace and quiet, and I had my book.

The Return to Childhood

I have noticed that in the past decade, although my childhood Christianity has not returned, more and more I am driven by the social issues so important in my Quaker early years. They were always there. In 1988, I wrote *Homosexuality: A Philosophical Inquiry*. Contrary to popular supposition, I did not write it because I was working out personal issues. I am the Platonic form of heterosexual—perhaps not a wise comparison given the sexual practices of the Greeks. Indeed, I think it is because I do not feel conflicted about my answer that I had the confidence to tackle the issue. I doubt I could do the same for a book about "mother love." I wrote the homosexuality book because, in my adolescent years in England, homosexuality was much discussed, particularly by Friends, who were at the fore of those arguing that homosexuality should not be a crime. Apart from anything else, sending a homosexual to prison was a bit like sending a drunkard to a brewery. I felt that sociobiology had something to contribute to the discussion. Hence my book.

Today, I am still much driven by those early Quaker concerns. I make this crystal clear in my little book on hatred. Already I have told of its dedication to those wonderful people of my childhood. Continuing, my preface:

> I was raised a Quaker in the years after the Second World War. Quakers don't have the usual trimmings of religion—preachers, churches ("steeple houses" as we called them of old), or creeds and dogmas and that sort of thing. However, to conclude that Quakers have no strong beliefs is to make a major mistake. They could give St. Paul a run for his money. Above all, for me, being a Quaker meant being part of a community with my fellow human beings. We were never very good at literal readings of the Bible, but my goodness we took the Sermon on the Mount seriously. "Ye have heard that it hath been said, An eye for an eye, and a tooth for a tooth: But I say unto you, That ye resist not evil: but whosoever shall smite thee on thy right cheek, turn to him the other also" (Matthew 5:38–39). And: "Ye have heard that it hath been said, Thou shalt love thy neighbor, and hate thine enemy. But I say unto you, Love your enemies, bless them that curse you, do good to them that hate you, and pray for them which despitefully use you, and persecute you" (Matthew 5:43–44). (Ruse 2022, xi)

That is our role in life and how we serve our Lord. Loving other human beings. Quakers talk of the "inner light," that of God in every person, and that resonates to this day.

But why then so much hatred: war and prejudice? That is the theme of my book. It is part of my lifelong drive: to show the pertinence of Darwin's theory of evolution through natural selection. On problems back then, equally on problems today.

I have just finished what I suspect will be my swan song: *Charles Darwin: No Rebel, Great Revolutionary* (forthcoming). It is a rewriting of my *Darwinian Revolution* (1999b). Except it isn't. Having told the reader that now it is a history of ideas, using history to throw light on philosophical ideas that engage us today, in my introduction I say:

> I will use already-expressed ideas [about the history of the Darwinian revolution] to push forward to my concerns now.

> Specifically, I shall ask about the relevance of Darwin's work towards an understanding of attitudes towards foreigners, especially including immigrants; towards an understanding of the nature (if they exist) of racial differences, and how these (real or otherwise) affect society's attitudes towards African-Americans; towards an understanding of sexual orientation, whether it is a matter of nature or of choice; and, finally, towards an understanding of the nature and status of women. Recently, it has become evident that there is still huge prejudice against Jews. After I have discussed beliefs about foreigners and attitudes towards race, I add a short codicil addressing this issue. Overall, I shall look at Darwin's work against its background, at our thinking today and the extent it has been shaped by Darwin's work, and whether Darwin himself had any idea of the ways in which his findings and theories would be an integral part of our thinking today. (Ruse 2022, 3)

That is my latest work. As mentioned earlier, I have a discussion in this book of the science-religion interface. The way I approach Darwin practically demands that. This is not a stand-alone. Over the years, my thinking about the Darwinian revolution has changed—I would say "matured." As I just intimated above about what drives me, I have gone from what was obviously the first step, exploring the nature and facts of the Darwinian revolution—so I am not criticizing myself here—to trying to unpack social issues—their nature and their problems—and looking to see if a knowledge of the Darwinian revolution can help here. If anything, my Quaker background has grown in importance rather than faded. That truly makes me feel that my life's work has significant moral implications. I am not just building the Eiffel Tower out of matchsticks. Fun to do but now time to get serious and to deal with grown-up problems. I am already dealing with grown-up problems. Understanding the science-religion relationship is one of them. I would not have testified in Arkansas had things been otherwise.

One thing is worth noting, connected with my many decades of fascination with the question of values in science: more and more my (social constructivist) thinking about science has been shaped by my conviction that to grasp the nature of science, certainly science in any historical perspective, is to be aware of how greatly understanding has been—still is—shaped by what linguists refer to as rival "group metaphors"—akin in some sense to Kuhn's "paradigms." One such group metaphor is *organicism*, the world as an organism. It is very much a

Greek vision—it is the theme of Plato's *Timaeus* (Cooper 1997). It was revived and reenergized by the Romantics—leading figure Goethe, leading philosopher Schelling—at the end of the eighteenth century. The other group metaphor is the world as a machine, *mechanism*. It was this that drove the Scientific Revolution. Descartes is its champion. It is the dominant metaphor for today's science (Ruse 2021a).

The metaphors are a major theme of *Charles Darwin: No Rebel, Great Revolutionary* (forthcoming). The double helix is the triumph of machine thinking. However, there are still those attracted to organicism. Enthusiasts for the Gaia hypothesis make up one group. Agriculture is also a favored area with debates over global warming and the propriety of using GM foods. A new book on the topic is *The New Biology: A Battle between Mechanism and Organicism*, coauthored with science educator Michael Reiss (Reiss and Ruse 2023). One fascinating offspring was seeing the importance of this divide for modern theological thinking. The vitalism of Henri Bergson, the evolutionary interpretation of Christianity of the Jesuit Teilhard de Chardin, Alfred North Whitehead, and process theology make much more sense if one approaches them with the organicist group metaphor in mind. Interestingly, although Michael Reiss is an Anglican priest, it was I who wrote the chapter on this topic. In light of what I have said about myself, not so surprising. Overall, my strong interest in social issues, like anti-Semitism, is couched in terms of the divide—something that throws much light on the different positions people espouse. For instance, in the case of homosexuality, much depends on how we categorize human beings—machines or organisms? The same is true when we think about how God regards human behavior.

Favorite?

One question remains. Which is my all-time favorite book? It will surprise you to learn that I have not yet mentioned it. In 2014, I spent a semester at the Institute for Advanced Research in Stellenbosch in South Africa. North of Cape Town, the university to which it is attached is embedded in the Afrikaans wine-growing area. It is truly the most beautiful place on Earth. I had total freedom about what I was to do, and so I determined to write something on global warming. The next day, on my first trip to the library, I discovered that all the books are in Afrikaans! What was I to do? I love nineteenth-century British and American novels and poetry—Jane

Austen, Charles Dickens, George Eliot, Emily Dickinson, Walter Whitman. Lewis Carroll too! I had long thought that, as a retirement project, I might dig into them and try to find traces of Darwin and his revolution. I saw that I now had the perfect opportunity. Nothing to distract me—good meals were provided—and, thanks to the internet, I could download a novel or a poem in seconds. More than this, OUP was putting together bibliographies—on Emily Dickinson for example. About a hundred-plus references, divided into topics—God, evolution, death, love, and more. I had done one on Darwin, so in return I was given six months of access to all the bibliographies. (No need to go hunting in library holdings.) In a week, I could become an expert on sexual selection in Dickens's novels. Start with Eugene and Lizzie in *Our Mutual Friend*.

> "Undraw the curtains, my dear girl," said Eugene, after a while, "and let us see our wedding-day."
>
> The sun was rising, and his first rays struck into the room, as she came back, and put her lips to his. "I bless the day!" said Eugene. "I bless the day!" said Lizzie.
>
> "You have made a poor marriage of it, my sweet wife," said Eugene. "A shattered graceless fellow, stretched at his length here, and next to nothing for you when you are a young widow."
>
> "I have made the marriage that I would have given all the world to dare to hope for," she replied. (Dickens 1952, 496)

It was the most exhilarating time of my whole career, especially when my Lizzie flew over for two months. She would spend the days exploring and, in the evening, we would eat supper at one of the many sidewalk restaurants. (Did I tell you it was wine country?) We even had a week's trip to Zimbabwe, including a safari—cameras, not guns. We saw warthogs, who promptly made an appearance in my next book.

I leave it as an exercise for the reader as to how D. H. Lawrence tied Bergsonian vitalism to some very rude sexual practices.

> But still the thing terrified him. Awful and threatening it was, dangerous to a degree, even whilst he gave himself to it. It was pure darkness, also. All the shameful things of the body revealed themselves to him now with a sort of sinister, tropical beauty. All the shameful, natural and unnatural acts of sensual voluptuousness which he and the woman partook of together, created together, they had their heavy beauty and their delight. Shame, what was it? It was part of extreme delight. It was that part of delight

of which man is usually afraid. Why afraid? The secret, shameful
things are most terribly beautiful. (Lawrence, 1915, 113)

One starts to understand why Lawrence is read so avidly by adolescents, and then goes entirely unread as we become adults.

Darwinism as Religion: What Literature Tells Us about Evolution (2017g). Note how I cannot stay away from the science and religion topic. I argued that literature, novels and poetry, shows how Darwinian ideas offer contrary explanations to Christian claims. Love of neighbors is not some alien directive forced upon us. As Pip shows in *Great Expectations*, it is something that is part of the nature of beings for whom sociality is the direct consequence of natural selection. This all-important (to me) book brings together, for the first and only time, my public work on Darwin and my private love of Dickens. How special for me is *Darwinism as Religion*! Never to be repeated. I have opened up once. I owed that—to give them a good book—to the generosity of the folk in Stellenbosch. Never again. Private is private.

And now, it is time to turn to the readings from my books: *Reading Ruse: Michael Ruse on Darwinism, Science, and Faith*. To get in ahead of those others, if my readers are not critical, I shall think they are not taking me seriously. And they have missed the opportunity to have a lot of fun at my expense.

— PART II —
The Readings

3

Atheism, Belief, and Faith

Chapter 3, Reading 1

The Arkansas Creationism Trial Forty Years On[1]

THE CONCEPT OF FALSIFIABILITY is something that has been talked about a great deal by scientists and others recently. It's an idea that has been made very popular by the Austrian-English philosopher Karl Popper. Basically, the idea of falsifiability is that there must be, as it were, if something is a genuine scientific theory, then there must, at least conceivably, be some evidence that could count against it. Now, that doesn't mean to say that there's actually going to be evidence. I mean, one's got to distinguish, say, between something being falsifiable and something being actually falsified. But what Popper argues is that if something is a genuine science, then at least in the fault experiment, you ought to be able to think of something that would show that it's wrong.

For example, Popper is deliberately distinguishing science from, say, something like religion. Popper is not running down religion. He's just saying it's not science. For example, you take, say, a religious statement like God is love, there's nothing in the empirical world that would count against this in a believer. I mean, whatever you see—you see, for example, a terrible accident or something like this, and you say, "Well, God is love. It's free will," or, for example, the San Francisco earthquake, you say, "Well, God is love; God is working his purpose out. We don't

---

1. Ruse 2021a. Reprinted with permission of Springer.

understand, but nothing is going to make me give this up." Now, with science, you've got to be prepared to give up.

Let's take evolutionary theory, for example. Suppose, I mean, contemporary thought on evolutionary theory believes that evolution is never going to reverse itself in any significant way. In other words, the dodo, the dinosaurs are gone; they are not going to come back. Suppose, for example, one found, say, I don't know, somewhere in the desolate north up in Canada, suppose one found evidence in very, very old rocks, say, of mammals and lots and lots of mammals and primates, this sort of thing, and then nothing for what scientists believe to be billions of years, and then suddenly, mammals come back again.

Well, that would obviously be falsifying evidence of evolution theory. Again, I want to make the point, you've got to distinguish between something actually being shown false and something being in principle falsifiable. I mean, the fact that you've got no contrary evidence doesn't mean to say that you don't have a theory. I mean, it could be true. What one means when one says that science has got to be tentative is that somewhere at the back of the scientist's mind, he, or increasingly she, has got to be prepared to say at some point, "Well, enough is enough; I've got to give this theory up." It doesn't mean to say you are going to be every Monday morning sort of requestioning your basic principles in science, but it does mean that if something is scientific, at least in principle, you've got to be prepared to give it up.

In 1957, the Soviet Union launched an artificial satellite that orbited the planet, Sputnik. It was the height of the Cold War and was recognized at once as a huge propaganda success for the Russians. Appalled, America set about responding and, in the postmortems following Sputnik, it became clear to all that American science education, particularly at the school level, was in dreadful shape. Money and resources were poured into organizations formed to improve such education, and in 1958, the American Institute of Biological Sciences founded the Biological Sciences Curriculum Study (BSCS) to tackle issues of high school biological education.

At the time, a number of southern states had anti-evolution laws, but by the mid-sixties, educators in these states wanted to get on board with the new direction in science, and so the BSCS books were adopted. The state of Arkansas, which had an anti-evolution law on its books, fought back, and, thanks to counter resistance by evolutionists, backed by the American Civil Liberties Union (ACLU), the case—Epperson

v. Arkansas—went to the Supreme Court. The anti-evolution law was struck down as a violation of the First Amendment separation of church and state. The premise: "The overriding fact is that Arkansas' law selects from the body of knowledge a particular segment which it proscribes for the sole reason that it is deemed to conflict with a particular religious doctrine; that is, with a particular interpretation of the Book of Genesis by a particular religious group" (Epperson v. Arkansas 1968, 103). The conclusion: "The state has no legitimate interest in protecting any or all religions from views distasteful to them" (Epperson v. Arkansas 1968, 107).

That seemed to be that. But not quite. The biblical literalists, formerly known as "fundamentalists," now more commonly as "creationists" (or "scientific creationists"), fought back. They had a formidable weapon. In 1961, two literalists, John C. Whitcomb, a Princeton-trained biblical scholar, and Henry M. Morris, a hydraulic engineer, coauthored *Genesis Flood: The Biblical Record and Its Scientific Implications* (1961). It became the bible (if one might use such a metaphor) of the literalist movement. Pushing the doctrine of "young Earth creationism," the authors claimed that every word of the Bible, read literally, is supported by modern science. The focus is on Noah's flood. Geology shows the Earth is recent, and that at some point it was covered with water; the fossil record shows that evolution is untrue and is more consistent with the pre- and post-flood biblical accounts of animals; and much, much more along the same lines.

Finally, things came to a climax in 1981. A young creationist, who was also a lawyer, had written up a proposed bill, insisting on balanced treatment between evolution and creation science, and in the legislature of the state of Arkansas he found takers. It was proposed and passed at record speed, taking only one Friday afternoon, when most had left or were eager to leave. Bill Clinton had been governor of the state from 1978 to 1980, when, for the first and last time not minding his fences, he was kicked out of office. He returned in 1982, and continued as governor until 1992, when he defeated incumbent G. H. W. Bush and became president of the USA. In the interregnum, from 1980 to 1982, was a man (Frank D. White) who was as surprised to find himself governor as he was unfit for the post. Unreflectively, he signed the bill, and on March 19, 1981, Arkansas Act 590 became effective.

As with Epperson v. Arkansas, the ACLU swung into action, preparing to bring suit against the law on account of its unconstitutionality. It lined up an impressive number of Arkansas religious leaders as plaintiffs, the lead being the Reverend William McLean, a United Methodist minister, whose name therefore became part of the subsequent trial and judgment—McLean v. Arkansas. (Actually, technically, McLean v. Arkansas Board of Education.) As is its wont, the ACLU looked for help from a prominent law firm, and the New York firm of Skadden, Arps, Slate, Meagher & Flom came on board, pro bono, giving the free support of a rather junior female partner and a number of (very sharp) even younger associates. (No one in the New York world of law is a disinterested altruist. This was very good publicity for the firm.) Everyone headed for trial, which took place in the first week of December 1981.

This is where I came in, in the early fall of 1981. Why me? I was not an American (not then, in 2000 I moved to a job in the States and ten years later became a citizen) and was not particularly distinguished. I was a (full) professor, fairly young (forty-one), in the Philosophy Department of a university in Guelph, Ontario, Canada. It was not a major established university, having been founded only fifteen years previously. . . . I had, however, the background, the talents, and the eagerness that the ACLU was looking for in its search for expert witnesses to testify at the trial. Expectedly, certain names kept coming up again and again, and the eventual witnesses practically chose themselves. There was Langdon Gilkey, professor at the Chicago Divinity School, and the leading Protestant theologian in the country. There was George Marsden, evangelical historian of religion, then at Wheaton College, later at Notre Dame. There was—this was a foregone conclusion—Stephen Jay Gould at Harvard, evolutionist, and one of the best-known scientists in America because of the monthly column he wrote for the science magazine *Natural History*. There was Francisco J. Ayala, Spanish born, former priest, now one of America's most distinguished evolutionary geneticists. And there were more, including Arkansas schoolteachers. (Missing was Carl Sagan, the most famous scientist of the day. He had been a little hoity-toity when first approached. Later, as the approaching trial started to gather publicity, he offered his services. But it was too late.)

But why me in this august group? Obviously, my name had come up, so I was not entirely unknown, and there was reason for this. I was one of a number in the 1960s (prominent member, David Hull from Chicago) who had kick-started the modern subfield of the philosophy of biology, leading to my writing an introduction to the area, *The Philosophy of Biology* (1973). Also, like many in the 1960s, I had been much intrigued by *The Structure of Scientific Revolutions* by Thomas Kuhn. It was not so much that I was taken by his thesis of change—more on this in a moment—but that I was excited by his demand that philosophers of science take seriously the history of science. So much so, that I took my first sabbatical (1972–73) in Cambridge, England, working in the university library, immersed in the Darwin archives. This led to my writing *The Darwinian Revolution: Science Red in Tooth and Claw* (1979). . . . In a way, it is the complement in the history of science—the history of evolutionary biology particularly—to my *Philosophy of Biology* in the philosophy of science—the philosophy of evolutionary biology particularly. It is a full overview of the revolution, making use of twenty years of archival research by Darwin scholars, including myself.

The point is that this preadapted me to take on the creationists. It was not so much that I had done much work on the creationist literature—although I had started work on this and by the time of the trial had a manuscript of what came out next year as *Darwinism Defended: A Guide to the Evolution Controversies* (1982). I should say that the manuscript was circulated to both sides and became a major source for the state in my cross-examination. What I had done is much work on the kinds of arguments that the creationists used. Many of these arguments were not that new and were around (and answered) at the time of Darwin. I knew the ropes. . . .

I had background preparation, I had the kind of personality that made me a natural for this sort of thing, and I was eager to do it. Not just the publicity—although most of my relatives and friends would say it was all about the publicity—but because I really do have moral concerns. As someone raised a Quaker, for all that my beliefs were long gone, I worried a lot about whether what I was doing was worthwhile, serving my fellow humans. You might say that of course being a teacher means you are serving your fellow humans, and increasingly over my life I came to see that. I really enjoy the scholarship, but I do take teaching seriously and have done my share and more.

Then the Arkansas trial came along and I saw a real chance of getting up and fighting what I believe are wrong and socially dangerous beliefs. It is not me and my pals who are against abortion, against homosexuals, and don't think women should be ordained. I should say that combined with this was the fact that my fellow philosophers wanted nothing to do with any of this. They thought it vulgar and misplaced to get into the witness box. Philosophers are not like other men. (Some had more legitimate concerns. David Hull was gay at a time when homosexuality was still much in the closet. He didn't want that coming out and being used, publicly, to discredit him, in a court trial.)

I got roped in, and, in the fall of 1981, went off down to New York City to be deposed before the trial. It was then that I discovered that the lawyers for Skadden Arps were by no means convinced that I should be a witness. It was not so much me as generally a prejudice against philosophers. We tend to go on and on about arcane topics that no one can or wants to understand, and on top of that we are so very arrogant. Convinced to a person that we are the brightest people on campus, we don't take instruction very well. You soon learn that lawyers are less concerned about the truth than about winning and this can lead to some very tense times. On top of this, of course, why a philosopher? Obviously, you need scientists, and theologians need hardly more justification. Educators are a must, and if you want to round things out with a historian or like person, why not? But why a philosopher? As it happens, this had nothing to do with my merits. The creationists rather forced it on the plaintiffs. It is a big mistake to think that creationists are necessarily stupid—before he changed track, Gish had published in the *Proceedings of the National Academy of Science* (Gish et al. 1960)—and they certainly do their homework. They knew full well that the biggest thing to hit the philosophy of science in the past half century had been Thomas Kuhn's *The Structure of Scientific Revolutions*. (Before Popperians who are reading this essay throw it down in uncontrollable rage, as belittling the status of their hero, note what I am saying and more importantly what I am not saying. I am not saying *The Structure of Scientific Revolutions* was the most important book or the most profound book or the longest lasting book. I am talking about immediate attention and controversy, and Kuhn's book wins hands down.)

The creationists had studied structure with great care and they knew full well the central concept and its supposed implications. Paradigms! Those conceptual frameworks within which scientific thinking is embedded. And what is the biggest mark of a paradigm, that which makes it so different and so controversial? That commitment to paradigms and changing from one to another is not simply a matter of reason and evidence. Paradigms require a kind of commitment to be found in religion or politics. People do change from being, say, a Catholic, to being a Protestant. Luther did! And people go the other way. John Henry Newman, for example. But the change from one to the other is not simply a matter of sitting down and saying, "I prefer consubstantiation to transubstantiation" or "I'm into justification by faith rather than good works." These may be important factors but in the end they are not decisive. Change needs almost a Kierkegaardian leap of faith. Creationists seized on this and argued that Darwinian evolution and creation science are different paradigms, with the supposed implication that one is as good as the other, and you cannot impose choice from without. At this point, you go beyond rationality and so that is it. There is no justification in education for preferring evolution over creationism. Balanced treatment is not only the fairest moral way forward, it is sanctioned by strong (and fashionable) philosophical argument.

How were our lawyers—as I will now feel free to call them—to counter this? They too were bright and had done their homework. They knew full well that when Kuhn came onto the scene, and started to pick up steam in the mid-sixties, the person and the group most immediately and strongly in opposition were Karl Popper and his merry men. Above all, as spelled out in his *The Logic of Scientific Discovery* (first published in English in 1959), Popper stood for rationality and, above all, he found it in science. What separates science from all else is the demarcation criterion of falsifiability. Even the best science is constantly putting itself to the test of the empirical evidence and, if it cannot handle this, it falls. No matter how prestigious. The way that Newtonian mechanics—the best and most fruitful science ever—had had to give way before Einstein and the other physicists of the twentieth century. Kuhn is wrong. Call them paradigms or whatever, if they are part of science, they must be falsifiable. Science is not like religion. And if you doubt that, go and look at the book edited by Imré Lakatos and Alan Musgrave (1970), the report on a conference earlier in the decade, where the philosophies of Popper and Kuhn were spelled out and the two sides went at each other, trying to show the flaws in the position of their opponents. The urgent need of a philosopher became obvious and the argument that the philosopher must make was no less obvious. The Kuhnian strategy must be countered and Karl Popper showed the way! I became part of the team that descended on Little Rock, Arkansas, in early December 1981.

The Trial

This was what was at stake:

On the side of creation science the claim was:

1. Sudden creation of the universe, energy, and life from nothing
2. The insufficiency of mutation and natural selection in bringing about development of all living kinds from a single organism
3. Changes with only fixed limits of originally created kinds of plants and animals
4. Separate ancestry for man and apes
5. Explanation of the Earth's geology by catastrophism, including the occurrence of worldwide flood
6. A relatively recent inception of the Earth and living beings

On the side of evolutionary science, the claim was:

1. Emergence by naturalistic processes of the universe from disordered matter and emergence of life from nonlife
2. The sufficiency of mutation and natural selection in bringing about development of present living kinds from simple earlier kinds
3. Emergence by mutation and natural selection of present living kinds from simple earlier kinds
4. Emergence of man from a common ancestor with apes
5. Explanation of the Earth's geology and the evolutionary sequence by uniformitarianism
6. An inception several billion years ago of the Earth and somewhat later of life (Ruse 1988a, act 590)

Expectedly, both plaintiffs and defense made much of Popper. I opened this essay with what I said to our side early in the morning, and under cross-examination we came back to it again and again. But really, that was easy. We had a party line and stuck to it. Evolutionary theory can be falsified and creation science cannot be. This is a good example of the sort of thing that went on under cross-examination. . . . I left Arkansas after our side had finished testifying. In a way, I felt a bit sorry for the creationists. We had such a stellar cast (I am not talking about me). They really had to scrape the barrel. No Langdon Gilkeys or Steve Goulds for them. Judge Overton handed down his ruling in early January 1982, and it was unambiguous. Evolution is science. Creationism is religion.

Teaching the latter violates the First Amendment separation of church and state. No balanced treatment for the kids of Arkansas. The points that the judge made were all fairly obvious and expected. No one in the real world ever accused him of misreading things or getting into dubious convoluted arguments. Again, at the risk of seeming unduly immodest, my testimony was at the heart of his ruling.

> It is guided by natural law;
>
> It has to be explanatory by reference to natural law;
>
> It is testable against the empirical world;
>
> Its conclusions are tentative, i.e., are not necessarily the final word; and
>
> It is falsifiable.

I had—and still have—huge admiration for Popper as a voice of rationality in the 1930s and 1940s, at a time when the world was in dire need of voices of rationality. That is by far my overwhelming emotion when I think of Karl Popper. As far as the philosophy was concerned, it wasn't so much that I was in favor of falsifiability or against falsifiability. It was rather that it was never really a topic of mine. In the philosophy of science, I was working on theories and their construction—people like Hempel and Nagel were more central to me. Quite apart from the fact that I am not a physicist, so Popper's work was not really my flavor. Then, when I worked on the history of science, the philosopher I had in my targets was Kuhn, as I tried to show that the Darwinian revolution could not have been as Kuhn hypothesized. There was no abrupt switch from one position to another—incommensurable paradigms—but a general gradual change, with Darwin's thinking incorporating much that he had learned from the non- or anti-evolutionists. To this day, I say that Darwin was a rebel, not a revolutionary. The one exception to my lack of real interest was that already-mentioned paper on Popper on evolutionary biology. Popper had said that Darwinian theory is not real science but a metaphysical research program that could not be falsified—apart from anything else, he claimed that natural selection is a tautology so obviously is not empirical (Popper 1974). The first paper I ever had accepted—a presentation at the first meeting of the PSA in 1968 in Pittsburgh—was on that topic. I guess I was interested in falsifiability in a minor way, right through practically until the Arkansas trial. I thought then as I think now that falsifiability is important and it is a mark of genuine science, although I was not then (nor am I now) convinced that that is all there is to be said on the topic of demarcation. Overton got me right. Falsifiability is important but there are other factors too. In Arkansas I was not selling my birthright for a mess of pottage, or, more prosaically, the chance to get involved in an exciting and very public event.

Let me conclude this essay—more a memoir!—by taking up the effect by the Arkansas trial on my subsequent professional career as a scholar. I could not other than be struck, at a kind of meta level, at what was going on here. Why the hostility to evolution, especially to Darwinian theory and its mechanism of natural selection? Simplistically, because it goes against the Bible. Yes, but no one hates the Copernican theory even though, supposedly, the Sun stopped for Joshua. In any case, it is not a generic hatred. Creationists admit these days that the ark would not have been big enough to carry all the species of animal extant today. Their ploy is that the ark carried "kinds," and after the flood these evolved into the different species we have today. And how did the evolution occur? Natural selection! The Creationist Museum in northern Kentucky has a better display and discussion of natural selection than the Field Museum in Chicago, three hundred miles to the north (272).

After the trial, I got to know Gish well as we appeared often together on TV talk shows, and a constant theme of his complaint about Darwinism was that it was really just as much a religion as creationism. (We were offstage, so he was quite happy making those judgments of creationism.) For a long time, I resisted his suggestion, but then came to realize that he had a point. It is not so much that Darwinian theory is religion. That is perfectly good science in its own right. It is rather that people take Darwinism and use it as the basis for a form of secular humanism. Just think, we have Darwin Day, celebrating Darwin's birth. So, also, we have Jesus Day. We call it Christmas. We don't have Copernicus Day or Newton Day or, for that matter, Dawkins Day. (I suspect he would be embarrassed, but not that much.)

This insight set me on a thirty-year journey, trying to show exactly how Darwin's theory is turned into a religion. I wrote a book showing that both creationism and Darwinism (construed in this sense) are into eschatology, world systems about meaning and end times (Ruse 2005). Creationists are providentialists, thinking we can do nothing without the saving grace of the blood of the Lamb, and so we must prepare for the end trying to obey his commands. Darwinists are progressionists, thinking we must improve things through our own efforts, if we are to bring Jerusalem down here on Earth. Both sides are into heaven—a secular version at least for the Darwinists—but they have different prescriptions on how to achieve it. In the lingo of theology, creationists are premillennialists, thinking Jesus will come before things are put right. Darwinists are postmillennialists, thinking Jesus (in a metaphorical sense) will come later when we have put things right. More recently, I wrote a book on Darwinism and literature, showing how fiction and poetry show that folk worked through such Christian themes as origins, God, the status of humans, sin, sex, salvation from a Darwinian perspective (Ruse 2017). I followed this with a book on war, showing how Christians and Darwinians took different stances on all the moral issues that such conflict entails (Ruse 2018).

What was fascinating was how, topic after topic, I found parallel treatments. Both creationists and Darwinists obsessed about the special place of humans, for example, determined to find that everything revolves around us—made in the image of our providential God as opposed to the climax of a progressive process of evolution. Showing that there is a lot more than just science going on here, the scientific theory of Darwinian evolution explicitly eschews such progress. Humans are different, but the science does not say we are better. In fact, the opposite. In the immortal words of the paleontologist Jack Sepkoski: "I see intelligence as just one of a variety of adaptations among tetrapods for survival. Running fast in a herd while being as dumb as shit, I think, is a very good adaptation for survival" (Ruse 1996, 486).

Chapter 3, Reading 2

God and Humans[1]

Who Is God?

HOLD ON, YOU MIGHT say. Before we start to talk about God, tell us why we should talk about God. This is supposed to be a book on atheism. Shouldn't we start there, and bring in God only as needed? The trouble with this approach is that atheism is, as it were, the default position. Atheism says that there isn't a God. So why argue about that? I don't think there is an invisible giraffe in the middle of my sitting room, but I am not going to bother to mount an argument unless and until someone starts to claim that there is one and that it matters. If someone tells me that I am going to go to everlasting hell and damnation unless I believe in the giraffe, then I am certainly going to get on the job! I don't think it is irreverent—and if it is, I will chance it—to suggest God is a bit like that giraffe. Until I have heard about God and why I should take him seriously, I will remain silent and untroubled. The burden of proof is on those who believe. So first let us make the pro-God case. Then we can turn to the anti-God case. (I rely heavily on McGrath 1995, 1997; Gunton 1997; and, above all, on Taliaferro and Meister 2010).

Given that Christianity—and Judaism and Islam, for that matter—is a religion that makes God absolutely central to its world picture, you might think that it would at least be clear on the notion or concept of

---

1. Ruse 2015c. Reprinted with permission of Oxford University Press.

God. Boy, would you ever be making a big mistake. As soon as you start to ask, you find two thousand years of unbroken debate and that is still going on. As the skeptics realized very early on in the game, the problem for Christianity is that it suffers from divided loyalties. It looks to two very different traditions for guidance and understanding—Athens and Jerusalem, the Greek tradition and the Jewish tradition. The question is whether these can be brought together in a harmonious fashion. Christians generally say that they can, but this is for us to decide. And this we can do only when everything is out on the table. Because in a way it is the more basic, and it is certainly the earlier, let us start with the God of the Jewish tradition—the God of the patriarchs and the prophets and (later) the evangelists—and then move to the God of the Greek tradition—the God of the philosophers.

Who Is the God of the Jewish Tradition?

This is the God of the all-defining and all-important work for Christians, the Holy Bible. This work falls into two parts: the Old Testament and the New Testament. What is taken by Christians as foundational is that the God of the Old Testament—the Jewish Bible—is also the God of the New Testament—the exclusively Christian part of the Bible. Nothing makes sense if these are not the same deity, and there are repeated passages in the New Testament, from the mouth of Jesus and others, affirming the identity and continuity. Marcion of Sinope (85–160) argued that there are two Gods, one for each Testament, and for his pains, he was excommunicated. One asks, therefore, whether there is a thread that runs through the Bible, linking the various conceptions of God, and the answer seems to be that there is. We find a story of God, of humans, and of the relationship between the two. Humans are persons, that is to say, beings with feelings, thoughts, and a sense of identity. This is true also of the deity. God is above all an intensely personal being. He is not a being like a human with a physical form, although sometimes in the Old Testament he is portrayed that way; he is a person like you and me. He is not a rock or a plant or even an animal. We humans are made in the image of God, and as persons we reflect God.

God is creator, absolutely and completely. He made the universe out of nothing. He is not the Platonic demiurge who shapes the materials he finds. He is a creator who cares for his creations—for the physical world, for the world of animals and plants, and, above all, for us humans, male and female. "Happy are those whose help is the God of Jacob, whose hope is in the Lord their God, who made heaven and earth, the sea, and all that is in them" (Ps 146:5–6 NRSV). Interestingly, although God is usually referred to by the masculine pronoun and in the New Testament frequently referred to as "Father," it is an important part of the Christian conception of God that he has no sex; in some sense, he is beyond or without it. What is important, whatever the language, God is Father not just in the sense of creator, but also in the sense of ongoing care and concern. God loves his children and wants the best for them.

Not that God is soft. We learn in Proverbs: "Those who spare the rod hate their children, but those who love them are diligent to discipline them" (Prov 13:24 NRSV). This may not be one of the Ten Commandments, but it is certainly a rule God follows. When humans transgress, they must be punished. Adam and Eve sinned, so they were cast out of Eden. God found that the world was given over to wickedness, so, except for Noah and his family, he drowned the lot. Dazzled by the beauty of Bathsheba, David sent her husband Uriah the Hittite to his death so he could have the woman for himself. God punished him by killing their firstborn. This said, God is amenable to reason and to the possibility of changing his mind. Abraham argues with God over Sodom and Gomorrah, hoping to find just a few righteous people and thus save the cities. It is not God's fault that he fails to find even a few.

Of course, none of the favoritism compares to that shown to Abraham and his descendants. God makes a deal with him. Make me your God, and I will shower blessings on you, most importantly making you the father of a great nation. For all that the Genesis story implies that there is one God and he did all of the creating, at this somewhat later point, as already noted, there is no implication that God—Yahweh—is the only God. But he is going to be the God of Israel and anathema to those who stand in his way. This does not mean that the "chosen people" never suffer—one thinks of the agonies in Egypt, for instance—but it does mean that God, the Lord, is ever mindful of their fate and their needs, and that when the time comes, he will deliver.

What about the God of the New Testament?

No one reading the Bible can miss how the picture of God changes through the various books. Not only is there a move to God being the only God but also a more refined and concerned being starts to come into view. The picture of a tribal God starts to fade, and God is seen with more overreaching feelings for the whole of his creation. Moreover, as we move through the Old Testament toward the New Testament, the thinking gets (shall we say) more sophisticated, as God (through the prophets especially) makes it very clear that we are not to think of ourselves as in any sense in the same league as God. In the book of Job, God reproves Job for pretending to understanding on a par with his. "Where were you when I laid the foundation of the earth? Tell me, if you have understanding" (Job 38:4 NRSV). Again in Isaiah: "For my thoughts are not your thoughts, nor are your ways my ways, says the LORD. For as the heavens are higher than the earth, so are my ways higher than your ways and my thoughts than your thoughts" (Isa 55:8–9 NRSV). Saint Paul echoes this: "But we speak God's wisdom, secret and hidden, which God decreed before the ages for our glory" (1 Cor 2:7 NRSV).

Nevertheless, the personhood of God remains central, and this holds true through the New Testament, although obviously with the appearance of Jesus and then the increasing role of the Holy Spirit (or Holy Ghost), things do get more complex. God gets less and less focused on the Jews exclusively and starts to extend his care and compassion to all peoples. Perhaps relatedly, he does start to do things more at a distance, through Jesus particularly. But it is the same God, and it is a personal God. The parables of Jesus make no sense unless we assume this. The father in the parable of the prodigal son, for instance, cares very much for his lost son and rejoices on his return. At the same time, he is sensitive to the worries and needs of the older son, who behaved and yet apparently gets no real affection. More than this, although he is now preaching a doctrine of love and compassion, God is still fully prepared to show his wrath when he is crossed or otherwise upset. Think of the parable of the fig tree. "A man had a fig tree planted in his vineyard, and he came seeking fruit on it and found none. And he said to the vinedresser, 'Look, for three years now I have come seeking fruit on this fig tree, and I find none. Cut it down. Why should it use up the ground?' And he answered him, 'Sir, let it alone this year also, until I dig around it and put on manure. Then if it should bear fruit next year, well and good; but if not, you can cut it down'" (Luke 13:6–9 NRSV). Is this really any different from a modern-day televangelist saying to God: "Give them another chance, Lord, and then if they refuse to come to you, strike them down with hellfire and destruction"? The God of the New Testament is a father, but like real-life fathers, there is not only unconditional love but also serious expectations and obligations. If you really love your father, then you had better show it. Dad is picking up the tab for your college fees. Don't spend four years drinking and playing video games.

Who Is the Philosophers' God?

It would be an unfair caricature to suggest that the God of the Jewish tradition is that grandfatherly fellow in the Michelangelo painting of the creation of Adam. But it wouldn't be quite that unfair, because the emphasis is totally on the personhood of God. He is an individual who creates and who relates to his creation in an almost physical way. We turn now to a very different picture of God, a God who, one can state with some authority, does not have a beard. The trouble is that he is not clean shaven either, and therein lies the start of our difficulties. God on the classic theistic position of the philosophers is not one of the chaps at all. God is not a person like Michael Ruse or Richard Dawkins or Queen Elizabeth the Second—or like the God sketched in the passages just given (B. Davies 2004, 27).

The problems start with the Greeks, although one should not really blame them because the answers come from the Greeks also. The difficulty is that when Jesus died on the cross, when Jesus went up to heaven, there was no religion of Christianity. Those who came after, first Peter and Paul, and then the great theologians—the church fathers and their successors—had to develop and articulate the faith. Agreed that God was the creator out of nothing, a being who had started everything simply because he wanted to—a being, therefore, who is worthy and demanding of worship by us—the question now is how to flesh this out, and it is here that the philosophers and the theologians turned to the Greeks, initially Plato via the Neoplatonists (particularly for Augustine) and later Aristotle (particularly for Aquinas). Plato's theory of Forms is the crucial model, the theory that shapes, infuses, drenches classical theistic views of God—although often as filtered or modified by later philosophers, Aristotle and the Neoplatonists, Plotinus (204–70) in particular. Remember that we have this supersensible world—ideal but in a way far more real than our physical world—filled with eternal entities, all bound together and getting their being from the supreme entity, the Form of the Good. In Plotinus's language: "A nobler principle than anything we know as Being; fuller and greater; above reason, mind and feeling; conferring these powers, not to be confounded with them" (Plotinus 1992, 5.3.14). The classical theistic view of God takes in all of this and ends with an entity that is eternal, in some sense self-sufficient, perfect, and sustaining everything else. Plato's Form of the Good may not be a creator out of nothing, but it is very much a sustainer in the sense that everything else gets its being from Good. For Plotinus, it is the eternal soul "attached to the Supreme and yet reaching down to this sphere, like a radius from a centre" (Plotinus 1992, 4.1.1).

It is the same with the philosopher's God—he doesn't come into being, he doesn't go out of being, he just is, and he is the source and reason for the existence of everything else. He uniquely exists necessarily; all else is contingent. Hence, this is not the God of deism, where God stands back after he has finished his job. The work is never over—although in a paradoxical sort of way, it never began either. Note that this does not mean that God is identical with his creation (pantheism) any more than Pythagoras's theorem is identical with Euclid's axioms, but it does mean that God—although himself outside time—is there for all of material existence, and that without him, all would collapse into nothingness. He is immanent.

God is loving; in fact, he is all loving. He is as well all powerful (omnipotent) and all knowing (omniscient), although God doesn't so much have the properties as that he himself is the properties (Wainwright 2010). Plato, again, argued that things of this world have their properties because of the Forms, because, remember, they "participate" in the Forms. The Forms are the properties, and God is love and power and knowledge. This leads to one of the more puzzling aspects of classical Christian theology. You might think (with Richard Dawkins) that with everything he has to do, God is pretty complex. This is to misconceive the kind of thought pattern followed here. God is not like a member of Mensa, with an IQ of 170. God is not a person. The traditional position is that God is the ultimate simple! He is the properties, but in an important sense, the properties are all one. In the words of Saint Anselm: "God is life, wisdom, eternity, and every true good.—Whatever is composed of parts is not wholly one; it is capable, either in fact or in concept, of dissolution. In God wisdom, eternity, etc., are not parts, but one, and the very whole which God is, or unity itself, not even in concept divisible" (Anselm 2008, 72). I hardly need say that in some respect this makes sense only by thinking of Plato and the Good. It is the One from which all else stems or flows. Simple, therefore, is not opposed to complex, in the sense that "the instructions for putting together this bookshelf are unnecessarily complex." It is more an ontological claim about the nature of God being the ultimate, on whom all else depends (B. Davies 2010).

How Can We Know God?

All of this is starting to sound pretty remote. How on earth can we be expected to relate to—let alone worship—a being who is so very different from us? We may not always approve, but we know what worship is about when we are talking about persons, about humans. It was one of the great tragedies of the twentieth century that so many Germans worshiped Adolf Hitler, and we know it was one of the great tragedies precisely because we know what it is to worship—to adore, to venerate, to put our trust in. But how can we worship something that is not a person, that is eternal and unchanging, that has (or rather is) all of the properties rolled into one? It is a question not of being willing but of being able. It is at this point that, through his theory of analogy, Aquinas made one of his great contributions to theology (B. Davies and Stump 2012, 161). Let's distinguish three senses in which we might use words. First up is *equivocal*, where the meanings are quite different. Compare "John is a naturalist," meaning John doesn't believe in miracles, with "Mary is a naturalist," meaning Mary is a nudist. Not the same thing at all! Next is *univocal*, where the meanings are the same. Compare "John is a naturalist" with "Fred is a naturalist," meaning Fred doesn't believe in miracles either. The same word with the same meaning. And finally *analogical*, meaning similar in respects but not all respects. Compare "John is a naturalist" with "Alice is a naturalist," where in her case it means, inasmuch as she is doing science, she doesn't allow miracles, but that as a Christian, she doesn't rule out a priori the possibility of miracles. Not the same, but an overlap. Aquinas applied this trichotomy to the problem of speaking of and understanding God. When we say that God is love or God loves us, we are clearly not using the word in the same way as when we say that Romeo and Juliet were lovers, Romeo loved Juliet, and Juliet loved Romeo. Apart from anything else, not to put too fine a point on it, God does not want to have sexual intercourse with us. So God's love and human love are not univocal. But neither are they equivocal. If I had a hobby of stomping on small babies just for laughs, I don't think I would be very loving. But when I say that God is love, I am ruling that out for him also. God doesn't do that sort of thing for laughs. God is love. At which point you might be inclined to throw up your hands and declare defeat. The best we can do is say what God is not—he is not a baby stomper. God is ineffable. The Jewish philosopher Maimonides (1135–1204) argued this way, going so far as to say that negation was positively better than alternatives: "I do not merely declare that he who affirms attributes of God has not sufficient knowledge concerning the Creator . . . but I say that he unconsciously loses his belief in God" (Maimonides 1936, 87). However, Aquinas thought there was a better option, namely, the middle way of analogy. God's love is not identical, but we can still see that God merits worship. Human love and God's love are analogical. "In this way some words are used neither univocally nor purely equivocally of God and creatures, but analogically, for we cannot speak of God at all except in the language we use of creatures, and so whatever is said both of God and creatures is said in virtue of the order that creatures have to God as to their source and cause in which all perfections of things pre-exist transcendently" (Aquinas 1920–22, 1a.13.5).

What Do Christians Believe about Jesus?

Where does Jesus Christ fit into all of this? Not easily. Judaism may have begun in polytheism, but it became increasingly and stridently monotheistic. "Hear, O Israel: The Lord is our God, the Lord alone" (Deut 6:4 NKJV). Christianity, on the other hand, seemed to split not just into two but into three: "The grace of the Lord Jesus Christ, the love of God, and the communion of the Holy Spirit be with all of you" (2 Cor 13:13 NKJV). We can say two things with confidence about the second figure in the Trinity (McGrath 1997; cf. Rogers 2010). On the one hand, Jesus is purely human. He had a mother; he ate and drank like the rest of us—seems to have been quite approving of a glass of wine—he had friends and loved some more than others; he got mad at times; as far as we know, he never had sex, but he had close relations and friendships with women, and much more. He was also mortal in the sense that he could be and was killed. He was the "son of man." On the other hand, Jesus is God. He did not just turn up happenstance. He came for a purpose, to redeem us. There is the aura of miracle about him, from having no physical father, through being able to perform quite domestic miracles like turning water into wine at a wedding feast, to truly stupendous ones like raising Lazarus from the dead, and finally coming back to life himself. He was anticipated and had a preordained role to play. He was a great teacher, but more than that, he spoke with the authority of God about what we should think and how we should behave. He was kind and loving beyond normal human abilities, and in the end was prepared to make the supreme sacrifice for others—for us—when he could quite well have kept his mouth shut and stayed out of harm's way. He was the "son of God." The Holy Spirit is a little trickier and has rather amusingly been described as the Cinderella left behind when the other two sisters had gone to the theological ball, although whether one wants to describe either God or Jesus as "ugly" is another matter (Feenstra 2010). It is that which Jesus left behind to be with us always, helping and guiding. However, general Christian tradition is that it was around before the coming of Jesus and is often associated with breath, as when God breathed life into Adam and Eve. Early Christians were not always entirely sure that the Holy Spirit was indeed God or just an emanation of God, but opinion swung to its divinity. Very influential (as so often) was Augustine, who identified the Holy Spirit with love: "Scripture teaches us that he is the Spirit neither of the Father alone nor of the Son alone, but of both; and this suggests to us the mutual love by which the Father and Son love one another" (McGrath 1995, 8; quoting *De Trinitate*).

What Is the Status of Humans?

We are pretty important. One hesitates to say that we are all important, because it seems that God is aware of every sparrow that falls to the ground, but it is hard to escape the conclusion that we are the real point of the creation. However you read the creation stories of Genesis, they lead up to the appearance of humans, and then we uniquely are the beings made in the "image of God" (Vanhoozer 1997). Expectedly, there has been a lot of discussion about precisely what this means, but Augustine (again!) is definitive in saying that we humans uniquely are rational beings, able to enter into a relationship with God. The influence of Plato is the key here, because the Greeks' psychology (especially in the *Republic*) makes rationality not just one of the parts of the soul but the key, dominant part. Englishmen like me spend a lot of time discussing whether there is a place for dogs in heaven—could it be heaven without dogs?—but even if there is a divine hound lying at the foot of the creator, it will not have quite the same relationship with God as every one of us humans has or has the potential to have. With rationality goes morality. Indeed, it seems fair to say that without rationality, you cannot have morality. Someone with a damaged brain who is not thinking properly may do horrendous damage to people or to property, but, because such people are not in control of their senses, we do not judge them culpable. Likewise, in civilized countries at least, we do not judge children by the same standards we apply to adults because we realize that they are not fully formed human beings—in an important sense, they are not fully rational. Note, then, that this all rather presupposes that humans have free will. A falling rock may do terrible things to those in its path, but we do not blame the rock. Once released, it had no choice about the path it was taking. However, if I am bashing you and not the rock, I am to blame. I did have a choice about whether to harm you. We shall have to dig into the notion of freedom, but for now, we can leave it at the level of the ability to make choices and also the possibility of carrying out those choices. I may want to kill you, but if I am in chains, I am not free to carry out my desires.

What Is Original Sin?

If we are free, then we are free to do the right thing, to be good, or to do the wrong thing, to be bad or sinful or evil (Moser 2010). Later, we shall have more to say about right and wrong, good and evil—that is, about morality. For now, let us simply ask why on earth we would ever do wrong, be sinful? If we are the special creation of a good God, then surely, above all else, we ourselves are going to be good. We would never sin. To do so would imply that somehow God has opened the door to ill behavior, to evil. And that is simply impossible. The Christian has a ready answer to this paradox. It was better that God gave humans freedom than that he made humans like clockwork automata, going through the motions without any choice (Meister 2010). God himself is obviously free. What constraints could there be on him? Hence, here is a major instance of how and why we are made in his image, thinking now of the features we share with God (like rationality) rather than just the functional side, of being set over the creatures of creation as he is over us. But being free means that we could do wrong; otherwise, it is a chimerical form of freedom, like the freedom possessed by a hypnotized person who then goes through certain absurd motions on stage. Unfortunately, we humans did do wrong—Adam and Eve disobeyed the direct orders of God—and so sin entered the world. From then on, humans were in some sense tainted. We are born free but with a disposition to do bad things. Not all of the time obviously, but much of the time and many times. "Therefore, just as sin came into the world through one man, and death came through sin, and so death spread to all because all have sinned" (Rom 5:12 NRSV).

This, I should say, is the Augustinian line on original sin—one that obviously owes much to Paul—and it has been incredibly influential in the West, not the least because the great Reformers—Martin Luther (1483–1546) and John Calvin (1509–64)—were both much influenced by Augustine and bought into this perspective entirely. It reached its apotheosis in the famous sermon "Sinners in the Hands of an Angry God," preached in 1741 by the Puritan minister Jonathan Edwards (1703–58): "The God that holds you over the pit of hell, much as one holds a spider, or some loathsome insect over the fire, abhors you, and is dreadfully provoked: his wrath towards you burns like fire; he looks upon you as worthy of nothing else, but to be cast into the fire; he is of purer eyes than to bear to have you in his sight; you are ten thousand times more abominable in his eyes, than the most hateful venomous serpent is in ours" (Edwards 2005, 178). Later, we shall encounter other positions on the topic. One thing that this Augustinian viewpoint does is give an immediate reason for the coming of Jesus. He died on the cross for our salvation (Graham 2010). Through his suffering, he took on our sins and hence made possible our eternal happiness. "And now you have an extraordinary opportunity, a day wherein Christ has thrown the door of mercy wide open, and stands in calling and crying with a loud voice to poor sinners; a day wherein many are flocking to him, and pressing into the kingdom of God" (Edwards 2005, 183). It is not really necessary here to go into the many convoluted explanations as to how this all could be or what eternal salvation might mean exactly—although if I were asked, it would be having a new Mozart opera every night of the week and lots of fish and chips at the intermission. (Coupled, I might say, with never, ever, having to grade a student paper again.) The point is that Jesus suffered on our behalf, vicariously as one might say, and this opens up the prospect of never-ending bliss with our creator.

Of course, this all comes at a price. Although there have been universalists, believing that eventually all humans will be saved, generally there is agreement that even Jesus dying on the cross is not going to do much for the eternal prospects of Adolf Hitler. Humans can never be worthy of salvation, but we can do our bit. Unfortunately, there is considerable controversy over what doing our bit might entail. As noted in our history of atheism, the Catholic position has always stressed good works—helping the poor and so forth. These can never be enough, but they are a good start. "Come, you that are blessed by my Father, inherit the kingdom prepared for you from the foundation of the world; for I was hungry and you gave me food, I was thirsty and you gave me something to drink, I was a stranger and you welcomed me, I was naked and you gave me clothing, I was sick and you took care of me, I was in prison and you visited me" (Matt 25:34–36 NRSV). Protestants, on the other hand, as we also noted, have tended to put belief above works. Perhaps reflecting his own history of persecuting Christians and yet being saved, Paul put a big emphasis on commitment: "We hold that a person is justified by faith apart from works prescribed by the law" (Rom 3:28 NRSV). This does not mean that Protestants have no obligations to help others, but such help is done in gratitude for mercies shown rather than as a down payment on a ticket into heaven. (To think otherwise is known as the Pelagian heresy.)

Christians have had two thousand years to work on their religion. In discussing it, if one is not careful, one runs the risk of giving the unfortunate reader a lesson in the meaning of eternity. There are those who would say that that is precisely what is done by the great theologians, from Augustine through Aquinas and on to the greatest theologian of the twentieth century, Swiss thinker Karl Barth (1886–1968). Let us pause now, and turn from exposition of the claims to the matter of belief.

Chapter 3, Reading 3

Belief[1]

Why Should We Believe Any of the Claims of Christianity?

TRADITIONALLY, PHILOSOPHERS AND THEOLOGIANS make a distinction between beliefs based on revelation (revealed theology or religion) and beliefs based on reason and evidence (natural religion or theology). Starting with the former, this is all about the kinds of beliefs that you get from reading the Bible or from listening to the pope (at least when he is speaking authoritatively or ex cathedra). But more basic in a way than even the Bible or the pope is the ground of your conviction that these are the places to go for information. One speaks now of the medium or power of faith. So let's ask about that, for whatever the significance of faith in the ultimate scheme of things, Catholics and Protestants agree that it is central in the life of the Christian.

What Is Faith?

It has many dimensions. Some are psychological, involving commitment. We shall look at those later. Others, those on which we focus here, are more to do with belief, for essentially faith is the means by which we come to knowledge of God and of his ways, inasmuch as

1. Ruse 2015b. Reprinted with permission of Oxford University Press.

this is possible. In a sense, therefore, faith is like a telephone line to the divine, or to update things a bit, the theological equivalent of Skype. On September 18, 1998, Pope John Paul II, writing an encyclical letter to his bishops, *Fides et Ratio* (Faith and Reason), laid it out. "Underlying all the Church's thinking is the awareness that she is the bearer of a message which has its origin in God himself (cf. 2 Cor 4:1–2). The knowledge which the Church offers to man has its origin not in any speculation of her own, however sublime, but in the word of God which she has received in faith (cf. 1 Thess 2:13)." This is something given to humans by God, revealing the essential truths about God, his nature, his purpose for us, and how this is to be achieved. "As the source of love, God desires to make himself known; and the knowledge which the human being has of God perfects all that the human mind can know of the meaning of life" (John Paul II 1998, §7).

From the Protestant corner, John Calvin is pretty good on these matters. How do we know God exists? Because God put within all of us an awareness of his existence. "To prevent anyone from taking refuge in the pretense of ignorance, God himself has implanted in all men a certain understanding of his divine majesty. . . . Therefore, since from the beginning of the world there has been no region, no city, in short, no household, that could do without religion, there lies in this a tacit confession of a sense of deity inscribed in the hearts of all" (Calvin 1960, 43, 46). If faith is such a natural instinct put there by God, how then do you explain those without faith, like Baron d'Holbach or Bertrand Russell? Obviously, through original sin. Their vile nature leads them to distort or ignore the avenue to God. For the rest of us, though, faith is self-sufficient.

What about Reason and Evidence?

Belief in God through faith is what the Calvinist philosopher Alvin Plantinga (b. 1932), following the Dutch theologian Herman Bavinck (1854–1921), calls "properly basic": "The believer is entirely within his intellectual rights in believing as he does even if he doesn't know of any good theistic argument (deductive or inductive), even if he believes there isn't any such argument, and even if in fact no such argument exists" (Plantinga 1981, 42; see also Bavinck 1951). What is perhaps not the contradictory but more the complement to faith? It is knowledge obtained through reason and the senses, in the area we are discussing as the subject matter of natural theology. Aquinas puts faith somewhere between knowledge, where we deal with what he characterizes as "something which is self evidently true" (or derivable from self-evident truth), and opinion (Aquinas 1920–22, 2.1.1). This rather implies that faith is not up to the same high standard as knowledge, but such an inference is a bit misleading. Famously, Aquinas believed that we can get knowledge of God and his attributes through a number of arguments (of which more in a moment), but he didn't believe that this made faith superfluous. It was much more a matter of showing that faith could be given a rational basis. Anselm put the point well when he said, *Credo ut intelligam* (I believe so that I may understand) and when he spoke of *fides quaerens intellectum* (faith seeking understanding) (see Anselm 2008). (No prizes for guessing that these phrases go back to Augustine.) The point is that no one thought that the arguments were uniquely going to lead you to belief or that they would be the basis of one's belief. It was more that they fleshed out the beliefs given through faith.

Aren't we letting the believer get away with too much here? Faith trumps everything, and reason and evidence just play catch-up? So much for being made in God's image! Hardly. Whatever the actual position of Aquinas on the faith and reason issue, he and his fellow philosophers did the heavy lifting on the God and reason and evidence front. And this is very much the Catholic tradition to this day—a tradition that has stimulated responses part in rejection and part in parallel, for Protestants have been somewhat divided on the issue of natural theology. Perhaps because of the emphasis on justification by faith, there has long been a thread of Protestant suspicion about the whole appeal to reason and evidence. Influenced by Calvin, the most noted recent representative being Karl Barth, these skeptics reject natural theology with distaste and scorn. They go so far as to say that natural theology is not just wrong, but somehow demeaning of God, because it elevates reason above him.

The Danish writer Søren Kierkegaard (1813–55) argued that faith backed with reason is faith emasculated. Faith is genuine only when it carries a whiff of risk, a "leap into the absurd" (Kierkegaard 1944). Countering this, however, for many Protestants, natural theology is a vital part of their belief system. Since the Reformation, for historical as much as theological reasons, this has been very much an English tradition (Ruse 2003d). Elizabeth the First (1533–1603), known still as "Good Queen Bess," was faced with a division between the would-be Catholics (backed by powerful Spain), focusing on the authority of the church, and the Calvinist Protestants just back from exile on the Continent during the short but awful reign of Catholic "Bloody Mary" (1516–58). These latter focused on the unadorned word of God as found in the Bible. The queen and her church forged a path down the middle—the Elizabethan Compromise—and one of its distinctive features was an emphasis on natural theology, something that obviously meshed nicely with the growing importance of science in that period.

The great empiricist philosopher John Locke (1632–1704), writing in the years after the English Civil War and concerned about all of the new rival sects that showed undue "enthusiasm," made the definitive case for reason measuring faith: "Nothing that is contrary to, and inconsistent with, the clear and self-evident dictates of reason, has a right to be urged or assented to as a matter of faith, wherein reason hath nothing to do" (Locke 1959, 2.28.10). This opened the way for a succession of English philosophers and theologians down to the present to proclaim the glories of their Lord through reason and evidence. Best known was Archdeacon William Paley (1743–1805), who wrote a series of texts used not just by the young Charles Darwin but by undergraduates at his alma mater (Cambridge) even into the twentieth century. What are the fruits of all of this effort? What are the various arguments that have been made for the existence of God? Let us find out. Brian Davies gives a good overview of the arguments (2004); Russell Re Manning puts natural theology in historical and theological context (2013).

What Is the Ontological Argument?

The ontological argument is not the earliest of the arguments—that honor probably goes to the argument from design, to be found in the thinking of Socrates—but it is the one that usually comes first. It strikes almost everyone as irritating, altogether too clever by half, and frankly not very convincing at all. After that, people divide. For most folk, that is enough. Richard Dawkins is scathing, not just about the argument but about the foolishness of those who take it seriously (2006). It is not worth wasting the time. For a minority, the irritation is more a matter of knowing that something important is being said, of knowing that it is almost certainly not the conclusion to which the argument points and that it is going to be the very devil to dig out precisely what is of value. I won't say that belonging to this second group is a necessary and sufficient condition of being a philosopher, but it helps. For all that he rejects it, it is interesting to note the respect with which Dennett treats the argument, as opposed to Dawkins's contemptuous dismissal.

By a reductio ad absurdum argument, Anselm sets out to show that this cannot be. We start with a definition: God is that than which no greater can be thought. The question now is whether God exists. Deny that he does and assert that he is just a mental fiction. Then think of another entity, like the first entity, but with the attribute of existence. Clearly, this second entity is better than the first entity. But this cannot be because the first entity is greater than anything else. Hence we have a contradiction, so it must be the case that God exists! The second version comes some several hundred years later from Descartes in his *Meditations*. "Certainly, the idea of God, or a supremely perfect being, is one that I find within me just as surely as the idea of any shape or number. And my understanding that it belongs to his nature that he always exists is no less clear and distinct than is the case when I prove of any shape or number that some property belongs to its nature" (Descartes 1964, 63). What's he saying here? Basically that I have the idea of God as a being with all perfections. Existence is a perfection. Therefore, God exists!

What Is the Causal or Cosmological Argument?

What Anselm and Descartes are saying is that God doesn't just exist. By his very nature, he had to exist. This point is crucial to understanding the next argument to which we turn, the causal or cosmological argument, one that goes back to the Greeks, Plato and Aristotle, and played a major role in medieval thought, particularly that of Saint Thomas Aquinas (Ruse 2003d). Again we find an argument with somewhat different forms, but for our purposes, it is enough to focus on the central inference. Everything has a cause. There must therefore be a cause of the world. This is, or call this, God. It is either that or we cannot break or stop the chain of causation, which seems to imply that it is infinite, which simply doesn't make sense. "But if the chain were to go back infinitely, there would be no first cause, and thus no ultimate effect, nor middle causes, which is admittedly false. Hence we must presuppose some first efficient cause—which all call God" (Aquinas 1920–22, 1a.2–3).

Proceed with caution here. Prima facie, you can drive a horse and carriage through the cosmological argument. If everything has a cause, then what caused God? What caused that "first efficient cause"? Actually, if we are going to take science seriously, we know that what caused everything was the big bang. This may be many things, but it is certainly not God. No one is about to worship an explosion. However, if you start to dig into Aquinas's writings, as well as those of others who have supported the argument, you soon see that they are ahead of you here. For a start, they are not really thinking of a cause that just so happens to come at the beginning of a chain. Aquinas, as a Christian, thought that revelation shows that the universe has a beginning. But as a good Aristotelian, he was quite prepared on philosophical grounds to accept the possibility of an infinitely old universe.

It is more a cause in the sense of what keeps everything going, sort of orthogonal to the chain of time. It is the answer to what has been called the fundamental question of philosophy: "Why is there something rather than nothing?" So all of the talk about the big bang is irrelevant. For a follow-up, everybody knew that you had to stop the chain with God, or you are stuck with an infinite chain of causes. The claim is that God is something that doesn't need a cause. He stops the chain dead. . . . It is here that the insight of the ontological argument is crucial. Clearly, God needs no cause because he is in some sense necessary. He doesn't need a cause any more than two plus two equals four needs a cause. He just is. In the language of the philosophers, God has a mode of existence that is entire unto himself, to which they give the name *aseity* (Hick 1961).

What Is the Teleological Argument?

Better known as the argument from design, this argument focuses on Aristotelian final causes, the ends that seem to control and explain so many features, especially so many biological features. The existence and nature of the hand, for instance, is clearly a matter of embryology—first, there were buds, and then there were fingers and thumbs—but it does not seem to be simply a matter of embryology. To understand the hand fully, we need to know what it is for—what is it intended to do? What is its purpose? To which the answer comes that it serves the end of grasping, as the nose serves the end of smelling, and the teeth of biting and chewing. Note that in a sense there is a forward reference—the teeth exist to chew, even if I live off fluids all of my life—but there is nothing spooky here, with little men in the future pulling strings affecting my present. Intention is the keyword, meaning that this is what hands and noses and teeth are for, even if they never actually get used. Brushed aside is Lamy's suggestion that features might first appear without design and only then find their function. How could something as intricate as the eye, something that is so good for seeing, just appear by chance?

Intention implies a consciousness. I intend to write three thousand words today. My computer and its software intend nothing at all. This is the key to the teleological argument. Final causes seem to imply that someone intended them to work, someone designed the system, and in the case of the physical world, that someone was God. Paley's exposition is the classic. The eye is like a telescope. Telescopes have telescope designers and makers. Therefore, the eye must have a designer and maker. God, the supreme optician. Moreover, you can make an inference about the nature of the designer. How can you doubt God's intelligence and concern? Denial of this kind of reasoning? "This is atheism."

What Is the Anthropic Principle?

An argument that in respects seems to be a corollary of the argument from design is the argument from law. Here one suggests that the very existence of the laws of nature points to the deity, and if one can bring in the nature of these laws, one has a correspondingly stronger argument. One enthusiast for this position starts with the (somewhat tenuous) biblical evidence: "As long as the earth endures, seedtime and harvest, cold and heat, summer and winter, day and night, shall not cease" (Gen 8:22). He then makes the key abduction: "Law implies a lawgiver. Someone must think the law and enforce it, if it is to be effective." He adds an elaboration about the nature of law and hence of the lawgiver: "Scientific laws, especially 'deep' laws, are beautiful. Scientists have long sifted through possible hypotheses and models partly on the basis of the criteria of beauty and simplicity. For example, Newton's law of gravitation and Maxwell's laws of electromagnetism are mathematically simple and beautiful. And scientists clearly expect new laws, as well as the old ones, to show beauty and simplicity. Why? The beauty of scientific laws shows the beauty of God himself" (Poythress 2010, 20).

What Is the Moral Argument?

The moral argument for the existence of God was made popular by the great eighteenth-century German philosopher Immanuel Kant (1949), and it appeals to the moral world. . . . Humans are social animals, and much of our thought and behavior is dedicated to getting along with our fellows—finding partners, having children, making friends and enemies, getting a job, playing a role in the civic arena, and much, much more. A lot of what we do starts with some fairly basic or gut emotions. I work hard to send my kids to college because I love them. I pay my taxes because I am scared of what would happen if I didn't. I join the army during a war because I love my country and I don't want to be judged a coward by my friends. But there is more to all of this than just emotions. There is reasoning—I start a plan to pay college tuition, I work out exactly how much I owe the government, I decide which branch of the military I want to join—and entering into this reasoning there is morality, our sense of right and wrong.

Of course, there is a lot of complexity about the exact nature of morality—the calls of obligation—and sometimes over such things as abortion, there are major differences between people. But whatever the exact nature of moral dictates (philosophers call this part of the business substantive or normative ethics), apart from psychopaths and like misfits, no one denies the existence of morality. And this point is not denied by the fact that from Friedrich Nietzsche on, many Continental thinkers have been much given to pronouncing on its end. It does not take much digging to show that they are not denying morality as such, but more the hypocrisy that infects so many contemporary claims about moral behavior. Most of the naysayers would be very uncomfortable roommates, because every little peccadillo would be subjected to endless critical analysis.

Is Morality Objective?

The big question, then, is where does morality come from and what is its status? (Philosophers call this part of the business metaethics because it is about morality, substantive morality, that is.) It seems to be a generally accepted aspect of what we might call the phenomenology of morality that it doesn't just seem to be a matter of opinion or of how people may sometimes behave. . . . The point being made is that somehow morality seems to be objective (Mackie 1977; cf. Ruse 1986b). "Rape is wrong" is not up for grabs in the way that "I don't much care for broccoli" and "I'd rather be a philosopher than a sociologist" are matters of taste—although, as it happens, even the most atheistic of philosophers think of their calling as divine destiny. "Rape is wrong" is binding on all of us. . . . Morality is a bit like the laws of nature in that it exists outside me and I am subject to it. (Of course, it isn't always the case that I follow it, but that is another matter.) Or in a stronger way, morality is a bit like the laws of mathematics, because morality seems to be about the world, not of it. David Hume made this point very clearly. The world is about matters of empirical fact. Morality is about matters of obligation. . . . In the lingo of philosophers, apparently one ought to be a moral realist, meaning one ought to believe that moral claims (substantive ethics) really exist as things in their own right. Morality is binding on us humans. And yet, morality is not simply a matter of the way that (empirical) things are. It has its own being.

Some people have thought—Plato obviously—that the analogy between mathematics and morality strikes deep. They both refer to entities in the supersensible world of pure rationality. Unfortunately, even if this is so, mathematics doesn't have the sense of obligation that morality has. So where does morality come from, and what is the authority behind its binding nature? The obvious answer is that it is to be found in the will of God. Morality is what God wants of us. Conversely, this gives us an inference-to-the-best-explanation proof for the existence of God. Morality must have a cause or foundation. It is not to be found in this world. It is not to be found in the world of the Forms. Hence, the most reasonable explanation is that it is the will of God. Hence, God must exist.

John Henry Newman in his *A Grammar of Assent* makes the case powerfully: "If, as is the case, we feel responsibility, are ashamed, are frightened, at transgressing the voice of conscience, this implies that there is one to whom we are responsible, before whom we are ashamed, whose claims upon us we fear" (Newman 1870, 109). The only being who would fit this bill is God. Anything less will not do. "These feelings in us are such as require for their exciting cause an intelligent being; we are not affectionate towards a stone; we do not feel shame before a horse or dog; we have no remorse or compunction on breaking merely human law; yet so it is, conscience excites all these painful emotions: confusion, foreboding, self-condemnation; and on the other hand it sheds upon us a deep peace, a sense of security, a resignation and a hope, which there is no sensible, no earthly, object to elicit" (110).

Chapter 3, Reading 4

The Unraveling of Belief[1]

TODAY, LIFE EXPECTANCY IN England is around eighty years—a bit more for women than for men, a bit more in the more prosperous south of the country than in the north. If you go back five hundred years or so, however, life expectancy is less than half of that—around thirty-five years, to be exact. That, however, is a little misleading, for up to a third of children didn't make it past the first year, and a tenth or more of women died in childbirth. If you got past these hurdles, you might make it to around forty-five. Of course, as today, some people lived much longer, but they were the exceptions. There is no big mystery here. As the seventeenth-century English philosopher Thomas Hobbes said, life tended to be "solitary, poor, nasty, brutish, and short" (Hobbes, *Leviathan*, ch. 13, as quoted in Ruse 2019b). He said it in the context of the Civil War, but it applied in peacetime too.... When Christianity was functioning properly, life made a lot of sense. For here and now and for the future.

The Story Comes Apart

What went wrong, if that is the right term? It is no great secret. The three *r*'s: the Renaissance, the Reformation, and the (Scientific) Revolution. Before we attempt an integrated picture, let's take them in turn.

---

1. Ruse 2019c. Reprinted with permission of Oxford University Press.

The Renaissance, starting at least a century before the full activity of the sixteenth century, saw an invigorating of many aspects of human life and culture—in the arts, in music, in politics, and more. For us, all important is the discovering afresh the glories of Greece and Rome, particularly in their extant writings, something made more accessible by a parallel interest in the languages of Greek and Latin and the desire to know them and use them effectively. The paradigm, of course, was Erasmus of Rotterdam, most particularly his preparing of high-quality editions and translations of the Bible. . . . The Reformation meant many things. What it certainly did not mean was an end to Christianity and the sense of purpose. Anything but! In many, many respects, people like Martin Luther and John Calvin were more sincere than the Roman prelates and priests they were displacing. There was a whole new emphasis on the truth of the Bible, rather than on the authority of the church. Although not every reading was literal—Luther, for instance, was not at all keen on the Epistle of James, because he thought there was too much emphasis on good deeds—there was a drive to a simpler, more straightforward understanding, unencumbered by all the metaphorical interpretation that Scripture had acquired. Above all, there was a hatred—that is not too strong a term—of the idea that good works at all influence or determine our future fate. In a return to Augustine, there was a total emphasis on faith and grace. Accept Jesus, and you have a chance of being in. Don't accept Jesus, and you are certainly out. Even more certainly out was purgatory. Apart from its lack of biblical support, it was seen by the Reformers as a trick to con the faithful into generous support of the church. With the emphasis on faith and faith alone—*sola fidelis*—paying for Masses and the like was simply a waste of time.

Showing how artificial it is to separate the Renaissance from the Reformation, the emphasis on the Bible and de-emphasis of the church meant that laypeople needed direct access to the Holy Word. Thus, people like Luther set about offering editions of the Bible in the vernacular, and to do this, the groundbreaking work of people like Erasmus—who never left the mother church—was essential. Printing helped, too, of course. No longer were copies of the Bible handwritten with the consequent opportunities for mistakes. As with the Renaissance, though, there was a downside, if not at first, then later. In the West, no longer did you have just one version of Christianity—there was always the Eastern version—but now you had Catholic and Protestant and, before long, Protestant and Protestant and Protestant. Did the bread and wine still have Jesus present, as is claimed by Lutheran consubstantiation, or is it just symbolic, as it was for Calvin? Can you take up arms, as is insisted by Lutheran and Calvinist, or must you be a pacifist, as is insisted by the Anabaptists? With so many options on offer, thoughts of jettisoning it all started to lurk. At least, thoughts of a world without Jesus.

There was one major striking thing that happened during the Scientific Revolution, and turning to it guides us toward a unifying explanation of what was happening in the sixteenth and seventeenth centuries (Ruse 2017i). Aristotle had divided causation into a number of categories, the most important of which were what we would call "efficient causes" and what even Darwin—especially Darwin—referred to as "final causes." On the one hand, we have the physical (or mental) things that make other things happen. I hammer on a nail, and there is sound, and the nail is firmly embedded in the wood. On the other hand, we have purposes or intentions. I hammer the nail in order to build a house, an end that I value. Note that these two kinds of causes, for all that one refers only to the past and the other brings in the future, are not entirely symmetrical. . . . A big question now was whether final causes were legitimate or should be done away with. Are they a sign of intellectual weakness? Increasingly, the feeling was that this was true in the physical sciences at least. Francis Bacon likened final causes to vestal virgins, decorative but sterile (Blane 1819, 69). The great French thinker René Descartes (1596–1650) argued that ontologically, God created two basic substances: *res extensa* and *res cogitans*, things extended and things thinking (Descartes 2014, 2.001). The mark of the material world is that it has spatial dimensions. It is completely inert, unthinking, basic.

The root metaphor changed from that of an organism to that of a machine. It didn't happen overnight, and we find some embracing both metaphors. . . . Increasingly, however, the machine metaphor ruled supreme. The philosopher-chemist Robert Boyle is always good on these things. Using the same analogy as that of Kepler, he argued that the world is "like a rare clock, such as may be that at Strasbourg, where all things are so skillfully contrived that the engine being once set a-moving, all things proceed according to the artificer's first design, and the motions of the little statues that at such hours perform these or those motions do not require (like those of puppets) the peculiar interposing of the artificer or any intelligent agent employed by him, but perform their functions on particular occasions by virtue of the general and primitive contrivance of the whole engine" (Boyle [1688] 1996, 5:394–95). Likewise, across the

Channel, Descartes chipped in that we should think of our body "as a machine created by the hand of God, and in consequence incomparably better designed and with more admirable movements than any machine that can be invented by man" (1964, 41).

At a metaphysical level, a machine is just as much into final cause—what came to be called teleology—as an organism. The vacuum is for cleaning the house (down the road) no less than an eye is for seeing (down the road). The crucial difference, and Hall sees this (1983), is that the teleology of the organism is internal—often associated with Aristotle—meaning that it does not appeal outside the organism for value. The leaf is of value to the tree, whether or not humans exist, whether or not God exists. The teleology of the machine is external—often associated with Plato—in that it gets its value only from humans or God. A vacuum is just steel and plastic until someone fashions it and uses it. And this means the possibility is opened of sloughing off the designer. If you don't find final cause talk very useful, then why bother at all with a designer or creator, at least in one's science? Not just spirits but God himself gets kicked out. In the words of one of the most eminent historians of the Scientific Revolution, God became "a retired engineer" (Dijksterhuis 1961, 491).

"An organized being is thus not a mere machine, for that has only a motive power, while the organized being possesses in itself a formative power, and indeed one that it communicates to the matter, which does not have it (it organizes the latter): thus it has a self-propagating formative power, which cannot be explained through the capacity for movement alone (that is, mechanism)" (Kant 2000, 246). And yet they are simply not acceptable in proper science. Uneasily, Kant concluded that they must have only a heuristic role, a point he conceded in a paradigmatically convoluted paragraph:

> The concept of a thing as in itself a natural end is therefore not a constitutive concept of the understanding or of reason, but it can still be a regulative concept for the reflecting power of judgment, for guiding research into objects of this kind and thinking over their highest ground in accordance with a remote analogy with our own causality in accordance with ends; not, of course, for the sake of knowledge of nature or of its original ground, but rather for the sake of the very same practical faculty of reason in us in analogy with which we consider the cause of that purposiveness. (Kant 2000, 247)

This didn't stop Kant from being rather nasty about biology, concluding that it would never live up to the physical sciences: "We can boldly say that it would be absurd for humans even to make such an attempt or to hope that there may yet arise a Newton who could make comprehensible even the generation of a blade of grass according to natural laws that no intention has ordered; rather, we must absolutely deny this insight to human beings" (Kant 2000, 271).

That, of course, was a challenge far too good to be ignored by an ambitious new researcher in the life sciences. A few decades later—in the 1830s—a Cambridge-educated young man by the name of Charles Darwin, fresh off a five-year voyage as ship's naturalist circumnavigating the globe on *HMS Beagle*, went right at the problem and showed that in biology, he rightfully could claim the mantle of Newton (Browne 1995, Ruse 2018b). Put him in context before introducing him. There were very few atheists at the beginning of the nineteenth century and not many agnostics. Deism of a stronger or weaker form was the general choice of those who had real doubts about the truth of Christianity.

The big problem was final cause. There was no convincing naturalistic explanation. What about evolution? In the *Critique of the Power of Judgment* ([1790] 2000), Kant considered the possibility seriously; the analogies (homologies) between unrelated organisms—humans, apes, whales, even bats and birds—were strong evidence in its favor. Final cause was the block. Nevertheless, by the end of the eighteenth century, people were becoming evolutionists. Darwin's grandfather, Erasmus Darwin, was a well-known example. Although expectedly there were many detractors, he had significant influence; it may well be the case that the German translation tipped the aged Kant. Although adaptations still could not be explained—the (false idea of the) inheritance of acquired characters (later called Lamarckism) went some way but by no means far enough—since organic evolution was seen as an epiphenomenon of ideas of social and cultural progress, that for most was reason enough. This did not mean that people jettisoned God. Rather . . . there was continued movement in the direction of deism, a God who works through unbroken law. Here was a case where, apparently, science supported religion. Evolution is as much a support for deism, with its god of law, as it was later taken to be a refutation of theism, with its god of miracles.

What was needed was for a professional scientist, probably British and soaked in natural theology, including—especially including—the central status of design, to get captivated by and converted to an evolutionary perspective. Such was Charles Darwin. On the *Beagle* voyage—from 1831 to 1836—he first lost his Christian faith, becoming a deist (because he could no longer believe in miracles), and then edged toward evolution, partly because of the fossil evidence but most strongly because of the idiosyncrasies of organic geological distributions.... Soon after his return to England—spring of 1837—Charles Darwin became an evolutionist, and then he searched frenetically for eighteen months to find a cause, the kind of equivalent of Newtonian gravitation. He soon spotted that selection might be the key. It was in the world of the farmyard or the bird or dog fancier, and he even came across its extension (in an 1809 pamphlet by a well-known breeder) into nature.

> A severe winter, or a scarcity of food, by destroying the weak or unhealthy, has all the good effects of the most skillful selection. In cold and barren countries no animal can live to the age of maturity, but those who have strong constitutions; the weak and the unhealthy do not live to propagate their infirmities, as is too often the case with our domestic animals. To this I attribute the peculiar hardiness of the horses, cattle, and sheep, bred in mountainous countries, more than their having been inured to the severity of climate. (Sebright 1809, 106)

Darwin took careful note of this passage, and even though he could not quite see the full import grasped that if something like this went on long enough, we would get full-blooded species. In a private notebook, he wrote of "excellent observations of sickly offspring being cut off so that not propagated by nature.—Whole art of making varieties may be inferred from facts stated" (C. Darwin 1987, C 133). Darwin had little problem in generalizing back again and going from the struggle to its consequences. Nigh uniquely in the history of science, we can document the very moment of insight. This is from a private journal (which explains its somewhat staccato style):

> Even a few years plenty, makes population in Men increase & an ordinary crop causes a dearth. Take Europe on an average every species must have same number killed year with year by hawks, by cold &c.—even one species of hawk decreasing in number must affect instantaneously all the rest.—The final cause of all this wedging, must be to sort out proper structure, & adapt it to changes.—to do that for form, which Malthus shows is the final effect (by means however of volition) of this populousness on the energy of man. One may say there is a force like a hundred

thousand wedges trying force into every kind of adapted structure into the gaps of in the economy of nature, or rather forming gaps by thrusting out weaker ones. (C. Darwin 1987, D 135)

Darwin had gotten his cause. What he had to do now was to fold it all into a coherent evolutionary theory, which he did over the next five years or so. Then he sat on his ideas for the next years until nigh the end of the 1850s. More later for possible reasons why. Finally, spurred by the arrival of an essay from a younger naturalist—Alfred Russel Wallace—Darwin was moved to action, and *On the Origin of Species by Means of Natural Selection, or the Preservation of Favoured Races in the Struggle for Life* appeared at the end of 1859. Darwin's style is casual, but the theory is carefully and professionally constructed (Ruse 1999b, Ruse 2008a, Richards and Ruse 2016). The aim was to show that natural selection, like Newtonian gravitational force, is what the philosophers called a *vera causa*, a true cause. There is some ambiguity about the precise meaning of this term, with some of an empiricist bent, like the physicist John F. W. Herschel (1830), insisting that one had to find analogies with everyday experience. Others of a rationalist bent, like the scientist and then philosopher and historian of science William Whewell, insisted that one had to show the cause explained in many different areas. This form of argument was what Whewell called a "consilience of inductions" (Whewell 1840). No need for direct experience.

Darwin covered both options! He opened with a discussion of human selection of organisms for our ends—farm animals and crops, from sheep to turnips, and organisms shaped by us for pleasure, like birds and dogs. He made much of the fact that pigeons have such variety and yet clearly all come from common stock. Then, having postulated that new variations are constantly appearing in populations—not uncaused but undirected—he was ready for his key inferences. These were put in (quasi- or proto-) deductive form. As the philosophers pointed out constantly, this was the form of the gravitational theory of Newton. First, the Malthusian element to a struggle for existence:

> A struggle for existence inevitably follows from the high rate at which all organic beings tend to increase. Every being, which during its natural lifetime produces several eggs or seeds, must suffer destruction during some period of its life, and during some season or occasional year, otherwise, on the principle of geometrical increase, its numbers would quickly become so inordinately great that no country could support the product.

> Hence, as more individuals are produced than can possibly survive, there must in every case be a struggle for existence, either one individual with another of the same species, or with the individuals of distinct species, or with the physical conditions of life. (C. Darwin 1859, 63–64)

Then, second, to natural selection:

> Let it be borne in mind how infinitely complex and closefitting are the mutual relations of all organic beings to each other and to their physical conditions of life. Can it, then, be thought improbable, seeing that variations useful to man have undoubtedly occurred, that other variations useful in some way to each being in the great and complex battle of life, should sometimes occur in the course of thousands of generations? If such do occur, can we doubt (remembering that many more individuals are born than can possibly survive) that individuals having any advantage, however slight, over others, would have the best chance of surviving and of procreating their kind? On the other hand, we may feel sure that any variation in the least degree injurious would be rigidly destroyed. This preservation of favourable variations and the rejection of injurious variations, I call Natural Selection. (C. Darwin 1859, 80–81)

The all-important point is that this cause—natural selection—points to adaptation, to final cause. The eye is created as it is in order to see. The flower to attract pollinators. The fangs of the snake to kill. The instincts of the nest-building bird to promote and continue life. It is not just change but change of a particular cause:

> Under nature, the slightest difference of structure or constitution may well turn the nicely-balanced scale in the struggle for life, and so be preserved. How fleeting are the wishes and efforts of man! how short his time! and consequently how poor will his products be, compared with those accumulated by nature during whole geological periods. Can we wonder, then, that nature's productions should be far "truer" in character than man's productions; that they should be infinitely better adapted to the most complex conditions of life, and should plainly bear the stamp of far higher workmanship? (C. Darwin 1859, 83–84)

"Of far higher workmanship"?! Strong echoes here of Darwin having had Paley on design drilled into him at Cambridge. Natural selection wasn't just any old mechanism.

Now, having first argued that this final cause force of natural selection could lead to the tree of life, Darwin moved into the final phase of his theory, as he argued that he could explain phenomena right across the life sciences. Social behavior, the fossil record (paleontology), geographical distributions (biogeography), anatomy and morphology, systematics, embryology, vestigial organs. Why is there a progressive fossil record? Evolution through natural selection. Why the Linnean system? Because it reflects the tree of life. Why are there such similarities between the organisms of very different adult forms? Because selection only works on the adults. This all done, Darwin was ready for his final famous passage:

> It is interesting to contemplate an entangled bank, clothed with many plants of many kinds, with birds singing on the bushes, with various insects flitting about, and with worms crawling through the damp earth, and to reflect that these elaborately constructed forms, so different from each other, and dependent on each other in so complex a manner, have all been produced by laws acting around us. . . . Thus, from the war of nature, from famine and death, the most exalted object which we are capable of conceiving, namely, the production of the higher animals, directly follows. There is grandeur in this view of life, with its several powers, having been originally breathed into a few forms or into one; and that, whilst this planet has gone cycling on according to the fixed law of gravity, from so simple a beginning endless forms most beautiful and most wonderful have been, and are being, evolved. (C. Darwin 1859, 489–90)

In the *Origin*, Darwin said virtually nothing about our own species, Homo sapiens. We know that he always included us in the story. Indeed, in his private notebooks, the first clue that we have that Darwin is now working with natural selection is an application to humans, and not just to our bodies but to our mental abilities. In his *Autobiography*, written toward the end of his life, Darwin reaffirmed this belief: "I am inclined to agree with Francis Galton [Darwin's half cousin] in believing that education and environment produce only a small effect on the mind of any one, and that most of our qualities are innate" (C. Darwin 1958, 43). From the first to the last, Darwin was (in today's terms) a human sociobiologist or evolutionary psychologist. In the *Origin*, however, Darwin wanted to get his main theory into the public domain. So, not to appear cowardly, he simply said that "light will be thrown on man and his history" (C. Darwin

1859, 488). No one was fooled, and at once Darwin became known as the author of the "monkey theory" or the "gorilla theory."

Probably, if he had been a free agent, he would have stayed on the sidelines of the human evolution debate—Darwin was not fond of controversy—but in the 1860s, Wallace turned to spiritualism, arguing that only supernatural forces could explain human evolution. Horrified, Darwin again put pen to paper, and in 1871, he published *The Descent of Man, and Selection in Relation to Sex*. Much of the early part of the book is predictable, as Darwin argued that his theory applies to us. He spent considerable time on the evolution of morality, arguing that cooperation is a major positive adaptation for human beings:

> It must not be forgotten that although a high standard of morality gives but a slight or no advantage to each individual man and his children over the other men of the same tribe, yet that an advancement in the standard of morality and an increase in the number of well-endowed men will certainly give an immense advantage to one tribe over another. There can be no doubt that a tribe including many members who, from possessing in a high degree the spirit of patriotism, fidelity, obedience, courage, and sympathy, were always ready to give aid to each other and to sacrifice themselves for the common good, would be victorious over most other tribes; and this would be natural selection. At all times throughout the world tribes have supplanted other tribes; and as morality is one element in their success, the standard of morality and the number of well-endowed men will thus everywhere tend to rise and increase. (C. Darwin 1871, 1:166)

Evolution after Darwin

A popular account of the fate of Darwin's theory has an almost immediate conversion of scientists to evolution and an equally great reluctance to accept natural selection (Bowler 1988, 2013). This mechanism came into its own only in the twentieth century, when it had the backing of Mendelian—later molecular—genetics. Even now, there are serious concerns. Well, the first part of this story is true, at least. Almost overnight, like the emperor's new clothes, when Darwin said something evolves, people said they had known that all along. The second part of the story really isn't true, although some things—other than a desperate religious or quasi-religious desire to avoid embracing natural selection—do give a cover for the myth. It is true that for the rest of the nineteenth century, many active biologists downgraded selection, mainly because it didn't speak to the kinds of issues with which they were concerned. No one could see selection at work on the dinosaurs, for instance. When you are doing systematics, as Darwin himself discovered when doing a massive study of the barnacles, adaptation gets in the way! You are looking for homologies to discern underlying relationships. However, expectedly in the areas where selection could help—especially when dealing with fast-breeding organisms like insects—it was prized (Kimler and Ruse 2013; Ruse 2019a, 2018b). Most notably, shortly after the *Origin* appeared, a sometime traveling companion of Wallace, Henry Walter Bates, proposed a selection-based theory of mimicry, arguing that some butterflies and moths—nonpoisonous—mimic other butterflies and moths—poisonous. The biggest predators are birds, and they learn to avoid the poisonous forms. The nonpoisonous forms slip in under cover, as it were. They have to be uncommon—as they are—otherwise the birds would ferret out the deceit and all would fail. A few years later, a German-Argentinian naturalist, Fritz Müller, found another form of mimicry, which today bears his name, as Bates's discovery bears his name. Much insightful selection-based work was done on industrial melanism—where insects became darker to be better camouflaged against trees that, thanks to the Industrial Revolution, were already becoming darker.

When Mendelism was discovered at the beginning of the new century, it was indeed extended and developed (Provine 1971, Ruse 1996b). This was thanks particularly to Thomas Hunt Morgan and his students at Columbia University, who in the second decade of the century produced the "classical theory of the gene." This, as we still believe today, located the units of heredity in real time and space on the chromosomes in the nuclei of cells. The theory was generalized to populations—random variation (mutation) being acted upon by selection—and so was the beginning of modern evolutionary thinking, labeled neo-Darwinism (or just Darwinism) in Britain and the synthetic theory of evolution in America.

What of natural selection? No one ever claimed it was the only force for change. Darwin was ever a Lamarckian, although now, with the coming of Mendelism, that was long gone. Fisher trained as a physicist, and he always regarded populations like gases, with molecules/genes dashing around in the cloud and selection (like pressure) working on the whole. Wright worked for years with the US Department of Agriculture on breeding shorthorn cattle. He tended to think in terms of small, isolated groups that change and then come back into the big population with new variations to spread.

The Loss of Meaning

"What is Darwinism?" asked Charles Hodge, principal of Princeton Theological Seminary and leading Calvinist theologian of the mid-nineteenth century. His conclusion: "It is atheism" (Hodge 1874, 177). This is not true, actually. At least, it is not true in the eyes of many Christians and more than a few nonbelievers, including myself. Some people of faith, indeed, have embraced it with enthusiasm. A good example is the late-nineteenth-century High Anglican Aubrey Moore. He wrote:

> Science had pushed the deist's God farther and farther away, and at the moment when it seemed as if He would be thrust out altogether, Darwinism appeared, and, under the guise of a foe, did the work of a friend. It has conferred upon philosophy and religion an inestimable benefit, by showing us that we must choose between two alternatives. Either God is everywhere present in nature, or He is nowhere. He cannot be here, and not there. He cannot delegate his power to demigods called "second causes." In nature everything must be His work or nothing. We must frankly return to the Christian view of direct Divine agency, the immanence of Divine power from end to end, the belief in a God in Whom not only we, but all things have their being, or we must banish him altogether. (A. Moore 1890, 99–100)

Both believers and nonbelievers embraced evolutionary thinking, Darwinian thinking even. This rather suggests, which is true, that the main reason driving people to nonbelief was not Darwinism as such (Benn 1906). Biography after biography of nineteenth-century figures tell us that the real reasons for leaving Christianity were almost always theological (Budd 1977).

On the one hand, there was the corroding force of German higher criticism, arguing that the Bible is just a collection of fallible old human-inspired and written documents. Stories like the resurrection were not literally true but were fairy tales concocted to meet eschatological expectations. At the same time, there was a feeling that some of the events and interpretations were unacceptable to the point of total rejection. To think that Jesus turned a few loaves and fishes into food for thousands is to turn him into an employee of a grocery store—Whole Foods in America or Sainsbury's in England. Even worse were claims about substitutionary atonement. The very idea that someone has to suffer dreadfully to buy off God is in itself repellent. To think that Jesus could do it on my behalf is simply grotesque. No morality here.

On the other hand, people like Hume were having an effect. Natural theology was falling out of favor. The Danish theologian Søren Kierkegaard argued that Christianity demands a leap of faith, and proving the existence of God rather undercuts this (Kierkegaard 1944). The greatest British theologian of the nineteenth century, John Henry Newman, whose spiritual path took him from an Anglican evangelical childhood, through the very high church—the Oxford Movement—and on over to Rome, where he ended as a cardinal and prince of the church, was very wary of design arguments. He wrote: "I believe in design because I believe in God; not in a God because I see design." He continued: "Design teaches me power, skill and goodness—not sanctity, not mercy, not a future judgment, which three are of the essence of religion" (Newman 1973, 97).

Darwin was typical. On the *Beagle* voyage, because he found he could not believe in the biblical miracles, he lost his belief in Jesus as Savior. He continued as a deist right through the writing of the *Origin*, finally fading to agnosticism in the 1870s. He was explicit. What really upset him was the thought that nonbelievers would go straight to hell: "I can indeed hardly see how anyone ought to wish Christianity to be true; for if so the plain language of the text seems to show that the men who do not believe, and this would include my Father, Brother and almost all my best friends, will be everlastingly punished. And this is a damnable doctrine" (C. Darwin 1887, 87).

If the theory of evolution through natural selection had a more nuanced relationship to Christianity, what indeed was this relationship? It was something as crucial to our discussion as it is possible for a topic to be. It undercut the belief in a loving and caring God. It may not have made atheism mandatory, but with the naturalistic explanation of final cause, it made nonbelief possible. In the immortal words of Richard Dawkins, it was now possible to be an "intellectually fulfilled atheist" (Dawkins 1986, 6). Moreover, God, even if he does exist, seems now to be totally indifferent to our well-being and status. Thanks to the struggle for existence, life truly is a Hobbesian nightmare. Far from creating us in his image, God didn't do it and simply couldn't care less about us. Like Aristotle's unmoved mover, he may not even know of us. Unlike

Aristotle's unmoved mover, there is no reason to think him perfect and no reason to strive to emulate him. . . .

It is not that God permits evil and even occasions it, as one might take the message of Job, but that God is as likely to be friendly as hurtful. In the world of the struggle for existence and natural selection, everything, including us humans, is simply the product of the blind forces of nature—no rhyme, no reason, no meaning or Meaning. Forget about eternal bliss and that sort of thing. You are not going to get it down here, and you are not going to get it up there. Truly, in the words of Camus, nigh a century later, life is absurd. It is this aspect of Darwinism—note, not just evolution—that makes it so all important in our discussion. Life has no Meaning. Get over it.

Chapter 3, Reading 5

Why Atheism?[1]

FAMOUSLY, THE EARLY CHRISTIAN thinker Tertullian (AD 155–240) asked: "What has Athens to do with Jerusalem?" He was arguing that the Christian faith should avoid the snares of the false pagan philosophy of the Greeks—Plato and Aristotle, particularly. Later Christian thinkers, above all Augustine and Aquinas, were to disagree strongly with this position, but they (as do we) agreed with Tertullian that it is to Greek and to Jewish thought—seen in harmony or seen in opposition—that we turn to discover the nature of Christian monotheism, and even more to discover the nature of (let us use the oxymoron) Christian atheism. In respects, it does seem that Tertullian has a point. Nonbelief simply does not come as an option in either the Jewish contribution to the Bible, the Old Testament, or the Christian Bible, the New Testament. "The fool hath said in his heart there is no God." Although Anselm famously quoted this passage from the Psalms (14:1), stating his case for the ontological argument, general agreement is that this was not truly an avowal of nonbelief. Rather, it was a denial of the God of the Jews. There were lots of people like that—the already mentioned followers of Baal, for example. And there was certainly much hostility to the devotees of alien deities. But there was no atheism, or even agnosticism, in the senses we are using the words. . . .

---

1. Ruse 2019d. Reprinted with permission of Cambridge University Press.

Why was this? To get at God, as it were, there seem to be two paths. On one hand is the path . . . of faith, meaning that, in some sense, psychologically you are overwhelmed by the conviction of God's existence like Saul on the road to Damascus. On the other hand is the path Anselm is about to pursue, where you try to use reason and evidence to prove the existence of God. Using the conventional terms of "revealed theology" (meaning belief on faith) and "natural theology" (meaning belief on reason and evidence), there is very little of the latter in any part of the Bible. Exceptionally, in the Psalms we learn: "The heavens declare the glory of God; and the firmament sheweth his handywork" (19:1 KJV). Passages elsewhere, notably Paul speaking in the Areopagus (Acts 17), reveal hints of natural theological reasoning. Generally, the very attempt to prove (or deny) the existence of God gets short shrift. The Jews were not into that sort of thing. It was faith or nothing, and nothing was not an option. Jesus made that clear. "Then saith he to Thomas, reach hither thy finger, and behold my hands; and reach hither thy hand, and thrust it into my side: and be not faithless, but believing. And Thomas answered and said unto him, My Lord and my God. Jesus saith unto him, Thomas, because thou hast seen me, thou hast believed: blessed are they that have not seen, and yet have believed" (John 20:27–29 KJV).

In the spirit of this kind of thinking . . . belief in the existence of God is always somewhat at a distance, that little bit in the corner that you can't get out. It is obviously true that many people of faith don't have this worry—"I know that my redeemer liveth, and that he shall stand at the latter day upon the earth" (Job 19:25 KJV)—but it is equally obviously true that many people do have doubts and sincere believers can wrestle with these throughout their lives. Indeed, paradoxically, it can be that which makes faith so vital. Followers of natural theology would tend to disagree. They would argue that reason and evidence can prove definitively the existence of God. You can empty all the corners. "God exists" is true or not true. Forget all the worries about morality and meaning.

Faith and Reason

To answer this sturdy argument, three points are pertinent. First, in the Christian tradition, faith has always trumped reason and evidence. With reason, Thomas Aquinas is taken to be the greatest natural theologian of all time. Yet he makes no bones about where he stands on the faith/

reason divide. "The truth of the intelligible things of God is twofold, one to which the inquiry of reason can attain, the other which surpasses the whole range of human reason" (Aquinas 1975, 7). Aquinas asserts definitively that faith is the more important—else the ignorant and stupid and lazy would never get knowledge of God. The recent pope Saint John Paul II stood right in this tradition: "The results of reasoning may in fact be true, but these results acquire their true meaning only if they are set within the larger horizon of faith: 'All man's steps are ordered by the Lord: how then can man understand his own ways?' (Prov 20:24)" (John Paul II 1998, 16).

Second, the natural-theological proofs may be found wanting. This is a major item discussed in this element. Full knowledge of God may not be so easily available as you first thought. Note that, here, revealed theology is in a somewhat stronger position. The critic can go after revealed theology, for instance arguing that it is all a matter of psychology, wishful thinking, and belief in God has no stronger basis than belief in winning the lottery. This is true, but that is hardly going to stop the believer from believing. After all, he or she has already foresworn reason and evidence, so reason and evidence are not going to be definitively effective now. In any case, by next week you may know that you did not win the lottery. God belief will be authenticated only after death, when there is going to be no one around to laugh at you for your naivety.

Third, if you do go the route of natural theology, then you open the path to atheism. The person of belief might turn from God because of the horrors of the Holocaust. But you certainly cannot make them turn from God because of the Holocaust, and it is as likely that they will reaffirm their belief in the Christian God because of the Holocaust. Only in the overall Christian eschatological scheme of things can one make sense of the Holocaust. Don't mistake me. I am not trying to slide in at the beginning of this element that that makes Christians horrible people. I don't see that as necessarily or universally true at all. What I am saying is that the Holocaust for a person of faith is not the knockdown argument that a black swan is to the biologist who believes that all swans are white. I am also saying that if you go the route of natural theology—reason and evidence—then you do open yourself to refutations for the existence of God. Nothing in the corners of the tin to shield you. So atheism is now firmly on the table.

The Greeks

Neither Plato nor Aristotle was an atheist. They certainly knew of atheists and Plato for one disapproved of them. He wanted them locked up, fed only by slaves, and buried outside the city walls. Talk about a moral issue! This is about on a par with being a child abuser. Although neither Plato nor Aristotle was given to dancing around stark naked ("sky-clad") or cutting the sacred mistletoe or calling down the moon, in the sense of pagan as someone outside the Abrahamic religions, that is obviously where they fall. Neither was into the polytheism we associate with ancient Greece—gods on Mount Olympus fighting and copulating and feasting and so forth. Both thought that sort of thing not just wrong but rather common and vulgar. Plato had his theory of Forms, supposing that there is a rational world of universals or archetypes that our material world copies in some sense—"participates in." Just as our world is ordered, with the Sun being the prime force illuminating and giving sustaining existence to all else, so in the world of the Forms the Good is the prime force illuminating and giving sustaining existence to all else. Aristotle likewise had his unmoved mover, the totally perfect being toward which all else strives.

There are similarities between Plato's Form of the Good and Aristotle's unmoved mover—and incidentally, not contingently, with the Christian God (Ruse 2017i). All are outside time and space, perfect, unchanging, and the cause of all else. However, Aristotle's unmoved mover does the only thing such a perfect being can do, contemplate perfection, meaning think only of itself (!). It has therefore no knowledge or interest in anything else, certainly not the things of this world. Plato's Good is very different, for it does have concern for the rest of existence—not the rest of creation, for Plato like Aristotle (and very much unlike the Christian) did not think the world was created. It existed always—eternal. However, Plato's Form of the Good was a designer—it was this that made the world (universe) as it is, and it was this that strove to make everything within the world as good as possible. There is debate about whether the Good-as-designer—what Plato called the "demiurge"—did an actual act of designing in space and time, or if (as most think) it was more a principle of ordering.

Either way, Plato (probably drawing on earlier thinkers, especially Socrates) started the Western tradition of natural theology, for he argued that the design-like nature of our world points to an external intelligence that planned the way that things are and function. The eye, for seeing, did not come about by chance. It was intended to be that way, thanks to the benevolent forethought of the demiurge. For Plato, all physical existence shows design—inanimate objects as well as organisms—and this was true also of Aristotle. However, given the indifference of the unmoved mover—no designer, it—the principle of ordering had to be a "vital" force within, rather than an external intelligence. Also, because earlier in life he was a practicing biologist, Aristotle always thought more of functioning—what he spoke of as being guided by "final causes" as opposed to regular "efficient causes"—in the world of organisms than in the whole physical world. Whatever the differences, however, there is not much atheism about Plato and Aristotle.

The Christians

Which makes it hardly unexpected that the great Christian philosophers theologians, notably Augustine, Anselm, and Aquinas, picked up on essential elements of Greek philosophy and incorporated them right into their world systems (Ruse 2015a). Augustine's God bears remarkable similarities to the Platonic Form of the Good, no surprise since he was much influenced by the Neoplatonist Plotinus. In the *Confessions*, Augustine homes right in on the key points. Necessary: "For God's will is not a creature but is prior to the created order, since nothing would be created unless the Creator's will preceded it. Therefore God's will belongs to his very substance." Outside space: "No physical entity existed before heaven and earth." Outside time: "Your 'years' neither come nor go. Our years come and go so that all may come in succession. All your 'years' exist in simultaneity, because they do not change; those going away are not thrust out by those coming in. . . . Your Today is eternity" (Augustine 1991, 228). In some sense, as with the Good, the Christian God does not exist contingently—like the objects of this world—but necessarily. Hard as it is to imagine, there might indeed have been a world without Michael Ruse. It is impossible that there be a world without God.

This point leads to the most notorious of the proofs of God—the ontological argument of Anselm, which asserts his being straight from his definition. God is defined as "that than which none greater can be conceived" (Anselm 2008, 114). Suppose, with the fool, we say that God does not exist. We run into a reductio ad absurdum. "God cannot be conceived not to exist.—God is that than which nothing greater can be conceived.—That which can be conceived not to exist is not God." In the *Summa*, Aquinas offers a neo-Aristotelian, teleology-drenched picture of all of nature, although he is not at all averse to using Neoplatonic notions in his thinking too. . . . The fifth argument is a version of Plato's teleological argument, the argument from design.

We see that things that lack intelligence, such as natural bodies, act for an end, and this is evident from their acting always, or nearly always, in the same way, so as to obtain the best result. Hence it is plain that not fortuitously, but designedly, do they achieve their end. Now whatever lacks intelligence cannot move toward an end, unless it be directed by some being endowed with knowledge and intelligence; as the arrow is shot to its mark by the archer. Therefore some intelligent being exists by whom all natural things are directed to their end; and this being we call God.

Notice that, like Augustine and Anselm, Aquinas assumes that God exists necessarily. He must or we run into the obvious objection: "What caused God?" God for these great thinkers has no cause and has no need of a cause. Through a combination of faith and reason—remember, the first was always prior—right through the medieval period the basic, Christian monotheist position made good sense.

Atheism

What then of the atheists who so disturbed Plato? Most notably there were the atomists who argued that the universe is empty space filled with little balls of matter—atoms—that buzzed around aimlessly. Every now and then they collided and stuck together. Gradually over time these chunks of matter got bigger and bigger, and since there was infinite time and space—just like monkeys typing Shakespeare—every now and then something functioning appeared. "Friends, Romans, countrymen." Working ears and eyes. So it all came together, without rhyme or reason, without purpose or intention. Leucippus and his student Democritus (around the fifth century BC) were the early atomists, followed by Plato's contemporary Epicurus (341–270 BC)—who much influenced the Roman poet Lucretius (94–55 BC).

Even in offering an alternative, meaningless scenario, Plato and later thinkers saw this kind of thinking as a threat to societal stability.

What price ethics and the rule of law when all is simply a matter of chance, without rhyme or reason? Overall, though, the main objection—as Plato makes very clear in the *Phaedo*—is that it is all so implausible. It is all very well to talk about infinite time and space—who can grasp those concepts? In the real world, Murphy's Law prevails—if it can go wrong, it will go wrong. Piles of junk simply don't jump up and start functioning. They just don't.

The Modern Age

What changed things? There were no new atheists in the Middle Ages. Why do we have them now? Essentially, we have them because of the three *r*'s: the Renaissance, the Reformation, and the (Scientific) Revolution (Ruse 2019). The Renaissance brought a renewal of interest in the writings of the ancients. Works like *On the Nature of Things* had a whole new life and an eager audience. This did not mean that people at once became atheists, but the option was being presented anew. Similarly, the Reformation, the break with the Catholic Church by Luther and Calvin and others, hardly signaled a turn to nonbelief. If anything, the Reformers were more ardently Christian than the Catholic establishment. But the differences in beliefs and practices showed the way to thinking outside the loop, and this pointed the way to the possibilities of little or no belief at all. Finally, the Scientific Revolution was no clarion call to atheism. Copernicus, at the beginning, was a minor cleric who died in good standing. Newton, at the end, was deeply religious, in later life spending far more time on biblical interpretation than on physics. It did, however, put the Sun at the center of things, rather downgrading the special status of things on Earth, and, more important, it challenged Aristotelian final causes. Rather than thinking of the universe in organic terms, the new breed of scientists thought in mechanical terms, of the world as a machine. Of course, machines have purposes, but that part of the metaphor (in the physical sciences, at least) was downgraded and dropped. The new science simply thought of the world as in endless motion, governed by blind laws. God could still exist, but he was pushed out of scientific explanation. In the words of one of the greatest historians of the Scientific Revolution, he became "a retired engineer" (Dijksterhuis 1961, 491). It is significant that, for all his religiosity, Newton moved toward a form of deism, denying the Trinity, and thinking in terms of a world where God no longer interferes.

Note that this is deism, not agnosticism or atheism. And this, despite that through the eighteenth century, people like David Hume hammered away at the natural theological arguments for God's existence—especially the argument from design—and as we entered the nineteenth century, people (especially German scholars) were attacking the foundations of revealed theology. So-called higher criticism looked at the books of the Bible as though they were humanly written documents, and quickly and thoroughly they started to fall apart as authentically divine narratives. They appeared much more likely the parables and tales of people long ago, who wrote down these fables and then through traditional retelling considered them divinely inspired. The fly in the ointment, however, if we might be allowed an appropriate metaphor, was the world of organisms. They just didn't seem to be the products of blind laws. They were teleological, bound by final causes, and to one and all—theist and deist—this meant a designer.

Natural Selection

It was Charles Darwin, in his *Origin of Species* published in 1859, who threw a bomb into all of this (Ruse 1999b). A convinced evolutionist after a five-year voyage around the world on *HMS Beagle*, he was also a Cambridge University graduate who had been fed a steady diet of Paley. He knew of and was completely convinced of the design-like nature of organic features, what he called "contrivances" or "adaptations." The question was to find a natural—blind law-governed—solution. This he found in an equivalent to the selection that breeders practice to produce fleecier sheep and meatier cows and more beautiful and melodious birds. They pick from the best and breed only from them. Darwin quoted an eminent practitioner of the art: "It would seem as if they had chalked out upon a wall a form perfect in itself, and then had given it existence" (C. Darwin 1859, 31). In nature, argued Darwin, the key lies in the ongoing population pressures made much of by political economist Thomas Robert Malthus. More organisms are born than can survive and reproduce. Available space and food sets limits. There will therefore be a "struggle for existence," and even more a struggle for reproduction. Apparently new variations are always appearing in natural populations—not uncaused but not according to need (in other words, no teleology built in here)—and in the struggle, some of these variations will prove of value to their possessors and so there will be an ongoing winnowing, what Darwin called "natural selection." This will lead to change, and the point is that this process points to the creation of design-like organic attributes. This was not a chance discovery. It was something that framed the whole discussion.

At once people saw what was going on. This may have been basic science, but it was science with nasty implications for God belief. Until then, we had what philosophers call an "inference to the best explanation." Something had to explain adaptation/contrivance. Blind law does not do so. Ergo, God is the best explanation. Now, Darwin was saying that blind law can do the job. So the God explanation is less compelling. Adam Sedgwick, a professor of geology at the University of Cambridge, and an old friend and mentor of Darwin—not without a sense of humor, for he referred to himself as a "son of a monkey"—saw this full well. "I call (in the abstract) causation the will of God: & I can prove that He acts for the good of His creatures. He also acts by laws which we can study & comprehend—Acting by law, & under what is called final cause, comprehends, I think, your whole principle." Except of course the principle doesn't do the job. "I humbly accept God's revelation of himself both in His works & in His word; & do my best to act in conformity with that knowledge which He only can give me, & He only can sustain me in doing" (C. Darwin 1985–2023, 7:12041).

We will discuss in some detail the exact implications of Darwin's theory for the theism-atheism debate. But we can say this straight off. Richard Dawkins was right when he said that after Darwin, it was finally possible to be "an intellectually fulfilled" atheist (Dawkins 1986, 6). The stranglehold of the design argument was, if not broken, then at least loosened. And now, perhaps not surprisingly, we do start to see a growing number of nonbelievers making their positions known. Darwin was raised a Christian, a member of the Church of England (Anglican or Episcopalian), probably given his family's detestation of slavery at the evangelical end of things. At Cambridge, he intended to be a clergyman. This avocation faded on the *Beagle* voyage, and his thinking moved from theism to deism. He held to this right through *Origin*. "Authors of the highest eminence seem to be fully satisfied with the view that each species has been independently created. To my mind it accords better with what we know of the laws impressed on matter by the Creator, that the production and extinction of the past and present inhabitants of the world should have been due to secondary causes, like those determining the birth and death of the individual" (C. Darwin 1859, 488). Despite the fact that Darwin constantly toyed with the text, right through the six editions until 1872, this passage went unchanged, although it is true that, by then, Darwin's beliefs had faded. Never atheism. He would have thought that almost vulgar and associated with disgusting practices like birth control. To a correspondent (John Fordyce) in 1879, Darwin wrote: "In my most extreme fluctuations I have never been an atheist in the sense of denying the existence of a God.—I think that generally (& more and more so as I grow older) but not always, that an agnostic would be the most correct description of my state of mind" (C. Darwin 1985–2023, 7:12041).

What is interesting about Darwin is that he reflected almost all mid-Victorian intellectuals in that it was not science as such that pushed him toward nonbelief. It was the problems inherent in religion—Christianity—itself that made for the repudiation of childhood beliefs. For Darwin, totally unacceptable was the Pauline claim that nonbelief would lead to hellfire and damnation. "Knowing that a man is not justified by the works of the law, but by the faith of Jesus Christ, even we have believed in Jesus Christ, that we might be justified by the faith of Christ, and not by the works of the law: for by the works of the law shall no flesh be justified" (Gal 2:16 KJV). Darwin's beloved father and older brother were both nonbelievers. He could not accept that they were thereby condemned.

> I can indeed hardly see how anyone ought to wish Christianity to be true; for if so the plain language of the text seems to show that the men who do not believe, and this would include my Father, Brother and almost all my best friends, will be everlastingly punished. And this is a damnable doctrine. (C. Darwin 1958, 87)

What, then, was the chief effect of *Origin* on people's religious beliefs? It made nonbelief possible if one had other reasons for nonbelief. As important, notions like the struggle for existence made the very idea of a caring God seem difficult or absent.... God just doesn't care.

Creationism

This all had a chilling effect on the teaching of evolution, and mention of Darwin was removed from school biology texts—paradoxically at a time when the theory was making immense steps because of the synthesis of Darwinian selection with Mendelian genetics. Things were not to change until the late 1950s when, in response to the perceived Soviet superiority in science and technology thanks to Sputnik, American science education was radically updated and evolution was brought fully and openly into biology texts. To every action there is an equal and opposite reaction, and this led to a resurgence of evolution-denying biblical literalism. It was now given the name "creationism," or "scientific creationism" since supposedly it was as good science as it was theology (and thus could rightfully be taught in state-supported public schools). The bible of the new movement, *Genesis Flood*, was written by biblical scholar John C. Whitcomb and hydraulic engineer Henry M. Morris (Whitcomb and Morris 1961). It was the definitive statement of "young Earth creationism"—divine creation six thousand years ago, six days in which all was accomplished, a universal flood wiping out all except a few chosen ones on a specially constructed ark, and the rest of the story.

Chapter 4
Darwinism, Belief, and Religion

Chapter 4, Reading 1
The Origins of Religion[1]

OUR CONCERN IN THIS chapter, however, is with his second question—the origin of religion in human nature—and how it is to be answered in the light of Charles Darwin's theory of evolution. . . . Given that mine is a book about Darwin viewed from the perspective of philosophy, you might think that this is a topic to be skipped. Surely a question about origins is less a philosophical question and more an empirical question, to be answered by historians or sociologists or anthropologists, or perhaps even by evolutionists? For three reasons, it deserves our attention. First, whether or not it is truly a philosophical issue, it is certainly something that has captured the attention of philosophers—from David Hume in the Enlightenment to (most recently) Daniel Dennett. Second, not only did Darwin have things of interest to say on the topic, but today there is a huge amount of interest in the putative evolutionary origins of religion, so to omit all discussion would be unduly to truncate the overall picture of Darwinism today. Third, ultimately—as with just about everything—there are lurking philosophical questions worth considering. So let us plunge into the debate, starting with David Hume. Although he has many insights that cry out for an evolutionary interpretation, I do not think that Hume was an evolutionist. But he certainly set the terms of the debate.

1. Ruse 2008c. Reprinted with permission of Blackwell.

David Hume

Hume offered a "natural" history of religion: that is, he explained religion entirely in natural terms—no miracles or hand of God or any such thing ([1779] 1947). He started by suggesting that polytheism was the original belief state of humankind and that it came from a tendency to see life in all things, including the inanimate. Primitive man was worried about food and security and all such things and this led him to interpret the world as though it were full of animate beings.... After he got going with the god idea, Hume then supposed that some divinities started to gain importance over all others, until we went all of the way to monotheism. Hume noted how in the Middle Ages the Virgin Mary was on the way to being promoted to the very top rank in people's minds—until the Protestant Reformation took the gloss off her glory. Not that Hume was much enthused by the end results: he inclined to think that the more a god gets in charge, the less tolerant are its supporters of rivals, and this leads to all kinds of tensions and frictions. Also, a more exalted god, elevated above the everyday nature of life, starts to take on attributes that reason suggests are absurd. Showing his good Calvinist training, Hume was particularly scornful of the Catholic doctrine of transubstantiation—the idea that we might be eating our god.... One starts to get some sense of why David Hume made so many good believers very uncomfortable and why one critic (a fellow Scot) described him as God's greatest gift to the infidel. Hume drew a firm line between belief in a divinity and moral behavior. If anything, there seems to be an inverse ratio, with greater intensity of belief leading to greater inclination to commit moral atrocities. People reason that being onside with God gives them license to do what they will.

The Descent of Man

For all of my caginess about direct links between Hume and Darwin, in the matter of religion and its putative natural origins, the case for an immediate Humean influence is strong. Darwin turned to religion and its origins in *The Descent of Man*, and, while it is true that he did not footnote Hume (the *Descent*, being a more measured work, has footnotes, unlike the *Origin*, which was written at speed and intended to be an abstract of a larger whole), the naturalistic spirit is very much that of Hume. Years earlier, when Darwin was working toward natural selection and when he was thinking out his whole, overall position on evolution—including, especially, his position on humans (namely that we are as much subject to natural causes as any other organism)—he did read Hume's *Natural History of Religion*. Although Darwin was obviously going to add an evolutionary perspective, a direct influence from Hume is more plausible here than at any other place in the Darwinian corpus.

What is striking about Darwin's discussion of religion is how brief it is, especially when compared to the detailed discussion that he gave of the origins and nature of morality. There are at least two reasons for this. First, by the 1870s, in the eyes of Victorians like Darwin, the battle for religion was over and God had lost. The problem now was to maintain morality in the face of nonbelief. The novelist George Eliot spoke for many when, discussing "God, immortality, duty," she declaimed: "how inconceivable . . . the first, how unbelievable the second, and yet how peremptory and absolute the third" (Myers 1881, 63). Remember the similar passage from Kant that found its way into the *Descent* (C. Darwin 1871, 1:70). Darwin gave much more space in the *Descent* to the evolution of morality. Second, and more importantly, Darwin was not a David Hume or (to speak of his contemporary) a Thomas Henry Huxley. His own belief may have gone, but by nature and class he was instinctively against religion-bashing. He had grown up with and in the church. His family was respectable and often religious—if not Anglican then Unitarian. One of his closest personal friends was the local vicar. Darwin was certainly not going to conceal his beliefs, but equally certainly he was not going to flaunt them. Measured and brief or not, what did Darwin have to say?

> There is no evidence that man was aboriginally endowed with the ennobling belief in the existence of an Omnipotent God. On the contrary there is ample evidence, derived not from hasty travellers, but from men who have long resided with savages, that numerous races have existed, and still exist, who have no idea of one or more gods, and who have no words in their languages to express such an idea. (C. Darwin 1871, 1:65)

To which sentiment, in Humean style, Darwin added immediately: "The question is of course wholly distinct from that higher one, whether there exists a Creator and Ruler of the universe; and this has been answered in the affirmative by some of the highest intellects that have ever existed." He also added that, if we are thinking in terms of vague spiritual sentiments, then obviously savages and other primitive folks come under the religion blanket. "If, however, we include under the term 'religion' the belief in unseen or spiritual agencies the case is wholly different; for this belief seems to be universal with the less civilised races" (C. Darwin 1871, 1:65).

How did this sentiment or belief about unseen forces arise? Here Darwin started to sound very Humean. It is all a question of seeing spirits in inanimate objects, feeling or pretending or mistakenly believing that they are truly alive.

> The tendency in savages to imagine that natural objects and agencies are animated by spiritual or living essences, is perhaps illustrated by a little fact which I once noticed: my dog, a full-grown and very sensible animal, was lying on the lawn during a hot and still day; but at a little distance a slight breeze occasionally moved an open parasol, which would have been wholly disregarded by the dog, had any one stood near it. As it was, every time that the parasol slightly moved, the dog growled fiercely and barked. He must, I think, have reasoned to himself in a rapid and unconscious manner, that movement without any apparent cause indicated the presence of some strange living agent, and that no stranger had a right to be on his territory. (C. Darwin 1871, 1:67)

Like Hume, Darwin saw a move from primitive religion, through polytheism, and on to monotheism. And, completing the story still in the vein of the Scottish philosopher, Darwin saw religion as connected with vile superstitions and practices, which only the rise to reason could conquer and prevent. . . . At the end of the *Descent*, in summarizing, Darwin returned to the topic.

> The belief in God has often been advanced as not only the greatest, but the most complete of all the distinctions between man and the lower animals. It is however impossible, as we have seen, to maintain that this belief is innate or instinctive in man. On the other hand a belief in all-pervading spiritual agencies seems to be universal; and apparently follows from a considerable advance in man's reason, and from a still greater advance in his faculties of imagination, curiosity and wonder. I am aware that the assumed instinctive belief in God has been used by many persons as an argument for His existence. But this is a rash argument, as we should thus be compelled to believe in the existence of many cruel and malignant spirits, only a little more powerful

> than man; for the belief in them is far more general than in a beneficent Deity. The idea of a universal and beneficent Creator does not seem to arise in the mind of man, until he has been elevated by long-continued culture. (C. Darwin 1871, 1:394–95)

Darwin's position, therefore, was that religion is a natural phenomenon—or rather a phenomenon that can be treated naturally—and he saw it as something that had evolved. Given the time that Darwin spent showing that morality is something deeply connected to natural selection, it is noteworthy that he did not at all attempt this in the case of religion. One presumes that since he thought it false—at least not proven in its essentials, and false in many details—he did not think that religion could be promoted by selection. It does not give us insights into reality, or (even if false) help us better to survive and reproduce. Nor did Darwin want to argue for religion as a by-product of selection, or as something that might be promoted by selection but that lacked direct adaptive significance for survival.

For Darwin, religion seems to be almost accidental, and brought about by animal features or powers that are simply misdirected. When we see something moving, it normally makes sense to think that it is living. We make mistakes, and ultimately this leads into religion. About the only thing that can be said in its favor is that, in the case of civilized people, it does help to reinforce morality. "With the more civilised races, the conviction of the existence of an all-seeing Deity has had a potent influence on the advance of morality" (C. Darwin 1871, 2:394). Here, perhaps, Darwin did differ from Hume, opting to be at one with his fellow Victorians, who, despite disbelief, were hymning the virtues of Bible reading as a guide to and reinforcement of morality.

Darwinism and Religion Today

From the middle of the nineteenth century down to the present, much has been written on the subject of the putative natural origins of religion. Anthropologists and sociologists turned in a major way to religious practices and beliefs, in Western society and elsewhere in the world, in the present and in the past—the latter being sometimes discovered from evidence and sometimes created by lively imagination. Well known, perhaps "notorious" is a better term, are such comparative studies as James G. Frazer's *The Golden Bough* (1906–15), which sees religion as rooted in fertility cults, where the king must die and be reborn for the cycles of nature to move forward as they do; the implications of this for Christianity, a religion where the god dies and is resurrected, did not escape its author. Others also got into the business of supplying histories of religion and explanations of their attraction. Freud was deeply interested in biology and thought that religion could be explained in biological terms. Deaths and sacrifices are echoes of actual events in the past, embedded in human nature by Lamarckian processes—the inheritance of acquired characteristics. Hence, even if Freud's claims about religion were correct—and many disputed such controversial ideas as that Moses was killed by his followers—the Lamarckian mechanism was wrong (Sulloway 1979). It was not until the 1970s that we see a genuine revival of attempts to explain religion in terms of modern evolutionary biology. Great credit must go to Edward O. Wilson, whose sociobiological synthesis made religion a central object of study; he discussed it in detail in his Pulitzer Prize–winning *On Human Nature*. For Wilson, religion is apparently all a matter of group identity and sticking together.

> The highest forms of religious practice, when examined more closely, can be seen to confer biological advantage. Above all, they congeal identity. In the midst of the chaotic and potentially disorienting experiences each person undergoes daily, religion classifies him, provides him with unquestioned membership in a group claiming great powers, and by this means gives him a driving purpose in life compatible with his self interest.
> (E. Wilson 1978, 188)

Wilson does allow that there can be cultural selection between sects, but essentially we start with the biology and all else is on the surface.

> Because religious practices are remote from the genes during the development of individual human beings, they may vary widely during cultural development. It is even possible for groups, such as the Shakers, to adopt conventions that reduce genetic fitness for as long as one or a few generations. But over many generations, the underlying genes will pay for their permissiveness by declining in the population as a whole. Other genes governing mechanisms that resist decline of fitness produced by cultural evolution will prevail, and the deviant practices will disappear. Thus culture relentlessly tests the controlling genes, but the most it can do is replace one set of genes with another.
> (E. Wilson 1978, 178)

Although he (like Edward O. Wilson and Reynolds and Tanner) thinks that culture is crucial, one suspects that ultimately David S. Wilson (and they) see real change as genetic. Coming from a very different direction, Daniel Dennett agrees entirely that religion is promoted by selection, but he is not at all convinced that this selection is necessarily for the benefit of humans, nor is it essentially (or, truly, in any way) genetic (Dennett 2006). Dennett adopted a theory of Richard Dawkins that posits the existence of "memes," units of culture akin to genes, which compete for people's allegiances (Dawkins 1976). Rival memes, as it were, invade people's minds, and those that win are those that are selected to continue. Winning is not random, but a function of the features—the adaptations—that the memes have or promote. Successful advertising obviously is a paradigmatic example of memes at work: you buy and smoke Marlboro cigarettes because this makes you feel like a real man, even though, in fact, you are acting in ways highly detrimental to your health and well-being.

For Dennett, religion is a meme parasite, which has features that make it attractive, even if it is not necessarily that good for the possessor. So that no one should miss this point, he begins his book *Breaking the Spell: Religion as a Natural Phenomenon*, by introducing the reader to the lancet fluke (*Dicrocelium dendriticum*): this parasite corrupts the brain of an ant, causing it to strive to climb blades of grass, so that it gets eaten by a sheep or cow, and thus the fluke can complete its life cycle before its offspring are excreted and take up again with ants.

> Does anything like this ever happen with human beings? Yes indeed. We often find human beings setting aside their personal interests, their health, their chances to have children, and devoting their entire lives to furthering the interests of an idea that has lodged in their brains. The Arabic word islam means "submission," and every good Muslim bears witness, prays five times a day, gives alms, fasts during Ramadan, and tries to make the pilgrimage, or hajj, to Mecca, all on behalf of the idea of Allah, and Muhammad, the messenger of Allah. Christians and Jews do likewise, of course, devoting their lives to spreading the Word, making huge sacrifices, suffering bravely, risking their lives for an idea. So do Hindus and Buddhists. (Dennett 2006, 4)

As we might expect, Dennett is disdainful of the suggestion that religion has much to do with morality.

> I have uncovered no evidence to support the claim that people, religious or not, who don't believe in reward in heaven and/or punishment in hell are more likely to kill, rape, rob, or break their promises than people who do. The prison population in the United States shows Catholics, Protestants, Jews, Muslims, and others—including those with no religious affiliation—represented about as they are in the general population. (Dennett 2006, 279)

In other words, although they would both claim to be Darwinians, when it comes to religion and its virtues, David S. Wilson and Daniel Dennett are about as far apart as they could be.

Religion as By-Product

Turning now to the other side of the equation, what of those who think religion falls more into the spandrel category? Gould certainly thinks somewhat along these lines: the whole of human culture comes under this category for him (Gould 2002). But most would not be so sweeping. Apart from anything else, religion with its costs—devotion to others, celibacy, ritual physical disfigurement, sacrifice, and so forth—simply does not seem to be the sort of thing that would happen as a by-product. It is just too costly. More likely, in the opinion of some, is the idea that religion piggybacks, as it were, into existence and power by attaching itself to other things—real, powerful adaptations—and manages to exist because it cannot be stopped or because ultimately its costs are simply not that great. Student of culture Pascal Boyer inclines to the first option. For him, religion simply subverts or borrows features that our biology has put in place for good adaptive reasons, and for whatever reason it cannot be eradicated.

But what is it that allows religion to get its hold in the first place? Anthropologist Scott Atran inclines to the second option, that religion grabs something adaptively useful and exploits it. For him, the big question facing organisms like humans is other living beings—above all, other living beings as threats. In an argument reminiscent of Darwin and his dog—in fact, reminiscent of Hume's speculations—Atran suggests that what we have is a somewhat overeager projection of the living onto the inanimate. It used to be thought that the baroque nasal appendages of the titanotheres were a case of sensible evolution having taken a step too far. Perhaps the same is true of religion. Cuckoos exploit the innate mechanisms that their host birds have for raising their young. Religion does much the same for humans.

Richard Dawkins falls somewhere in this category. Predictably, he has little time for the supposition that morality and religion have much connection, quoting with relish the Old Testament stories of the behavior of the patriarchs. Of Abraham and Isaac he writes: "By the standards of modern morality, this disgraceful story is an example simultaneously of child abuse, bullying in two asymmetrical power relationships, and the first recorded use of the Nuremberg defense, 'I was only obeying orders.'" Since, for him, religion has only a negative value, he too inclines to the religion by-product thesis. Although he thinks the general principle more important than the details, he wonders if the adaptive advantages of obeying informed authority—"stay away from the cliff"—have been subverted by the religion memes—"do this, that, and the other, or you will go to hell" (Dawkins 2006, 242).

Serious Science?

What can we say about these various ideas and hypotheses? One thing for certain. They can't all be true! For every action there is an equal and opposite reaction. For every idea about the evolution of religion, there is an idea that takes exactly the opposite tack. You think, with both Wilsons, that religion evolved and helps humans? Then what of Dennett, who likens it to the fluke parasite—hardly the ultimate model of warmth and friendliness. Or consider Atran, who thinks that religion is a case of adaptation going over the top. As he says explicitly: "Supernatural agency is an evolutionary by-product, trip-wired by predator-protector-prey detection schema" (2004, 54). You think, with Reynolds and Tanner, that religion evolved to help the individual? Then read David S. Wilson, who is convinced that it is groups all of the way. You think, with Edward O. Wilson, that religion is a function of biological evolution? Then look at Dennett, who thinks that it is culture—simple, albeit not very pure, culture. You argue with David S. Wilson that religion is the ultimate support of morality? Then take up Dennett (not to mention David Hume), who thinks that religion does absolutely nothing for morality, or, if it does, its effects are almost invariably negative. Amusingly, Dennett quotes the Nobel physicist Steven Weinberg: "Good people will do good things, and bad people will do bad things. But for good people to do bad things—that takes religion" (Dennett 2006, 279). Although Dawkins thinks Jesus is a step up from the figures of the Old Testament, he criticizes him for "his somewhat dodgy family values," and the death on the cross is "a new sado-masochism whose viciousness even the Old Testament barely exceeds" (Dawkins 2006, 250-51).

I do not want to argue that it is necessarily impossible or worthless to invoke Darwinian evolutionary theory in the explanation of the origin or nature of religion. I do rather argue that we have got to do a lot better than we have done thus far. Part of the issue, obviously, is with the general program of human sociobiology, or evolutionary psychology, or whatever you want to call it. As and when this improves overall, so biological explanations of religion as a part of human culture will get dragged along too. Part of the issue equally obviously lies in teasing out which causal explanation is generally better in these circumstances. Leave aside the individual selection/group selection discussion as something that has been considered in depth already, and take what Dennett relies on, the theory of memes, or memetics, or whatever it is called. It really is crude to the point of nonbeing, certainly to the point of nonhelpful. What is a meme? It is a chunk of culture analogous to a gene. As it happens, genes are hard enough to define, but we do have some idea of them as the smallest functioning length of DNA. But what is the smallest functioning length of culture? Is Catholicism a meme? Is the authority of the pope a meme? Is transubstantiation a meme? Why the authority of the pope, for example, rather than each and every one of the dogmas that he endorses? And what kind of theory do you have as memes clash and come together and sometimes fuse and sometimes break apart?

The point is not that Dennett is necessarily wrong in arguing that ideas sometimes have lives of their own, or that religions can be dreadful things that take over people's minds to their own detriment—if we do not think this way about Catholicism then most of us do about cults like Scientology—but that memetics is not very helpful in understanding what is going on. One is really just taking regular language and putting it in fancy terms. No new insights. No new predictions. No astounding claims that turn out to be true. More importantly, one is not really using Darwinian evolutionary theory to do any work. One could be a Lamarckian and make most of the claims that Dennett wants to make. In fact, given the way that memes can be transmitted from one individual to another, one would probably be more comfortable being a Lamarckian if one wanted to make most of the claims that Dennett wants to make.

And part of the issue here—part of the problem with Darwinian approaches to religion—is simply that so much of the scholarship is so crude. Let me praise David S. Wilson for wanting to use real examples to articulate and flesh out his thinking. But his discussion of Calvinism really will not do. If Calvinism was such a terrific booster of societies and helped them work so well, why did it so frequently fail to convince? Take the English (MacCulloch 2004). Henry VIII broke from the Catholic Church because he wanted to take a new wife and the pope forbad it. His son Edward VI was ultra Protestant. But when Edward died as a teenager his elder sister Mary came to the throne and, as an ardent Catholic, persecuted Protestants, many of whom fled to the Continent. By that time, the middle of the sixteenth century, the German Lutheran areas were torn by war and strife, and so these exiles headed for safer, quieter, Reformed (Calvinist) areas. When Mary died and her younger Protestant sister, Elizabeth, came to the throne, the Calvinists all flooded back. But generally, the English were not that keen on what Calvinism had to offer. They did not want the repressive morality and lifestyle of those who later came to be known as Puritans. So the Elizabethan compromise was worked out in the shape of an Anglican Church—to this day a funny bricolage of Catholic style and Protestant theology. But it was certainly not unsuccessful: the English saw off the Spanish and their armada. It is true that in the seventeenth century the Roundheads, the Puritans, won the Civil War and lopped off the head of Charles I, but within twelve years the Royalists, the more central Anglicans, were restored and the Puritans were out again.

The point I am making is that you cannot just isolate one bit of history, one place in time and space, and think you have the basis for a universal theory. You have got to spread your grasp much more broadly and confront the difficult cases, the examples that contradict your argument. This comment applies particularly to Americans—Edward O. Wilson and Daniel Dennett come at once to mind—who start with assumptions about the universal appeal and force of religion. By any measure, given its anti-Enlightenment obsession with religion, America is a very peculiar country, at least compared to the rest of the First World. It is very dangerous to argue about the need that humans have of religion if, in fact, a lot of humans really do not seem to need that much religion at all. England is a case in point. Most young couples want a church wedding. After all, they want things done properly. But do they spend their subsequent married life obsessed with the death of Jesus Christ on the cross as a payment for their sins? Such a claim would be laughable. They probably have some vague notion that the life hereafter will be a kind of extended weekend, with lots of telly and visits to the pub; but generally, to the lives of the average person in England, religion is about as relevant as the royal family. Significantly, the queen and her children are expected to observe the ritual practices as a proxy for the rest of the population. I am not saying that all writers—not even all American writers—on the biology of religion are equally provincial in thinking that their home society is the norm, but the issue is inadequately addressed. . . . It is hardly surprising that for Dennett religion is a parasite like the lancet fluke. For Dawkins, it is child abuse . . . but we must first ask the basic question about God and his existence—not whether God exists but whether a Darwinian account of origins shows that God exists or not. Whether or not he was writing sincerely, David Hume separated out a naturalistic account of religion's origins from its truth status, and this seems also to be Darwin's approach. Edward O. Wilson goes entirely the other way. Darwinism gives a naturalistic account of religion and that is end to religion as a reliable authority on what there really is. As it happens, Wilson thinks that the human psyche demands religion, and hence at this point he feels free to substitute his own kind of evolutionary humanism. But this is because Darwinism has already done its corrosive work.

Chapter 4, Reading 2

God[1]

WE HAVE TO WHITTLE down the discussion somewhat, so I am going to brush past a great deal of Christian theology, especially much that has to do with specific notions like the Trinity, important though they are. I am also going to ignore entirely much that makes Christianity vital and important to its practitioners—the churches, the music, the rituals, the fellowship, and all of that. But this must be. . . . I want to pick out four items or claims that are central to Christian belief—four items that the Christian takes on faith. If you do not believe in these, then you should not call yourself a Christian. First, that there is a God who is creator, "Maker of heaven and earth." Second, we humans have duties, moral tasks here on earth, in the execution of which we are going to be judged. Hence, God stands behind morality. Third, Jesus Christ came to earth and suffered because we humans are special, we are worth the effort by God. The usual way of expressing this is to say that we are "made in the image of God." We have "souls." Fourth and finally, there is the promise of "life everlasting." We can go to heaven, whatever that means.

I want to explain in a fair manner what is meant by Christianity in the terms of the four points introduced in the last paragraph. I want also to show that you can hold these, if you so wish, in the light of modern science—if you prefer, in the face of modern science. In other words, the Christian's claims are not refuted by modern science—or

---

1. Ruse 2010a. Reprinted with permission of Cambridge University Press.

indeed threatened or made less probable by modern science.... I warn that that means that the Christian is not and cannot be offering a science-like answer. Certainly, the Christian should not be offering such an answer. The noted Christian (Calvinist) philosopher Alvin Plantinga comments about naturalistic accounts of the nature of mind: "A theist may be able to learn a good bit from this; but fundamentally he will ask different questions and look for answers in a quite different direction" (Plantinga 1999, 19). Generalizing, this is precisely my claim. Although we must strive to understand science and its triumphs, questions such as that about ultimate origins are simply not attempted by science as we know it. My claim is also—and this is very important—that these questions are genuine questions.

Hence, I argue that it is open for others to attempt answers to these questions. However, these must be answers of a different type in some way: nonscientific answers. For this reason, it is not fair to criticize the religious person for not offering a science-like answer. As it happens, the Christian claims to be giving a faith-based answer, one that comes from a different source than the reason and empirical experience (through the senses) that yield science. You may think that if science is silent, we should also be silent. That is your opinion, and others have the right to disagree. This does not preclude arguing against the Christian on nonscientific grounds. That is another matter.

There is significant overlap between Protestants and Catholics on faith, especially inasmuch as the great Reformers—Luther and Calvin—looked back, as do Catholics, to Saint Augustine. Calvin, for instance, states explicitly that God has implanted in us a natural tendency to believe in and understand the nature of God. . . . Calvin worries a bit about primitive people. However, he reassures himself and his readers that "no nation so barbarous, no people so savage, that they have not a deep-seated conviction that there is a God." All humans have this sense or at least this capacity to come to realization of God. We are dealing with innate ideas here (the kind that John Locke was to deny in the next century): "That there is some God, is naturally inborn in all, and is fixed deep within, as it were in the very marrow. . . . From this we conclude that it is not a doctrine that must first be learned in school, but one of which each of us is master from his mother's womb and which nature itself permits no one to forget" (Calvin 1960, 1.3.3). Calvin adds that although we have this natural tendency, it is obscured by our sinful nature. Hence, the existence of unbelief and doubt.

If faith is indeed knowledge, then one important point follows immediately. We must eschew *fideism*, meaning an attitude that pits faith against reason—Tertullian (ca. 160–220 CE) notoriously (although possibly ironically) claimed that the resurrection is to be believed because it is "absurd," it is "impossible" (Roberts and Donaldson 1989, 3:525). Traditional Christian thinking is that faith cannot be unreasonable, in the sense of illogical. It might push you beyond reason, certainly beyond empirical evidence, but it cannot make you believe the contradictory. This is a point that will need more discussion later, but for now we are justified in making sure that the Christian's claims make at least some kind of sense. Otherwise it can hardly qualify as knowledge. . . . On this note, let us now turn to look at major planks of Christian belief. I shall not argue that you do not have to believe these things. That seems to me to be obvious. My inquiry is about those who do believe these things. Christians have faith. They believe that they have been told certain truths by God. What follows?

Necessary Being

Let us plunge right in with the first claim:

> I believe in God the Father Almighty, Maker of heaven and earth.

This is a claim based on the Old Testament: "In the beginning God created the heaven and the earth. And the earth was without form, and void; and darkness was upon the face of the deep. And the Spirit of God moved upon the face of the waters" (Gen 1:1–2 KJV). It relies likewise on words in the New Testament: "In the beginning was the Word, and the Word was with God, and the Word was God. The same was in the beginning with God. All things were made by him; and without him was not any thing made that was made" (John 1:1–3 KJV).

This claim is clearly intended to speak to one of the major questions we have seen to be unanswered by science: Why is there something rather than nothing? Does it do so, and does it do so in a way that does not infringe on the domain of science? In trying to answer this question, remember that I am going to do Christians the courtesy of assuming that we are not now dealing with absolutely literal readings of the Bible. I am assuming that we have moved beyond picturing God as he is depicted in Michelangelo's creation of Adam, that is, as someone who looks a little bit like Charlton Heston, dressed in a bedsheet. It is still legitimate and necessary to ask something about the nature of God. As just noted, even though the Christian God is not a scientific concept, our understanding cannot violate the rules of reasoning (or the findings of the senses). We would be beyond knowledge as we know it, and we would be violating things that science does hold dear. Here, therefore, we start to grapple with somewhat philosophical issues, meshing with points made in an earlier chapter. God cannot be a contingent being like us. If he were, then we would get caught up in sophomoric questions like "God caused the world, but what caused God?" (When I say that such questions are sophomoric, I do not mean that they are unimportant questions. I mean it is sophomoric to think that you thought them up and that no one previously has tried to answer them.) God has to be a necessary being. Or let me phrase this more carefully, because I do not want to assert even implicitly that God exists. If God exists, then God exists necessarily. One immediate consequence of this is that God (assuming he exists) is not something in time. He is not just very old and likely to go on living for some time yet. He is eternal. He is like the truths of mathematics: "two plus two equals four" did not start to be true, and it never will cease to be true. It is not like my dog Toby, with a beginning, a middle, and an end. Saint Augustine is quite clear on this. This is part of God's being.

For Calvinists particularly, given their obsession with God's absolute sovereignty, his freedom seems to be his most important feature: "the prerogative of divine sovereignty." Unsurprisingly, therefore, for Barth, God's very being is all bound up with just this: "Freedom in its positive and proper qualities means to be grounded in one's own being, to be determined and moved by oneself. This is the freedom of the divine life and love. In this positive freedom of His, God is also unlimited, unrestricted and unconditioned from without. He is the free Creator, the free Reconciler, the free Redeemer" (Barth 1957, 301). God's eternity follows from this sense of freedom, of God's being grounded in his own being. "God's eternity, like His unity and constancy," Barth writes, "is a quality of His freedom. . . . Eternity is God in the sense in which in Himself and in all things God is simultaneous, i.e., beginning and middle as well as end, without separation, distance or contradiction" (Barth 1957, 608).

What is to be said in reply? It is certainly the case that there are those who want to claim that God's existence is in some sense logically necessary. Anyone who subscribes to the validity of the so-called ontological argument, which derives God's existence from his very nature, thinks along these lines. Saint Anselm's version shows this clearly: God is that than which none greater can be conceived. Existence is part of this. Hence, it is contradictory to think that God does not exist. However, this argument is certainly not part of the mainstream of Christian thought. There is nothing in the Bible suggesting that God is logically necessary. In both the Old and the New Testaments it is true that, in some sense, the awareness of God is so strong that his nonexistence is unthinkable—Job in his trials never denies God's existence, nor do his comforters—but this is more as a matter of fact than of logical necessity.

After all, Anselm's ontological proof famously starts with a quotation from Psalms: "The fool hath said in his heart, There is no God" (Ps 53:1). This is not a sentiment the psalmist receives favorably, but neither is it an occasion to launch into a proto-form of Russell's theory of descriptions. Another point is based on the notions of potentiality and actuality. God has no potentiality; he is purely actual. Hence he exists out of his very essence, and this in some way makes him necessary. The third point picks up on participation and makes much the same point as the second. All of this is clearly bound up with an anticipation of Barth's claim about God being totally free. His being comes from himself, and hence he is dependent on no one and no thing for his nature or his actions.

God's Properties

Does any of this prove that God is the creator? I am not sure that it does. As opposed to Greek thought, which never, ever claimed that God (or the gods) created matter—for the Greeks, the stuff of the world was a given—it is central to Christian thinking that God created from nothing, ex nihilo. Together with the passages quoted earlier, the Bible harps continually on this theme. "By the word of the LORD were the heavens made; and all the host of them by the breath of his mouth" (Ps 33:6 KJV). "Through faith we understand that the worlds were framed by the word of God, so that things which are seen were not made of things which do appear" (Heb 11:3 KJV). I don't see that we have addressed this issue at all. What we do seem to have done is to have moved God into the position where he could be the creator. He starts to be the sort of being that the Christian creator must necessarily be. That is no minor thing and is probably enough for our purposes, for we are not trying to prove the truth of Christianity but to make room for it in the face of science. Theologically, it is still open for the critic to go after the Christian, asking exactly how a necessary being might be a creator. Christians from Augustine on (and before) have wrestled with this very issue. But this is a somewhat different matter from faulting Christianity because it conflicts with modern science.

However, we cannot stop here, for the Christian wants to claim more of God than that he just exists, although note that already we are starting to bring in some properties such as his total freedom. God had the power to create. In fact, for the Christian, because of his total freedom, he has total power—he is omnipotent. He also has total knowledge—he is omniscient. And he did not create just for fun. He did it out of pure love. He wanted beings in his own image whom he could cherish and care for, at the most extreme level. This means that he was prepared to intervene in his creation. "For us and for our salvation he came down from heaven, was incarnate of the Holy Spirit and the Virgin Mary and became truly human. For our sake he was crucified under Pontius Pilate; he suffered death and was buried." These are theological claims, and (you will be getting tired of my repeating this) they must be judged in this light. Christians believe these things on faith. But it is still incumbent on us to see how they measure up—or measure out—in the light of science. (In this paragraph, I have quoted the Nicene Creed [325 CE], which is somewhat later than the earlier version, the Apostles' Creed [ca. 180 CE], quoted at the beginning of the chapter. It stresses that Jesus came to earth for our salvation.)

Part of the problem here, of course, is understanding precisely what one means by such terms as "all knowing." The traditional Christian position, embedded in the thinking of Saint Thomas, has always been straightforward. We cannot know God directly, but we can know him analogically. When we say that God is our Father, we do not mean that it was his sperm that fertilized our mother's eggs. We mean that God has fatherlike qualities—he is the ultimate creator, and he provides for us, and he cares for us. Just remember, this is always a bit imperfect: "It should nonetheless be kept in mind that Revelation remains charged with mystery. It is true that Jesus, with his entire life, revealed the countenance of the Father, for he came to teach the secret things of God. But our vision of the face of God is always fragmentary and impaired by the limits of our understanding" (John Paul II 1998, 13–14). It is precisely this point that is stressed by many Protestants. Barth was a prime example. He wanted to stress above all the otherness of God, and this makes God rather distant, at least with respect to understanding. With regard to God's attributes, start running through the properties. All loving. One can certainly glimpse without contradiction what it would mean for a being to be all loving. Such a being would give unconditional love—freely, without hope of gain, all of the time, as much as could be given. Parents know what it is to give to their children. Teachers and professors know what it is to give to their students. Just keep multiplying this. Of course, what constitutes love in particular instances is often problematic—we all know parents (perhaps ourselves) who spoil their children with too much unconditional giving, but love never implied being soft or avoiding difficult decisions.

All powerful. There are Christians today who want to avoid or modify this attribute. Process philosophers, those influenced by the philosophy of Alfred North Whitehead, think that God is part of the creative process along with us (Whitehead 1929, cf. Cobb and Griffin 1976). They often invoke the notion of *kenosis*, that is, of God voluntarily giving up his powers, as when he let himself be crucified on the cross ("Let this mind be in you, which was also in Christ Jesus: Who, being in the form of God, thought it not robbery to be equal with God: But made himself of no reputation, and took upon him the form of a servant, and was made in the likeness of men: And being found in fashion as a man, he humbled himself, and became obedient unto death, even the death of the cross" [Phil 2:5–8 KJV]). For the process thinkers, God can help to direct things here on earth, but he cannot force anything (Ruse 2008b). This is still very much a minority position. Most Christians, especially those in the Augustinian tradition (which very much includes Calvinists) do not accept these limits (voluntary or not) on God. But one must tread carefully. Does the traditionalist mean that there are no limits whatsoever on God's powers? Can one think coherently of a being who has total control and can do anything? Every first-year undergraduate knows that there are pitfalls here. Could God make a stone that he could not lift? Could God make two plus two equal five?

Admittedly, not everyone has agreed on these points. Descartes notoriously thought that God could make mathematics do anything he wanted if he were so inclined. (This is known by philosophers as *universal possibilism*.) As someone with inclinations toward Platonism in mathematics, I am inclined to think that God is bound by the laws of mathematics (and of logic)—or, if you prefer to put it another way, God's omnipotence is not constrained by the fact that he cannot do the logically or mathematically impossible. However, if you warm toward universal possibilism, you might still want God bound by logic, even if you argue that in an age of non-Euclidean geometries (which seemed very non-possible to people before the nineteenth century) God has more freedom of action when it comes to mathematics. One suspects that in the end, full-blooded universal possibilism verges on the incoherent—to us at least. "If we cannot understand 'infinite power,' we also cannot understand and hence cannot believe or know, the proposition that God's power is infinite" (Plantinga 1980, 116). Certainly, if you want to be on-side with science, Christians are constrained in their thinking on these matters by the usual rules of reason. Even if you go so far as to think that God can do the logically impossible (even if he never would), you have to explain this to yourselves and to us in terms that do not imply the logically impossible.

The Problem of Evil

Raise now the venerable problem of evil. If God is all powerful, then he could prevent evil, and if he is all loving, then he wants to prevent evil; and yet evil exists. Hence God cannot have both of these predicates usually ascribed to him. Whether or not you think them adequate, the Christian has moves that avoid outright contradiction (Pike 1964, Hick 1978, Ruse 2000a). With respect to moral evil—the evil that led to Auschwitz—the Christian (following Augustine) invokes human freedom. It is better that humans have freedom, even though it led to the gas chambers, than that we be simply automata. With respect to physical evil—the child dying of leukemia—the Christian (following Leibniz) argues that it is all a matter of possibilities. Since God is constrained by the impossible, if he wanted to maximize the good things, necessarily he had to create a world containing physical evil. The good from being warned outweighs the pain from burning. Gravity is a good, even though people fall to their deaths from heights. You may or may not find these responses adequate. Notoriously, Voltaire was scathingly funny in his *Candide* about the Leibnizian argument. That is not our question now. Agreeing that the moves would let the Christian avoid outright contradiction, the question now is: Does science exacerbate the problem of evil? If you hold to modern science, does this make the problem of evil even more intense, to the point that reasonable Christian belief is impossible?

Many scientists and fellow travelers think that this is so. In particular, evolutionary biology, with its focus on a bloody struggle for existence, the force behind natural selection—the main cause of evolutionary change—mocks the idea of the Christian God. Richard Dawkins, predictably, has been eloquent on the subject. The definitive text is a letter that Darwin wrote on May 22, 1860, to his friend Asa Gray, a Harvard botanist and a deeply committed evangelical Presbyterian.

> With respect to the theological view of the question; this is always painful to me.—I am bewildered.—I had no intention to write atheistically. But I own that I cannot see, as plainly as others do, & as I shd. wish to do, evidence of design & beneficence on all sides of us. There seems to me too much misery in the world. I cannot persuade myself that a beneficent & omnipotent God would have designedly created the Ichneumonidae with the express intention of their feeding within the living bodies of caterpillars, or that a cat should play with mice. Not believing this, I see no necessity in the belief that the eye was expressly designed. (C. Darwin 1985–2023, 8:224)

Dawkins agrees wholeheartedly with this sentiment. Predators are designed for catching prey, prey for evading predators. If one wins, the other dies in the agony of skin-and-flesh-ripping fangs and claws. If the other wins, then the other dies in the agony of starvation. This is not just chance. This is the way that things are and must be. . . . "DNA neither knows nor cares. DNA just is. And we dance to its music" (Dawkins 1995, 133). The philosopher Philip Kitcher has recently recanted a lifelong willingness to see the possibility of Christian faith and practice (Kitcher 2007). In terms appropriate to a revival tent, he has confessed his sins and now agrees that if one subscribes to Darwinian evolutionary thinking, then it is simply not possible to commit to the Christian religion—at least, it is not possible if one wants to retain some modicum of integrity and to deserve the attribute of reasonable being. Unfortunately, although perhaps predictably, one suspects that a lot of Christians feel the same way. The Reverend Keith Ward, an Anglican priest and Regius Professor of Religion at Oxford University, seeks a warmer, friendlier form of evolutionary theory. We may not be able to get away from suffering and death, but at least we can tone them down a little: "On the newer, more holistic, picture, suffering and death are inevitable parts of a development that involves improvement through conflict and generation of the new. But suffering and death are not the predominating features of nature. They are rather necessary consequences or conditions of a process of emergent harmonization which inevitably discards the old as it moves on to the new" (Ward 1996, 87).

Before we get bowled over, let us look at these worries in the light of the Christian responses to the problem of evil. The questions for us to ask are about how these responses fare when faced with modern science, specifically with Darwinian evolutionary theory. Start with the free will defense against moral evil. Evolutionary theory is of course something that, like the rest of science, supposes that the world runs in a lawlike fashion. We need not dwell long on this point because we have already spoken to the fear that law in some sense precludes free will. I have taken a compatibilist position, arguing that free will demands law rather than fights it; but, even if you are not a compatibilist, you are still going to agree that the world is lawlike and that this applies to living things, including humans. The whole of child-rearing and education presupposes this. Why try to teach the multiplication tables to children unless you think that they are animals capable of learning them and of retaining them for more than five minutes? If you are a noncompatibilist, then you think that there is more to life than law, but we knew that already. The real question is whether Darwinian evolutionary biology in its own right in some sense makes the free will defense less compelling than otherwise. You might think so. A major cry about thirty years ago, repeated to this day, is that to apply Darwinian theory to human behavior is to suppose an illegitimate thesis about "genetic determinism." You are supposing that humans really are bound by laws, with their genes calling the shots, and that freedom is no more real than the hypnotized stage volunteer's illusion that he or she is free. Truly we are marionettes at the mercy of the controlling DNA (Lewontin et al. 1984).

In fact, as Daniel Dennett has pointed out, this simply is a travesty of the Darwinians' thinking on human nature. Ants are genetically determined. They are, to use a metaphor, entirely controlled by the tiny computers that they have for brains. They do not think. They just obey orders. And this is just fine for ants, but it would not be just fine for humans. The reason is simple. Ants have gone the route of having many offspring along with having little care or concern for individual well-being. The great thing about being genetically determined is that you do not have to waste effort on education and so forth. The really bad thing about genetic determination is that if something goes wrong, you do not have the ability to put things right. Dennett gives a lovely example. He tells of a wasp that brings food to its nest to provision its young. "The wasp's routine is to bring the paralyzed cricket to the burrow, leave it on the threshold, go inside to see that all is well, emerge, and then drag the cricket in. If the cricket is moved a few inches away while the wasp is inside making her preliminary inspection, the wasp, on emerging from the burrow, will bring the cricket back to the threshold, but not inside, and will then repeat the preparatory procedure of entering the burrow to see that everything is all right." This can go on and on indefinitely. "The wasp never thinks of pulling the cricket straight in. On one occasion this procedure was repeated forty times, always with the same result" (Dennett 1984, 11). This is the problem. A thousand ants are out foraging, and it starts to rain. A thousand ants are lost because the chemical (pheromone) trails that lead them home are washed away.

In the case of the ants, it does not matter much. There are hundreds of thousands more where the vanished ants came from. In the case of humans, it would matter very much. We have gone the route of having few children who demand much care. We cannot afford to lose even two or three every time it starts to rain. So how have we set about raising children in the face of adversities like rainstorms and predators and fellow humans? Here is where our brains are important. As Richard Dawkins says in *The Blind Watchmaker*, we have big onboard computers, and when we come to challenges, we can reason how to overcome them. This does not mean that we move from beneath the net of law, but that we have a dimension of freedom that the ants do not have. To continue with Dennett's examples, we are like the Mars Rover (a machine, take note). When it came to a rock, it did not stop, but had the computer-driven ability to reason how to get around the rock. Or to use an example of my own, ants are like cheap rockets that are aimed at the target, but if the target moves they miss it because they cannot change direction. Humans are like expensive rockets that have homing devices to track a moving target. Both rockets do what they do because of unbroken law, but the expensive rockets—a.k.a. humans—have a dimension of freedom that the inexpensive rockets—a.k.a. ants—do not. In other words, I suggest that far from Darwinian theory blowing holes in the free will defense, if anything it comes to its aid. (Technically, biologists say that humans are practicing K selection, whereas the ants are practicing R selection. Of course, everyone thinks that there has been a feedback process in evolution. As we got better at raising children, we could have fewer, and having fewer put more selective pressure on being better able to raise children.)

What about the physical evil defense that God cannot be expected to do the impossible? Bad things are inevitable as good things are maximized. Here again matters are a little more complex, and more paradoxical, than you might expect. As Dennett came to the aid of the free will defense, so Dawkins comes to the aid of the Leibnizian defense. He argues that we could not have a functioning organic world without adaptation—design-like features—and that the only way in which you can get such features is through natural selection. The intelligent design theorists' supposition of guided mutations or variations is just not science. Lamarckism, the inheritance of acquired characteristics, is false. Macromutations or large variations leading to instant new features simply do not create design-like effects. The second law of thermodynamics kicks in here—things run down and not up, creatively. Chance leads to mistakes and not to triumphs of design. This really leaves only selection.

For the Christian, the logic now is simple. God has created through unbroken law. There are fairly good theological reasons why he did things this way and not in one instantaneous flash of creative activity. If you are an Augustinian, thinking the thought of creation, the act of creation, and the product of creation are as one to God—because he stands outside time—then it is very natural to see God's creation as unfurling rather than pinned down to one moment. Augustine himself recognized this, because he spoke of God having created the seeds of life rather than life itself (McGrath 2009, 102). But what are the consequences of this creation through law? Adaptation is necessary. The only way you can get adaptation through law is by natural selection. Natural selection demands a struggle for existence. A struggle for existence necessarily involves pain and suffering. The virtues of having animals, humans particularly, outweighs the pain and suffering. Hence, God could not have done other than he did, given his all-loving nature. Pushing aside universal possibilism, none of this compromises God's freedom. Which is just as well, for, as we have seen, part of God's very being is his freedom. Indeed, this freedom is such a great good that, as part of his love, God gave humans freedom also. There are immediate theological problems here, especially for the Calvinist, who stresses God's sovereignty and total power. How much freedom can humans be allowed? Notoriously, Calvin (and his great inspiration Augustine) embraced the doctrine of predestination—God foreordaining all that will happen, especially the choice of the saved and the damned.

Miracles

God created heaven and earth. For the Augustinian, God always had this thought of creation, and so one might suppose that, in a sense, the existence of God and the existence of matter are a package deal. Although Genesis gives the impression of a definite moment of creation, a point at which presumably time began, in a deeper sense you never get God without his creation. Does not the belief in miracles fly in the face of science? What we mean by natural science is the attempt to explain the physical world, the world of experience, in terms of unbroken regularity, of law. It is true that back in the seventeenth century, some—Newton himself, prominently—were prepared to allow God some kind of intervening and adjusting role in the universe's running, but at least since the nineteenth century this has not been permitted. William Whewell, ardent though he was against evolution, admitted that his own appeal to the supernatural took him out of the realm of law. Of organisms he wrote: "When we inquire whence they came into this our world, geology is silent. The mystery of creation is not within the range of her legitimate territory; she says nothing, but she points upwards" (Whewell 1837, 3:588). The argument and the conclusion are obvious. Science demands unbroken law. Christianity posits miracles, violations of law. Hence, science and Christianity are in conflict. Even though Christianity may speak to science's unanswered questions, science cannot accept the legitimacy of the Christian faith.

My suspicion is that David Hume was right and that the evidence against miracles is always more plausible than the evidence for them (Hume 1999). Proving miracles gets you into the domain of science, and that is a sucker's game. Better simply to accept the miracles on faith and leave it at that. Also, one of the important things about miracles is that generally they are rare—they are miraculous! If too many miracles are called for, then the notion starts to get debased. It is true that some of the most common miracles are conceived in such a way that, even though God's specific action may be called for, they do not involve the breaking of natural law.

Chapter 4, Reading 3

Darwinism and Belief[1]

EVER SINCE CHARLES DARWIN published his landmark *On the Origin of Species* in 1859, no issue has fueled the science-and-religion debate more than his theory of evolution by natural selection. Indeed, that scientific theory and religious reactions to it have come to dominate and define the debate. Separately and together, we are called on to teach, write, or speak on this one issue more than on all the other issues in science and religion combined.

    Everyone knows families where the children look and act nothing at all like their parents and where their attitudes and lifestyles are so different that it causes friends to wonder if the hospital made a mistake and sent home some babies with the wrong parents. Then, one day, perhaps when everyone is a bit tired and the light is none too good, a gesture, a glimpse of a profile, a tone of the voice—and, oh my goodness, the child is not just a chip off the old block, he or she is a veritable cord. Christianity and Darwinism—meaning Charles Darwin's theory of evolution through natural selection—are a bit like that. And so are Judaism and Darwinism. Of course they are different. They could not be more so. Christianity and Judaism affirm that humans were created miraculously by a good God, who not only made the world for humans to live in but who continues to care for people right through their lives and beyond. Darwinism counters that humans are the end product of a long line of lower organisms,

---

1. Ruse 2017f. Reprinted with permission of Yale University Press.

right back to blobs and beyond. If the Genesis account shared by Jews and Christians says God made the first human from the dust of the ground, then people are modified mud; and if Darwinism says the unbroken laws of nature, forever cycling mindlessly, brought humanity into existence, then people are modified monkeys, with that simian line itself emerging from a mud-like ooze of organic molecules.

And yet, then suddenly the evolutionist working away in the lab or out in the field, perhaps collecting frogs or primroses or digging for fossils or counting fruit flies, stops suddenly, brought up short. There is a flash of something, and for those who had Christian or Jewish backgrounds, no matter how long ago or pushed down by newer memories, it all comes flooding back. Perhaps it is a flower's intricate design to capture insects, or it is the realization that it all started somewhere way back when and is incredibly beautiful and a bit frightening, or it is a sense that there is something about human nature that simply doesn't seem to be reducible to molecules or whatever. Through Darwinism the philosophical observer sees something older, something farther back in Western culture—something that might spring from a Jewish or Christian view of life.

Those who know a bit about the history of science should not really be surprised by this observation. They know that Darwinian evolutionary theory came from somewhere and that it didn't come from Japan or China or India or Africa or any other non-Western part of the world. It came from Europe, Western Europe, and although countries like Russia and obviously the more recent United States contribute to the story, it was places like Britain, Germany, and France that gave birth to such thinking.

And it didn't happen twenty-five hundred years ago when the Greeks began to think critically or a bit later when the Romans conquered much of Europe. It happened when Christianity was part of the very fiber of the culture of those countries. It happened when to make sense of the world and its inhabitants one turned to the stories of the Bible and to the philosophical schemes of Catholic thinkers like Augustine and Aquinas—later, Protestant Reformers like Martin Luther and John Calvin—and found what guidance and understanding was to be taken from these sources. By "guidance and understanding" in such a context, we want to suggest that there are at least three big questions of life that can be asked and that Christianity and its parent Judaism try to answer. This in itself is significant because it has not been obvious to every intelligent person at every time and in every place that there are these questions, let alone answers to them. When addressing these questions, this chapter focus on Christians because their beliefs most strongly influenced European thought when Darwinism arose during the 1800s, but the answers would be much the same for Jews of the period. For Islam, the answers have too little a tie to the origins of Darwinism to be of concern here, but as a theistic religion that posits the divine creation of humans, the questions are much the same and the answers are not too different.

First big question: "Where did everything come from?" Today, as in Darwin's day, this might strike most thinking Europeans and Americans as one of the obvious big questions of life. But many ancient Greek philosophers—Aristotle, for instance—never thought to ask it. To him, things just were. No one asks, "When did two plus two first start equaling four?" Or if people do, they show that they don't understand the meaning of mathematics. The equation "two plus two equals four" always was and always will be. That is how it goes in mathematics. And that is how it goes in the real world, in the eyes of Aristotle. Matter just is. Of course, it can be altered. But the basic stuff simply is no more a subject for inquiry about origins than are the truths of arithmetic and geometry. Many religions are much the same—Hinduism and Buddhism, to name two major examples, and also Confucianism, if that can be called a religion: the universe just is. And people just are.

Yet within Christianity, the question about origins is one that is asked, and it is one that is thought coherent and worthy of an answer. It is clear why: Christianity is based on Jewish thought, and Judaism is imbued with a historical consciousness. The ancient Hebrew people did think in terms of origins and beginnings, and Christians followed them in doing so. This is what Genesis is all about. God made the world and set it in motion. That is the answer to where things came from "in the beginning." Believers needed to know that is true. It is central to their theology.

Second big question: "What kind of world do people live in?" Some people may have so little imagination that they would never ask a question like this. They just take things as they are, one event after another. But most people have some curiosity, and of course there are all sorts of answers one might give, at different levels. One of us—the historian—after living in Seattle for two decades, only half-jokingly complained that it's "a world where it never stops raining!" while both of us have felt much the same after "summering" in England. Louis Armstrong can croon that it's "a wonderful world," while ancient Babylonians typically saw it as a dreadful place dominated by gods who were at best indifferent to humans and at worst malicious. Others might use words like "mystifying," "awe-inspiring," or "comforting." Responses like these get closer to some of the kinds of answers one finds in much modern religious thought, from mainline Protestantism through Reform Judaism to New Age mysticism.

More traditional Christians, those informed by natural theology, might go in a slightly different direction and say something like "We live in a world that is put together in a pretty clever fashion. It isn't just a random heap of bits and pieces. The rains fall, the crops grow, the animals and humans have food to eat, and so life goes on in an orderly, functional fashion." This is a very Christian sort of answer, and it is one that finds support in the Bible. It was also very much the way of Plato and Aristotle. It was Plato who drew the inference that such a beautifully functioning world could not be pure chance, there must be an intelligence (a.k.a. God) that lies behind it. Aristotle was more biologically minded than Plato. It was he who saw that it is in the living world of plants and animals that one really finds this intricate functioning, this design-like nature, which seems to be more than just random chance. Living organisms are such clever structures that there must be some meaning to them. And it is here that Christian theologian philosophers, from Thomas Aquinas to William Paley, stepped in to make sense of it all. To them, the world does seem like it is designed, because it is designed—by an all-powerful, all-loving being—the God of Christianity.

Third big question: "Where do humans fit into the scheme of things?" At least by their way of judging, by any relative measure, humans are way ahead of all other organisms, even ahead of—especially ahead of—those beings that seem most like them, the higher apes—gorillas, chimpanzees, orangutans. Humans may not be as fast as the cheetah or able to fly like the butterfly. They may not be as beautiful as a flower in the morning. But they have intelligence and the tools—in particular arms and hands, but also an upright nature—to do something with them. Humans also have a consciousness of their own existence and of a sense of morality and mortality far beyond that of any other kind of animal. They are not gods, but it is not surprising that the world's most widely accepted religious account of human origins maintains that people are made in the image of God and are given "dominion" over the earth and all its nonhuman inhabitants.

Now where does the theory of evolution—Darwinism in particular—fit into all of this? Well, to be quite frank, nowhere very much until after the so-called Scientific Revolution.... Then people started to speculate in a naturalistic way about physical origins....

And then there was the organic world. The focus on design implied stability. For example, the human or animal eye is just so perfect, it must have always been so, like a cleverly designed watch or, to use the eighteenth-century British natural theologian William Paley's analogy for it, a telescope. But there were pointers the other way. Perhaps most significant was an idea that went back to the medieval era, and through that, even farther back to some Greek philosophers. Could it not be that the whole of life might be placed in an upward-pointing "chain of being"? In it, the most primitive organisms are at the bottom and then, climbing up through the fish and reptiles and mammals, the humans are reached at the end. This is not a dynamic picture—it is a ladder fixed in eternity rather than an escalator moving in time—but obviously it does lend itself to speculations, and if change is in the air, so be it. For the late-eighteenth-century French evolutionist, it became an escalator. Backing this temporalizing of the chain were increasing numbers of embryological studies, where the sequence from the fertilized egg to the grown adult can be followed in detail—again giving rise to speculations about change in the broader context. Then, with more mining and the cutting of canals across the European countryside, the unearthing of fossils gave direction to animal development the deeper people dug, from more complex above to simpler below—a seeming temporal record of progressive change over time documented by Cuvier, William Smith, and others.

Above all, however, what truly triggered naturalistic speculations about organic origins was the new philosophy of progress. This is the idea that change for the better—in society, in well-being, in medicine, in science, in culture generally—can be achieved through human abilities and effort.... But for all this, gradually people were starting to think in evolutionary terms—although back then the word "evolution" referred more to embryological growth, and people tended to use such terms as "development" and "transmutation" for biological change over time, as in the transmutation of one species into another. So much for God specially creating each biologic kind at the beginning. Species might develop much like individual embryos—so much the grander and still potentially divine. A wholesale God beats a retail God, in the progressive mind. Even if humans were made in God's image, people began to think that evolution seemed more apt than stasis, though many who thought this way did not give much of an ongoing role to God beyond the creation.

As it happens, few of the early evolutionists, if they may be called that without undue anachronism, were atheists or even agnostics. Like Erasmus Darwin and his science-minded friends, such as Benjamin Franklin and Joseph Priestley, they might have trouble with the divinity of Christ. But they all thought him a wonderful role model, and behind him they all accepted the existence of God—a God who had created and set the world in motion and still watched over it, and us, with a beneficent eye—the God of deism or theistic Unitarianism. Nevertheless, they knew that they were going against central aspects of Christianity. At the time, they and their Christian critics were generally not greatly worried about literal aspects of the Bible—obviously an evolutionist cannot take the first chapters of Genesis as literally true, word for word, but from Augustine on, sophisticated Christians had been realizing that aspects of the Bible must be understood metaphorically, typologically, or allegorically. But those Christian critics were worried about progress, for this stood in direct opposition to what they took to be central, providence, which is the idea that sinners can do nothing to save themselves other than rely on faith in God's forgiveness.

Grant these differences. But what is fascinating are the similarities, most particularly the similarity that the evolutionists took so seriously: the three main questions that were of such concern to Christians. Above all, Darwinism is a story of origins. No nonsense about eternal existence or whatever.

It is telling about where everything came from, organic species in particular. Today, scientists tend to separate cosmogony, the origins of the universe, from evolution, the origins of species. Back then, however, it tended all to be part of one story. And this package deal may not have been the Christian answer, but it was an answer to the same question posed by the Christians: Where did everything come from? It is similar to the second question, about the nature of the world. For Christians, believing in a good God who was all powerful, the world (including the world of organisms) simply had to be well designed. God would not have done—could not have done—a botched-up job. The eye does not function as well as it does just by chance.

To be candid, before Darwin, early evolutionists did not have a ready answer to the question of design in nature. Their Christian critics said that blind law does not lead to complex functioning—it leads to rust and decay and disorder and things not working—and generally the evolutionists thought they had a good point. The eye is pretty convincing. How could something so functionally complex come about by chance? Early evolutionists pretty much had to tough this one out, hoping that a solution would emerge. As it did, in 1859 when Charles Darwin published his *On the Origin of Species*. Just as Newton had put forward a mechanism to explain the Copernican universe—the force of gravitational attraction—so Darwin put forward a mechanism to explain or account for organic evolution—natural selection. Darwin seized on the calculations of Thomas Robert Malthus, a turn-of-the-century (eighteenth to nineteenth) English cleric who obsessed about overpopulation. The rate of reproduction always outstrips the available food and space. This means that there will be a fight for resources, what Malthus called a struggle for existence. He was thinking about people, but the same reasoning would apply to competition among organisms of any given type or species.

In other words, all of the time there is going to be a natural winnowing of organisms, with (on average) those with locally beneficial variations—what Darwin called "adaptations"—being the winners. This process, using an analogy from the practices of animal and plant breeders, Darwin called "natural selection." The important thing is that this doesn't just lead to change, or evolution; it leads to change in a particular direction, toward ever more beneficial or efficient adaptations. More efficient eyes help their possessors in the struggle for existence—the same for grasping hands, sharp teeth, and everything else. In other words, Darwin gave an answer to the second big Christian question: What kind of world do we live in? Why is the world as well designed as it is? It was not the direct intervention of a good God but the end result of the slow, law-bound process of natural selection.

What about the third question, where human beings fit into the scheme of things. Let there be no doubt that the early evolutionists were as obsessed by the superior status of humans as anyone else. Progress leads up to human beings, they assumed. But Darwin made things a bit more complicated. Socially, he was as much in favor of progress as anyone. His maternal grandfather was Josiah Wedgwood, the commercial potter, one of the leaders of the Industrial Revolution. Darwin was not about to turn his back on upward change.

Biologically, he had little doubt that humans came out on top. The trouble is that natural selection seems to be relativistic. It is not a tautology—tautologies are things that are necessary and the struggle for existence is not that, for it might never have happened—but it does suggest that winning and losing are not an absolute thing but depend on the circumstances. Little food, and it pays to be small. A lot of food, and being big might be the way to go. And food-wise, humans are pretty high-maintenance organisms.

Compared to some other species, humans are not very strong or fast, so they have to make their way by cooperation, intelligence, and tool use. That means big brains, or at least that was how Darwin reasoned. Obviously, if the theory of evolution is to be believed, it worked. Otherwise humans would not be here. But it was not obviously going to succeed. Darwin solved the problem by appealing to what evolutionists today call "arms races." Lines of organisms compete against each other and adaptations improve. The prey gets faster and so the predator gets faster. Eventually intelligence emerges as it proves its worth in the races, and humans emerged on top. Darwinism, like Christianity, supplied answers to three fundamental questions that Westerners cared about, which made it appealing to many of them....

The Problem of Evil

Historically, one of the biggest problem facing the Christian is the problem of evil. How could a God who is all powerful and all loving permit pain and suffering? Why do bad things happen to good people? Expectedly, the Christian has a number of answers to these issues. Traditionally, one separates evil into two kinds. First there is moral evil brought on by human beings. Humans are made in the image of God but live in a fallen world of their own sinful choices. It is better—it is more human—to have free will and then choose Christ than to be bound, to be determined. This is so, even though it can lead to dreadful consequences. Second, there is natural evil, like the mutation causing Huntington's chorea or Tay-Sachs disease. Here the usual defense is a version of the argument made famous by the seventeenth-century German philosopher Gottfried Wilhelm von Leibniz. He pointed out that God's being all powerful does not mean God can do the impossible. God cannot make two plus two equal five. Making the world was a matter of balances. If God was going to warn us against the fire, then God needed a pretty powerful method of motivation—pain. On balance, the pain from burning for some outweighs the dangers from burning for all. Likewise with mutations. If God was going to have some way of making organisms, including humans, then in the end God needed mutations. On balance, the good ones leading to such things as intelligence outweigh the evils of the mutations that lead to pain and suffering. And how much can any suffering in this brief life matter anyway when balanced against eternity? All things lead to good in those that trust in God, Christians can affirm.

We do want to do two (related) things. First, we point out that the problem of evil is a Christian issue. It stems from the kind of God the Christian supposes. That is, an all-powerful, all-loving being. . . . Which brings us to the second point about the problem of evil. It may not be such a big problem for non-Christian religions, but what about under evolutionary biology, especially the Darwinian version? If Darwinism is so close to Christianity, is the problem of evil something with which it wrestles? Now, obviously in one sense it doesn't and can't. Science doesn't work like that. Science seeks to tell it like it is or appears to be based on testable observations and repeatable experiments. Which, of course, is true, but it doesn't mean that science cannot have something of interest, of pertinence, to say about the problem of evil. After all, Darwinian theory came into being creating and defining itself in relation to Christianity, and it attempts (perhaps because of or perhaps despite of) to answer the kinds of questions Christians find pertinent. As it happens, this suspicion is borne out. In particular, the strife and suffering caused by the evolutionary process have been taken by some as yet one more nail in the Christian coffin, yet one more reason why the problem of evil is simply an insuperable objection to the Christian's claims or at least to Christian natural theology.

Charles Darwin himself felt this, writing to his good friend the Harvard botanist Asa Gray, who was incidentally a sincere Presbyterian. "I had no intention to write atheistically. But I own that I cannot see, as plainly as others do, & as I should wish to do, evidence of design & beneficence on all sides of us. There seems to me too much misery in the world. I cannot persuade myself that a beneficent & omnipotent God would have designedly created the Ichneumonidae with the express intention of their feeding within the living bodies of caterpillars, or that a cat should play with mice" (C. Darwin 1860a, para. 3). A Calvinist Christian, Gray understood this perspective—only a few are selected, whether by God or nature—so it did not pose an insurmountable problem to his religious faith, but natural selection was identified early on as an issue by some Christian critics whose faith rested heavily on natural theology, such as the influential Princeton theologian Charles Hodge. "What is Darwinism?" Hodge asked in an 1874 book with that title. "It is atheism" (1874, 177).

Chapter 4, Reading 4

Darwinism as Religion[1]

"Naked came I out of my mother's womb, and naked shall I return thither: the Lord gave, and the Lord hath taken away; blessed be the name of the Lord" (Job 1:21 KJV). Well, natural selection has done its fair share of taking away. Can it also do some giving? Less metaphorically—although my quoting of the Bible was deliberate and, as you will see, contains a hook—if we do not take the religious route, and if natural selection seems to point to the bleak world of Thomas Hardy, is there any case for saying that natural selection can nevertheless contribute positively to an alternate world picture? A world picture that gives meaning to life? We shall see that there is. To tackle the problem properly, however, we need to make and follow a division made generally by those asking about the meaning of life.... Can a natural selection–governed or –inspired world picture give an objectivist understanding of meaning? Can natural selection impose upon us a set of rules for right conduct and point to a worthwhile end to which we should aspire and labor? Thinking of the contrary, our second question must be that if achieving an objectivist understanding proves impossible, and hence in some very real sense we are thrown back on ourselves, can natural selection contribute to such a subjectivist world picture? Can it help us to find meaning after all?

---

1. Ruse 2019a. Reprinted with permission of Oxford University Press.

Can Darwinism Be a Religion?

I will take these questions in turn, starting in this chapter with the objectivist approach. Let me now double back to my hook. Religion yields the paradigm of the objectivist approach. Can we put ourselves on the path to an answer by asking if, in some sense, a Darwin-inspired world picture can be or can function as a religion? In our opening lines, religion sets the pace for questions about natural selection. Is it worth continuing in this mode? Cutting at once to the quick, I argue that this is indeed the right way to go and gives the right questions to ask. From the time of the *Origin*, what I shall now by stipulation call Darwinism—restricting the term to a secular world picture inspired by Darwin's theory of evolution through natural selection—has existed and flourished. This holds down to—very much down to—the present (Ruse 2005a, 2017g, 2018b).

At once I make a caveat. I am now distinguishing Darwinism, as I use it, from Charles Darwin's theory of evolution through natural selection, as it was in the *Origin* and as it has matured into today's theory. I take without argument that this latter was and always was, and is now, a genuine scientific theory, with a proper causal understanding of the physical world (Ruse 2006). Increasingly, it has been backed by empirical evidence, both from nature and from the laboratory. I know there have been various scientific challenges—genuine and pseudo—to the supreme place of Darwin's theory. I assume, however, that these are either inadequate or, more often, readily incorporated into the main theory or paradigm. More important here is that in the light of the arguments, because Darwinian theory is genuine science, no more no less, on its own it cannot answer our questions about meaning/Meaning. If such questions are to be answered in the Darwinian world, then something must be added that makes this possible. If Darwinian thinking is to be turned from straight science into a kind of religion, ask about the new ingredient.

Many nonbelievers today, new atheists like Richard Dawkins, deny with indignation the charge that they are in the religion business. To be honest, it is hard to take too seriously the protestations of someone who begins a book with: "The God of the Old Testament is arguably the most unpleasant character in all fiction: jealous and proud of it; a petty, unjust, unforgiving control-freak; a vindictive, bloodthirsty ethnic cleanser; a misogynistic, homophobic, racist, infanticidal, genocidal, filicidal, pestilential, megalomaniacal, sadomasochistic, capriciously malevolent bully" (Dawkins 2006, 1). If those aren't the words of an Old Testament prophet, I don't know what are. We need not, however, quibble about terms. Evolutionists themselves openly tell us that they are in the religion business. . . . Evolutionary thinking simply does not cure a pain in the belly. Evolution, Darwinian evolution, has another function: to offer an alternative to the conventional Christian religion.

Thomas Henry Huxley was quite explicit that he was seeking a new religion to supplant the old, Christian religion. Even before the *Origin*—he had been primed about Darwin's theory—he wrote about seeing (conventional) religion and science forever at war. "Few see it but I believe we are on the eve of a new Reformation and if I have a wish to live thirty years, it is that I may see the foot of Science on the necks of her Enemies. . . . But the new religion will not be a worship of the intellect alone" (Desmond 1997, 253). "Extinguished theologians lie about the cradle of every science as the strangled snakes beside that of Hercules; and history records that whenever science and orthodoxy have been fairly opposed, the latter has been forced to retire from the lists, bleeding and crushed, if not annihilated; scotched, if not slain" (T. Huxley 1860b, 52).

Thomas Henry Huxley's grandson, the evolutionary biologist Julian Huxley . . . was even keener to make a religion out of his science. He wrote a book called *Religion without Revelation*. In the pattern of the older man, grandfather Thomas Henry, Julian did not want to rid the world of religion. He wanted to change it for secular purposes, God must go; but what remains of religion is vital: "If, finally, there be no reason for ascribing personality or pure spirituality to this God, but every reason against it; then religion becomes a natural and vital part of human existence, not a thing apart; a false dualism is overthrown; and the pursuit of the religious life is seen to resemble the pursuit of a scientific truth or artistic expression, as the highest of human activities" (J. Huxley 1927, 53–54). We need religion. Not the traditional religion. No prizes for guessing what the future author of the classic *Evolution: The Modern Synthesis* had in mind.

Prominent among today's evolutionists, Edward O. Wilson likewise sets about making a religion from his science. It is materialistic, or at least naturalistic, as it presents "the human mind with an alternative mythology that until now has always, point for point in zones of conflict, defeated traditional religion" (E. Wilson 1978, 192). Helpfully, Wilson tells us:

> Its narrative form is the epic: the evolution of the universe from the big bang of fifteen billion years ago through the origin of the elements and celestial bodies to the beginnings of life on earth. The evolutionary epic is mythology in the sense that the laws it adduces here and now are believed but can never be definitively proved to form a cause-and-effect continuum from physics to the social sciences, from this world to all other worlds in the visible universe, and backward through time to the beginning of the universe. (E. Wilson 1978, 192)

After this, we are hardly surprised to learn that "if this interpretation is correct, the final decisive edge enjoyed by scientific naturalism will come from its capacity to explain traditional religion, its chief competition, as a wholly material phenomenon. Theology is not likely to survive as an independent intellectual discipline" (E. Wilson 1978, 192).

Progress versus Providence

Take up now the need to identify the added ingredient that is to make all this possible. If we are to have a religion—secular or otherwise—we need an underlying metaphysic to hold it together. To make a picture. To confer meaning. This will be a kind of root metaphor. In the case of Christianity, although there are variations, we find our metaphysic, our root, in the idea of providence. A creator God, on whom we are totally dependent, who so loved us that for us he made the supreme sacrifice. What does evolution have to offer in its stead? If not providence, then what? . . . It is the idea that Darwin expressed at the end of the following poetic passage: "Imperious man, who rules the bestial crowd / . . . Arose from rudiments of form and sense, / An *embryon* point, or microscopic *ens*!" (Erasmus Darwin 1803, 1:11). And then he tied it in with a more general philosophy of progress, telling us that the idea of organic progressive evolution "is analogous to the improving excellence observable in every part of the creation; such as the progressive increase of the wisdom and happiness of its inhabitants" (Erasmus Darwin 1801, 2:247–48). This idea of progress—things getting better and better—in the cultural world and then reflecting into the biological world—thus conferring meaning as strongly as Christian providence—continued to get major and sympathetic attention in the years before Charles Darwin. It was the underlying theme of a very popular, pre-Darwinian, evolutionary tome, *Vestiges of the Natural History of Creation* (first published in 1844).

Then, in the 1850s, came the indefectible Herbert Spencer. For him, progress was everything. From an article two years before the *Origin*:

> Now we propose in the first place to show, that this law of organic progress is the law of all progress. Whether it be in the development of the Earth, in the development of Life upon its surface, in the development of Society, of Government, of Manufactures, of Commerce, of Language, Literature, Science, Art, this same evolution of the simple into the complex, through successive differentiations, holds throughout. From the earliest traceable cosmical changes down to the latest results of civilization, we shall find that the transformation of the homogeneous into the heterogeneous is that in which Progress essentially consists. (Spencer 1857, 445)

With acknowledgment to Spencer, Julian Huxley defined evolutionary progress as "increased control over and independence of the environment" (J. Huxley 1942, 545). . . . "The future of progressive evolution is the future of man. The future of man if it is to be a progress and not merely a standstill or a degeneration, must be guided by a deliberate purpose. And the human purpose can only be formulated in terms of the new attributes achieved by life in becoming human" (J. Huxley 1942, 577). One thing was that Huxley knew the enemy when he saw it. Thanks to its underlying metaphysic, Christianity leads to moral intellectual and physical laziness. "Divine Providence is an excuse for the poor whom we have always with us; for the human improvidence which produces whole broods of children without reflection or case as to how they shall live; for not taking action when we are lazy; or, more rarely, for justifying the action we do take when we are energetic. From the point of view of the future destiny of man, the present is the time of clash between the idea of Providentialism and the idea of humanism—human control by human effort in accordance with human ideals" (J. Huxley 1927, 18).

Wilson is open in his fervent belief in biological progress: "The overall average across the history of life has moved from the simple and few to the more complex and numerous. During the past billion years, animals as a whole evolved upward in body size, feeding and defensive techniques, brain and behavioral complexity, social organization, and precision of environmental control—in each case farther from the nonliving state than their simpler antecedents did." Adding: "Progress, then, is a property of the evolution of life as a whole by almost any conceivable intuitive standard, including the acquisition of goals and intentions in the behavior of animals" (E. Wilson 1992, 187). Wilson, Southern born and bred and saved from his sins as a teenager, has moved on from his early years. Out of respect and affection, compared to the Huxleys, he is a lot less hostile toward Christianity. Candor, nevertheless, forces him to admit that his heart now belongs to a rival. "I see no way to avoid the fundamental differences in our respective worldviews" (E. Wilson 2006, 3–4).

On the one hand, Christianity: "You are a literalist interpreter of Christian Holy Scripture. You reject the conclusion of science that mankind evolved from lower forms. You believe that each person's soul is immortal, making this planet a way station to a second, eternal life. Salvation is assured those who are redeemed in Christ." On the other hand, Darwinism: "I am a secular humanist. I think existence is what we make of it as individuals. There is no guarantee of life after death, and heaven and hell are what we create for ourselves, on this planet. There is no other home. Humanity originated here by evolution from lower forms over millions of years." We are fancy apes who have adapted rather well to life here on Earth. And this means that spiritual explanations and understandings are otiose. The same is true of behavior: "Ethics is the code of behavior we share on the basis of reason, law, honor, and an inborn sense of decency, even as some ascribe it to God's will" (E. Wilson 2006, 4).

Is There Meaning?

Evolutionists do not have identical thoughts about progress. T. H. Huxley rather lost faith toward the end of his life; the horrendous social conditions in Victorian cities and like phenomena shook his thinking. He was not against all thoughts of progress, but they were tempered. Christians likewise do not have identical thoughts about providence. Think of the difference on this topic between, let us say, the Roman Catholics and the Jehovah's Witnesses. And that is to omit the Calvinists! The important thing is that, given the notion of providence, meaning falls at once into place. We are to do our duty by God—praising him and following his orders. On its own, our efforts can never be enough. But God through his love and his mercy forgives us. "In my Father's house are many mansions: if it were not so, I would have told you. I go to prepare a place for you" (John 14:2 KJV). That is meaning or Meaning—all that we have and all that we could ever desire. What then about meaning or Meaning for the evolutionary progressionist?

Meaning for the evolutionist is found in the upward rise of the history of life—monad to man. We humans are in some objective sense the winners, the top of the tree, of more value than other organisms. This is a function of many things, but our minds, our consciousness, our intelligence, are the all-important factors. We are in some sense more complex than other organisms. This complexity, in some way, plays itself out by making us thinking beings with an ability to understand our world and with our own powers of choice, of deciding between good and evil. This readily translates into prescriptions. In the biological world, we are to keep up the evolutionary process, at least not letting it decline and perhaps helping it ever upward. In the social realm—for remember that biological progress is a child of cultural progress—we are to make for a better society for one and for all.

Why Progress?

I want to turn to a question that is becoming increasingly asked.... We have been talking about evolutionary progress. How does natural selection fit into all of this? To be honest, thus far, not much (Ruse 1996b). Based on some half-baked readings of physics—a lot of his thinking was based on half-baked readings—Spencer saw life naturally in a kind of equilibrium (Richards 1987). Then something disturbs this equilibrium, and there is turmoil and chaos. Finally, equilibrium is reachieved but at a higher point. Interestingly, into this potpourri of thinking—"dynamic equilibrium"—Spencer introduced the Malthusian population explosion leading to struggle. For Spencer, however, the struggle led not so much to failure and selection as to striving to succeed, with the winners passing on their better attributes through a Lamarckian process. Apparently, sharing his fellow Victorians' belief that an organism can produce only a limited amount of vital bodily fluid, Spencer argued that either it can flow out between the loins into making many offspring, or it can head up to the brain and make for ever greater intelligence. The consequence is that with the more intelligent producing ever fewer children, the Malthusian struggle slows down and stops. Spencer himself was so far advanced that he was a lifelong bachelor and had no offspring at all.

Darwin came naturally to progress, but he knew also that there is a worm in the bud. Most commercial enterprises fail. The number of bankruptcies far exceed the number of successes. Progress in the Darwinian world can never be the easy upward movement, a kind of naturalized providence, where all will come right, no matter. He wrestled with this in a private notebook:

> The enormous number of animals in the world depends of their varied structure & complexity.—hence as the forms became complicated, they opened fresh means of adding to their complexity.—but yet there is no necessary tendency in the simple animals to become complicated although all perhaps will have done so from the new relations caused by the advancing complexity of others.—It may be said, why should there not be at any time as many species tending to dis-development (some probably always have done so, as the simplest fish), my answer is because, if we begin with the simplest forms & suppose them to have changed, their very changes tend to give rise to others. (C. Darwin 1987, E 95–97)

Complexity gives rise to new opportunities, and increasingly Darwin saw that natural selection can aid organisms to take advantage of these—Spencerian-type thinking before its time. One is arguing that progress will necessarily occur because nature has a built-in drive to complexity. Increasingly, as grew his confidence in natural selection, Darwin dropped this kind of thinking. But not the belief in progress of some kind or another. The *Origin* does not discuss humans in detail. It still makes clear Darwin's belief in progress: "The inhabitants of each successive period in the world's history have beaten their predecessors in the race for life, and are, in so far, higher in the scale of nature; and this may account for that vague yet ill-defined sentiment, felt by many paleontologists, that organisation on the whole has progressed" (C. Darwin 1859, 345). By the third edition of the *Origin*, in 1861, Darwin was confident that he had the answer. Progress thanks to selection, rather than progress despite selection.

Today's Evolutionists and Progress

Today, Dawkins stands in this tradition. "Directionalist common sense surely wins on the very long time scale: once there was only blue-green slime and now there are sharp-eyed metazoan" (Dawkins and Krebs 1979,

508). The key lies in arms races.... Leave critical comment for a moment. Turn to another approach that claims to have found a selection-driven progressive process. Paleontologist Simon Conway Morris—a Christian but seeking an entirely natural explanation—argues that only certain areas of what we might call "morphological space" are welcoming to life-forms (Conway Morris 2003) . . . Again and again, organisms take the same route into a preexisting niche. The saber-toothed-tiger-like organisms are a nice example, where the North American placental mammals (real cats) were matched right down the line by South American marsupials (thylacosmilus). There existed a niche for organisms that were predators, with catlike abilities and shearing/stabbing-like weapons.

Darwinian selection found more than one way to enter it—from the placental side and from the marsupial side. It was a question not of beating out others but of finding pathways that others had not found. Conway Morris argues that, given the ubiquity of convergence, the historical course of nature is not random but strongly selection constrained along certain pathways and to certain destinations. Most particularly, some kind of intelligent being was bound to emerge. After all, our own very existence shows that a kind of cultural adaptive niche exists—a niche that prizes intelligence and social abilities. "If brains can get big independently and provide a neural machine capable of handling a highly complex environment, then perhaps there are other parallels, other convergences that drive some groups towards complexity." Continuing: "We may be unique, but paradoxically those properties that define our uniqueness can still be inherent in the evolutionary process. In other words, if we humans had not evolved then something more-or-less identical would have emerged sooner or later" (Conway Morris 2003, 196).

Hold, for a moment, final judgments about natural selection and progress, and turn to the final part of the discussion/meaning. The argument is straightforward. It has certainly seemed so to enthusiasts from Spencer to Wilson. Evolution is progress, which means that it is a value-laden phenomenon. Over time, value increases. Humans are the end point, which means we are the most valuable species here on Earth. Humans over warthogs over reptiles over fish over bacteria. Hence, meaning comes through humans, cherishing us, protecting us, keeping us from decline—biological or cultural—and if possible, helping us to advance even further—biological and cultural.

Chapter 4, Reading 5

Darwinism Explains Religion (?)[1]

DARWINIAN EVOLUTIONARY THEORY HAS always taken behavior seriously. From the beginning, and especially in the *Origin*, Darwin realized that what animals do is as important as what they are. Being fleet of foot can be as advantageous in the struggle for existence and the ability to digest (say) grass or meat. Darwinian evolutionary theory has always applied to humans, so human behavior has always been a topic of interest. Note therefore the crucial causal plank, the key method of inquiry. For the Darwinian, natural selection is the first-row key to everything. It may not always work but it is the first thing to be invoked. And, the important thing about selection is that it does not merely lead to change but to change of a particular kind, namely, in the direction of adaptation or contrivance. Selection produces things that function toward desired ends, such as the eye being used for seeing and the teeth being used for biting and chewing. Hence, for the Darwinian interested in human behavior, the key to understanding is adaptation, brought on by natural selection. However, the key is not all powerful. Complicating the picture is the fact that not all features of the living world are necessarily adaptive. Some occur by chance and some are by-products of selection. So a major part of the Darwinian's task is determining if something is adaptive and hence probably produced by selection, or if something is not adaptive and in which case what did cause

---

1. Ruse 2009a. Reprinted with permission of Prometheus.

it, if indeed there was an identifiable cause. This obviously applies very much to studies of human behavior.

Religion is a major factor in human behavior and culture and naturally it has attracted considerable Darwinian attention. The big problem, therefore, is whether or not it is adaptive and if so, in what way, and if not, why then does it exist. As always, it is best to start with Charles Darwin himself to set the background.

Darwin on Religion

Much ink has been spilled on the question of Charles Darwin's debt to David Hume. My own feeling is that, although clearly Hume's general empiricism was important, and although many things in Hume cry out for evolutionary understanding (I do not think that Hume was an evolutionist), we should not overemphasize the connections. There were other more immediate philosophical influences, John F. W. Herschel and William Whewell, for instance. Sir James Mackintosh in ethics, perhaps. (For more details, look at Darwin's *Autobiography* [1958].) However, one place where Hume did have a major influence was on Darwin's thinking about the natural origins of religion. We know that, as a young man, Darwin read Hume's *Natural History of Religion*, and when Darwin himself took up the problem of its origins, the Scottish skeptic's heavy accent can almost be heard up from the pages. Something to be justified or attacked, but [also] as something to be explained. Thinking that the most primitive form of religion is when savages believe in spirit forces, he asks about its origins. Apparently it is all a question of seeing spirits in inanimate objects, feeling or pretending or mistakenly believing that they are truly alive. Darwin did not think that explanations of religious belief bore on the truth or falsity of religion. Whether true or false, the important point is that religion as considered by the scientist be considered a natural phenomenon. He stressed this point again and again.

Darwin's position, therefore, was that religion is a natural phenomenon or rather, a phenomenon that can be treated naturally—and he saw it as something that had evolved. It is noteworthy that Darwin said little about religion and its relationship to natural selection. Here there is a major break from Darwin's parallel discussion of morality, which did get linked firmly to selection. Perhaps for all that he claimed not to be addressing the truth status of religion, because by this time in his life

Darwin had become an agnostic—certainly he thought that Christianity is not proven in its essentials and false in many details—he did not think that religion could be directly promoted by selection. For Darwin, religion seems to be almost accidental, and brought about by animal features or powers that are simply misdirected. When we see something moving, it normally makes sense to think that it is living. We make mistakes, and ultimately this leads into religion. The one concession that he was prepared to make is that, in the case of civilized people, religion does help reinforce morality: "With the more civilized races, the conviction of the existence of an all-seeing Deity has had a potent influence on the advance of morality" (C. Darwin 1871, 2:394–95). But even here, Darwin did not want to explore the matter in more detail.

Religion as Adaptive

In the hundred years after the *Origin*, there was much interest in putative natural origins of religion. But most discussion came from the newly developing social sciences rather than from biology. The growth of Darwinian studies of social behavior—what is often known as "sociobiology"—has changed all of that. As always in discussions of evolution it is natural selection that drives the thread of the investigation. Hence, let us be guided by biological categories—Darwinian categories, that is. Most importantly, we will expect a division between those who think that religion is something brought about directly by natural selection, and those (like Darwin) who think that religion is something of a by-product. Then among those who suppose selection as the cause, there will be division between those who think that religion is of direct adaptive advantage to humans and those who think that it might not be such a good thing to have—perhaps a product of something like sexual selection or perhaps adaptive for someone or thing other than humans (like parasites). There is also the possibility of division between individual and group selectionists—that is, between those who think that religion must be for the benefit of the individual and those who think that religion is of group worth, perhaps even to the detriment of the individual. Finally, there is the possibility of division between those who think that religion is essentially biological and those who think that culture is significant if not all important.

Starting with those who think that religion is selection produced and of value to humans, we find the grand old man of Darwinian social studies, of sociobiology, Edward O. Wilson. For him, religion is apparently all a matter of group identity and sticking together:

> The highest form of religious practice, when examined more closely, can be seen to confer biological advantage. Above all, they congeal identity. In the midst of the chaotic and potentially disorienting experiences each person undergoes daily, religion classifies him, provides him with unquestioned membership in a group claiming great powers, and by this means gives him a driving purpose in life compatible with his self interest.
> (E. Wilson 1978, 178)

Wilson has always been ambivalent about the comparative significances of individual- and group-selective processes. On the religion issue, he rather divides, thinking that it is something brought on by a group process, but surely with individual benefits also. More robustly individualistic are the physical anthropologist Vernon Reynolds and the scholar of religion Ralph Tanner. They are quite accepting of such hypotheses as that circumcision of males, a practice central to religion of Jews and others, is something that prevents disease. This is a practice that benefits individuals. Somewhat ingeniously, Reynolds and Tanner suggest that religions tend to divide into those that promote high reproductive rates—many Semitic religions—and those that do not, North European Calvinism, for instance. This is something echoing interests and concerns of Darwin in the *Descent*. There, the great evolutionist worried that the worthless Catholic Irish seem to have lots of children whereas the hardworking Presbyterian Scots have but few. This was a horrific reflection, seemingly negating the upward, progressive nature of the evolutionary process, a picture so dear to the heart of Darwin and his fellow Victorians. Darwin consoled himself with the reflection that the Irish do not look after their kids whereas the Scots do, and so on balance the Scots if anything do better than the Irish.

Groups and Memes as Units of Selection

Showing just how different people's thinking can be and yet still (in the eyes of advocates) be under the banner of Darwinism—that is, people who would think of themselves as working in the sociobiological mode—we have the biologist David Sloan Wilson and the philosopher Daniel Dennett. Wilson is openly committed to a group-selective analysis of religion, wanting to regard societies as akin to organisms and as strengthened by a sincere commitment to a religious doctrine. He ties this thesis strongly to morality, which he speaks of as having "both a genetically evolved component and an open-ended cultural component" (D. Wilson 2002, 119). Wilson analyzes the society that Jean Calvin founded in Geneva in the sixteenth century, listing the rules that governed this group: "Obey parents," "Obey magistrates," "Obey pastors," and on down the list to "No lewdness, and sex only in marriage," "No theft, either by violence or cunning," and so forth. Of this Wilson writes:

> To summarize, the God-people relationship can be interpreted as a belief system that is designed to motivate the behaviors [examples of which are listed just above]. Those who regard religious belief as senseless superstition may need to revise their own beliefs. Those who regard supernatural agents as imaginary providers of imaginary services may have under-estimated the functionality of the God-person relationship in generating real services that can be achieved only by communal effort. Those who already think about religion in functional terms may be on the right track, but they may have underestimated the sophistication of the "motivational physiology" that goes far beyond the use of kinship terms and fear of hell. Indeed, it is hard for me to imagine a belief system better designed to motivate group-adaptive behavior for those who accept it as true. When it comes to turning a group into a societal organism, scarcely a word of Calvin's catechism is out of place. (D. Wilson 2002, 105)

Although he thinks that culture is crucial, ultimately Wilson sees real change as genetic. Coming in a very different direction, the philosopher Daniel Dennett agrees entirely that religion is something promoted by selection, but he is not at all convinced that this selection is necessarily for the benefit of humans, nor is it essentially (or truly in any way) genetic. Dennett has adopted a theory of Richard Dawkins that posits the existence of "memes," units of culture akin to genes, which compete for people's allegiances. Rival memes, as it were, invade people's minds, and those that win are those that are selected to continue. Winning is not random but a function of the features—the adaptations—that the memes have or promote. Successful advertising obviously is a paradigmatic example of memes at work.

For Dennett, religion is a meme parasite that has features that make it attractive even if it is not necessarily that good for the possessor. So that no one miss this point, he begins his book *Breaking the Spell: Religion as a Natural Phenomenon* by introducing the reader to the lancet fluke (*Dicrocelium dendriticum*), a parasite that corrupts the brain of an ant, causing it to strive to climb blades of grass, that it get eaten by a sheep or cow, and thus the fluke can complete its life cycle before its offspring are excreted and take up again with ants:

> Does anything like this ever happen with human beings? Yes indeed. We often find human beings setting aside their personal interests, their health, their chances to have children, and devoting their entire lives to furthering the interests of an idea that has lodged in their brains. The Arabic word islam means "submission," and every good Muslim bears witness, prays five times a day, gives alms, fasts during Ramadan, and tries to make the pilgrimage, or hajj, to Mecca, all on behalf of the idea of Allah, and Muhammad, the messenger of Allah. Christians and Jews do likewise, of course, devoting their lives to spreading the Word, making huge sacrifices, suffering bravely, risking their lives for an idea. So do Hindus and Buddhists. (Dennett 2006, 4)

Religion as By-Product

What of those who think religion falls more into the by-product category? The late Stephen Jay Gould himself was one who thought along these lines. The whole of human culture came under this category for him.... But most would not be this sweeping. Apart from anything else, religion with its costs—devotion to others, celibacy, ritual physical disfigurement, sacrifice, and so forth—simply does not seem to be the sort of thing that just happened as a by-product. It is just too costly. More likely is the idea that religion, as it were, piggybacks into existence and power on the backs of other things—real, powerful adaptations—and manages to exist because it cannot be stopped or because ultimately its costs are simply not that great. Student of culture Pascal Boyer inclines to the first option. For him, religion simply subverts or borrows features that our biology has put in place for good adaptive reasons, and for whatever reason it cannot be eradicated:

> The building of religious concepts requires mental systems and capacities that are there anyway, religious concepts or not.

> Religious morality uses moral intuitions, religious notions of supernatural agents recruit our intuitions about agency in general, and so on. This is why I said that religious concepts are parasitic upon other mental capacities. Our capacities to play music, paint pictures or even make sense of printed ink-patterns on a page are also parasitic in this sense. This means that we can explain how people play music, paint pictures and learn to read by examining how mental capacities are recruited by these activities. The same goes for religion. Because the concepts require all sorts of specific human capacities (an intuitive psychology, a tendency to attend to some counterintuitive concepts, as well as various social mind adaptations), we can explain religion by describing how these various mind capacities get recruited, how they contribute to the features of religion we find in so many different cultures. We do not need to assume that there is a special way of functioning that occurs only when processing religious thoughts. (Boyer 2002, 311)

But what is it that allows religion to get its hold in the first place? Anthropologist Scott Atran inclines to the second option, that religion grabs something adaptively useful and exploits it. For him, the big question facing organisms like humans is other living beings—above all, other living beings as threats. In an argument reminiscent of Darwin and his dog, Atran suggests that what we have is a somewhat overeager projection of the living onto the inanimate. It used to be thought that the baroque nasal appendages of the titanotheres were a case of sensible evolution having taken a step too far. Perhaps the same is true of religion. Cuckoos exploit the innate mechanisms that their host birds have for raising their young. Religion does much the same for humans:

> Supernatural agent concepts critically involve minimal triggering of evolved agency-detection schema, a part of folk psychology. Agency is a complex sort of "innate releasing mechanism." Natural selection designs the agency-detection system to deal rapidly and economically with stimulus situations involving people and animals as predators, protectors, and prey. This resulted in the system's being trip-wired to respond to fragmentary information under conditions of uncertainty, inciting perception of figures in the clouds, voices in the wind, lurking movements in the leaves, and emotions among interacting dots on a computer screen. This hair-triggering of the agency-detection mechanism readily lends itself to supernatural interpretation of uncertain or anxiety provoking events. People

interactively manipulate this universal cognitive susceptibility so as to scare or soothe themselves and others for varied ends. They do so consciously or unconsciously and in causally complex and distributed ways, in pursuit of war or love, to thwart calamity or renew serendipity, or to otherwise control or incite imagination. The result provides a united and ordered sense for cosmic, cultural, and personal existence. (Atran 2004, 78)

Serious Science?

What can we say about these various ideas and hypotheses? One thing for certain: they can't all be true! For every action there is an equal and opposite reaction. For every idea about the evolution of religion, there is an idea that takes exactly the opposite tack! Science can break down for two basic reasons: the theories are no good, or the evidence is not supportive. Both of these reasons come into play with the sociobiological accounts of religion.

With respect to theory, straight Darwinism goes from strength to strength. The same cannot always be said of the ideas used to explain religion. Take the theory of memes. It really is crude to the point of nonbeing, certainly to the point of the unhelpful. What is a meme? It is a chunk of culture analogous to a gene. As it happens, genes are hard enough to define, but we do have some idea of them as the smallest functioning length of DNA. But what is the smallest functioning length of culture? Is Catholicism a meme? Is the authority of the pope a meme? Is transubstantiation a meme? Why the authority of the pope, for example, rather than each and every one of the dogmas that he endorses? And what kind of theory do you have as memes clash and come together and sometimes fuse and sometimes break apart? How is Mormonism a meme as compared to evangelical Christianity? Does Mormonism somehow include a lot of the evangelical Christianity meme, or are they separate memes? And so on and so forth. The point is not that Dennett is necessarily wrong in arguing that ideas sometimes have lives of their own, or that religions can be dreadful things that take over people's minds to people's own detriment—if not Catholicism, then most of us think this way about cults like Scientology—but that memetics is not very helpful in understanding what is going on. One is really just taking regular language and putting it in fancy terms. No new insights. No new predictions. No astounding claims that turn out to be true.

God?

Suppose that there is something to the naturalistic Darwinian approach to religion, its history, and its nature. How does this cash out philosophically? What does this tell us about God, his nature, and his existence? You might flip the argument entirely on its head, showing on non-Darwinian grounds that God does not exist and then setting forth on a naturalistic journey to explain why nevertheless so many people persist in believing that he is real. This is the tack taken by Dennett. He trots through the various arguments for the existence of God, following through with the standard objections. Then, God dismissed, Dennett is ready to give an argument about why we are deceived. It is hardly surprising that for Dennett religion is a parasite like the lancet fluke. What if you do not want to go down that path? Ask the basic question about God and his existence. Not the question about whether God exists, but whether a Darwinian account of origins shows that God exists or not. Was Darwin right in thinking that the reality or not of God is irrelevant to a naturalistic account of religion? Edward O. Wilson goes entirely the other way: Darwinism gives a naturalistic account of religion and that is an end to religion as a true description. As it happens, Wilson thinks that the human psyche demands religion, and thus he sees the place to move in with a kind of evolutionary humanism. But this is because Darwinism has already done its corrosive work:

> But make no mistake about the power of scientific materialism. It presents the human mind with an alternative mythology that until now has always, point for point in zones of conflict, defeated traditional religion. Its narrative form is the epic: the evolution of the universe from the big bang of fifteen billion years ago through the origin of the elements and celestial bodies to the beginnings of life on earth. The evolutionary epic is mythology in the sense that the laws it adduces here and now are believed but can never be definitively proved to form a cause-and-effect continuum from physics to the social sciences, from this world to all other worlds in the visible universe, and backward through time to the beginning of the universe. Every part of existence is considered to be obedient to physical laws requiring no external control. The scientist's devotion to parsimony in explanation excludes the divine spirit and other extraneous agents. Most importantly, we have come to the crucial stage in the history of biology when religion itself is subject to the explanations of the natural sciences. As I have tried to show, sociobiology

> can account for the very origin of mythology by the principle of natural selection acting on the genetically evolving material structure of the human brain. (E. Wilson 1978, 192)

If this interpretation is correct, the final decisive edge enjoyed by scientific naturalism will come from its capacity to explain traditional religion, its chief competition, as a wholly material phenomenon. Theology is not likely to survive as an independent intellectual discipline.

One suspects that Wilson is wrong and Darwin was right. The fact that you can give a naturalistic explanation of religion does not at once imply that religion is false. I can give a naturalistic explanation of my belief that the truck is bearing down on me, but it does not follow that the truck is not bearing down on me. It is true that if all you have is a naturalistic explanation, then (Dennett-like) you will probably not be eager to embrace religion. If you can show that religion is indeed a parasite on the mind, why take it any more seriously than the hucksters' email claims that their potions will increase your penis length? But for the traditional religious person—at least, for the traditional Christian religious person—religion has another source of epistemic power that email spam does not have: faith. This being so, then far from a naturalistic account being threatening, many expect such a naturalistic explanation of origins. God had to impart the information to humankind in some way, and why not through evolution? Nor would it be a counterargument that the explanation might make the arrival of religion rather less than edifying like a dog barking at a parasol blowing in the wind. The job is done.

What about something that often comes up in naturalistic discussions of the origins of religion, namely, the comparative issue? The Christians believe one thing, the Jews another, and the Muslims a third. Now, you might think that this is a pretty good argument against all of them. How can the Christian God be so loving and insist that we acknowledge and worship him, condemning to eternal damnation all of those Asians who grew up in ignorance? But even if you do accept the argument against God based on comparative religion, note that this has nothing whatsoever to do with evolution. It was an argument that moved the deists at the end of the seventeenth century. Moreover, evolution or not, the believer can continue to believe in the face of religious diversity—the Christians (or whatever) got it right and the others did not, and that is the end of matters.

If evolution is true, and it is, and if natural selection is the main mechanism, and it is, then the Darwinian approach to religion cannot be without merit. But it has far to go before it can command assent and respect.

5

Darwin, Darwinism, and Darwinian Thought

Chapter 5, Reading 1

The History of Evolutionary Thought[1]

THE IDEA THAT ALL organisms (including humans) are generated by natural means from other forms has ancient roots. Aristotle tells us that Empedocles (fifth century BCE) toyed with such thoughts. However, it was not until the eighteenth century and the Enlightenment that *evolution* (as we now call this idea of natural development) really started to gain a serious number of supporters. There are reasons both for the long delay and why the idea finally began to gain momentum.

The Early Days

The Greeks had no great religious objection to evolution, but their world picture did not have a place for any kind of significant developmental processes. Specifically, the Greeks thought that they had irrefutable reasons to reject ongoing, incremental organic change. They—particularly the philosophers Plato and Aristotle—thought that the world (especially the world of organisms) showed order and intention and, as such, was not something that could simply have appeared through blind, ungoverned processes of law. It certainly was not something that could have grown

---

1. Ruse 2009c. Reprinted with permission of Harvard University Press.

from simple beginnings to the complexity of today.... In an incredibly influential discussion, Aristotle in *De partibus animalium*, identified the factors at work here as "final causes" (1984a). These are causes that occur not just to produce or do something (the finger parts dry and make nails) but for the sake of some kind of purpose (the nails protect the finger ends). They show some kind of forethought or intention. For this reason, final causes cannot be reduced to blind, unguided law, as is demanded in evolution. The world, particularly the world of organisms, must in some sense have been designed rather than just produced under its own steam by natural processes. (Sedley 2008 is the definitive study.)

The Jews, and following them the early Christians, had religious reasons for the rejection of evolution. It goes against the creation stories of the early chapters of Genesis, which portray a world created miraculously by God and then peopled by him through divine fiat over a short time span. But do not think that religion as such was then and always an absolute bar to evolutionism. The church fathers (the major Christian theologians of the early centuries) worked toward an understanding of the biblical text that would allow interpretation, particularly in the face of advances of science. Saint Augustine was eager not to let ancient creation accounts stand in the way of modern thought. He himself, believing that God stands outside time, speculated in a kind of protoevolutionary fashion that the Divine had formed seeds of life that then sprang into full being when they were placed here on Earth (McGrath 2009, 102). However, one should not read too much into any of this. Like the Greeks, the Jews and Christians were simply not looking in the direction of evolution and would have thought final causes an unanswerable objection to significant developmentalism. As is well known, these kinds of causes became a foundation of one of the major proofs of God's existence, the argument from design, which moves from design here on Earth to the existence of the divine artificer.

Why, then, did evolution start its rise in the eighteenth century? The answer is simple. It was at this time that people started to challenge the Christian picture of world history—a providential picture of a world created by God, where humans are made in his image but have fallen and are able to achieve salvation only through his undeserved grace. Some began to argue that perhaps humans held their fates in their own hands and could progressively improve their own lots. It was this idea of progress—the belief that the world and its denizens are on a trajectory upward and that this upward rise is made possible by (and only by) the unaided efforts of the world's human inhabitants—that gave rise to the idea of organic evolution (Ruse 1996b). Enthusiasts for progress extended their thinking into nature and developed the idea of evolution—progressive change upward from the simple to the complex. They then read this idea back into human thought and social practice as confirmation of their beliefs about progress.

Similar ideas were to be found elsewhere, most notably in France. In his *Philosophie zoologique* (1809), the taxonomist Jean-Baptiste de Lamarck produced the first full-blown evolutionary theory—a picture of upward rise to our own species from the most primitive forms of life, which in turn had been produced from mud and slime through the actions of heat and electricity and other natural forces. Although the metaphysical idea of progress was the main factor behind the rise of evolutionary ideas, it is not true that there was no pertinent empirical evidence. Aristotle had noted that organisms of very different species seem to share common patterns or structures—what today are known as homologies—and the evolutionists were ready to interpret these as signs of common ancestries. Likewise, the successes of animal and plant breeders did not go unnoticed. But generally the evidence took a very secondary position. The fossil record, something that today many (if not most) people would invoke first as the proof of developmental origins, was less than helpful. As a systematic proof of progressive change, the gleanings from the rocks were meager indeed. In any case, counting against the empirical side was the fact that no one had any great understanding of what might have caused evolution. Most assumed some kind of vague, upwardly thrusting force or forces, but little more. Generally, everyone was committed to the folk belief that characteristics acquired in one generation could be transmitted immediately to the future generations—Lamarck was so enthused by this process that the inheritance of acquired characteristics has since become known as Lamarckism—but beyond this was silence. . . .

The ideology of progress was what counted, and it was for this reason that most people around 1800 would have regarded evolutionism less as a real science and more as a pretender, somewhat like animal magnetism (mesmerism) and the reading of character from skull shape (phrenology). Even judged by the standards of that time, evolution was what may fairly be called a pseudoscience. Obviously, Christian opponents of evolution disliked intensely the anti-providential underpinnings of the doctrine. But evolution was not associated with total nonbelief, atheism, or even what later in the nineteenth century Thomas Henry Huxley was to call agnosticism. Most evolutionists were deists who believed in God as unmoved mover, a being who had set the world in motion and now let it unfurl without need of miraculous intervention. For the deist, indeed, evolution was proof of God's power and intention rather than disproof. Everything was planned beforehand and went into effect through the laws of nature. . . .

Progress goes against the Christian doctrine of providence. Nevertheless, although Cuvier thought that there was evidence of Noah's flood, neither he nor other serious scientists wanted to make the case by simple reference to Genesis. Indeed, by the beginning of the nineteenth century, all were starting to realize that the Earth's history must be far older than the traditional six thousand years that one can work out from the genealogies given in the Bible. It is not that the Bible is false, but rather that it needs interpretation. Some solved the problem by thinking of the six days of creation as six long periods of time; others solved it by supposing that there were long, unmentioned gaps between the biblical days. God's creation therefore was a long, drawn-out process, but it was not evolutionary. . . .

The controversy was at an impasse, and not much had changed by the middle of the nineteenth century. On the one side were the evolutionists, committed to progress and ardent in their belief that organic development was the perfect complement to this ideology, with enthusiasm outstripping empirical knowledge. Confirming this pattern, in 1844 the Scottish publisher Robert Chambers wrote (anonymously) a highly popular work on evolution, *Vestiges of the Natural History of Creation* (1844), in which he argued that everything was in a state of upward becoming and that what happened in the social world mirrored what happened in the biological world. . . .

Charles Robert Darwin (1809–82) was sent to Edinburgh University to train in the family tradition of medicine. After two years he dropped out, bored with the lectures and revolted by the operations. Yet already Darwin had started to mix with scientists, especially naturalists interested in the living world. One of his acquaintances was Robert Grant, an anatomist and an avowed evolutionist. So, quite apart from his grandfather's work (the young Charles read Erasmus's major treatise, *Zoonomia*), evolution was an idea to which Darwin was introduced at an early age. It seems, nevertheless, that the youthful Darwin accepted in a fairly literal form the whole of Christianity, including the early chapters of Genesis, and that this was a factor in his redirected choice of a career: to be an ordained minister in the established Church of England. To achieve this end one needed a degree... from an English university, and so, in 1828, Darwin was packed off to Christ's College, Cambridge.

In 1831 Darwin (who continued mixing with scientists) got his big break. After he graduated, his career as a clergyman was put on hold through the offer of a lengthy voyage on *HMS Beagle*, just about to start on a surveying trip around South America. A major influence at this point was (vicariously) the Scottish geologist Charles Lyell, who at the beginning of the decade began publishing his massive *Principles of Geology* (there were three volumes; Darwin took the first with him and had the others sent out). Although he was no evolutionist, Lyell insisted that the physical world must be explained in terms of natural causes of a kind now still working. This had a great effect on Darwin, whose first systematic work was in geology, and prepared him to think about the organic world likewise less in biblical terms and more in terms of natural causes. In 1835, leaving South America, the *Beagle* sailed into the Pacific Ocean and visited the Galápagos Archipelago, a group of islands on the equator, far from land. Thanks to the governor, who pointed out that the giant tortoises indigenous to the archipelago were different from island to island, Darwin came to see that this held for the Galápagos fauna generally—the birds in particular, the finches and the mockingbirds, were peculiar to their specific homes.... Unlike earlier thinkers, what was of great concern to Darwin was the cause or causes of evolution. Without causes, he was no more than one among many evolutionists....

Darwin sat on his evolutionary ideas for fifteen years, during which time he turned to a massive study of barnacles. We are not quite sure why this delay occurred, although by this time Darwin had fallen sick with a mysterious ailment that was to plague him for the rest of his life, and so undoubtedly he was not relishing the huge debate that his ideas were bound to cause. Also a major factor must have been his reluctance to upset powerful science establishment figures (including Cambridge mentors) who had encouraged the young Darwin in his work. One of the things he did during the pause was to network with younger scientists, who could rally around him when he did go public. Finally, however, Darwin was pushed into action when, in the middle of 1858, he received a short essay by Alfred Russel Wallace, a collector in the Malay Peninsula, that had virtually the same premises and conclusion that he had discovered some twenty years earlier....

In the *Origin*, Darwin writes:

> Can it . . . be thought improbable, seeing that variations useful to man have undoubtedly occurred, that other variations useful in some way to each being in the great and complex battle of life, should sometimes occur in the course of thousands of generations? If such do occur, can we doubt (remembering that many more individuals are born than can possibly survive) that individuals having any advantage, however slight, over others, would have the best chance of surviving and of procreating their kind? On the other hand we may feel sure that any variation in the least degree injurious would be rigidly destroyed. This preservation of favorable variations and the rejection of injurious variations, I call Natural Selection. (C. Darwin 1859, 31)

Now, with some minor problems brushed away, Darwin was ready to present the second part of his theory. For a good two-thirds of the *Origin*, Darwin took the reader through the various branches of biological science—instinct, paleontology, biogeography, classification, morphology, embryology—and showed that phenomena in these branches are explained by evolution through natural selection, and, conversely, these various branches point to and support the mechanism of evolution through selection. . . . Paleontology raised some difficulties. It had its good points—for instance, that as we go down the record, increasingly we find organisms that seem to have features midway between features of extant organisms that are widely different. However, against the positive, there were gaps in the fossil record, and, even worse, the fossil record began abruptly at the start of what we . . . would call the Cambrian; there was no record before that. . . .

Darwin . . . never concealed that he was working in a God-backed mode, and there are frequent unself-conscious references to the creator. "Authors of the highest eminence seem to be fully satisfied with the view that each species has been independently created." This was not Darwin's position. "To my mind it accords better with what we know of the laws impressed on matter by the Creator, that the production and extinction of the past and present inhabitants of the world should have been due to secondary causes, like those determining the birth and death of the individual" (C. Darwin 1859, 488).

After Darwin

Before the *Origin* was published, evolution rode on the back of the doctrine of progress, it was opposed by the idea of final cause, and its status was akin to that of mesmerism and phrenology—it was a pseudoscience. How did Darwin change things? As far as progress was concerned, Darwin himself was certainly a cultural progressivist and saw evolution itself as progressive. . . . However, he also saw that the link between progress and evolution was something that brought down the status of the latter, and he realized that natural selection is a mechanism that denies the inevitability of biological progress. In other words, conceptually Darwin broke the link between the two, although, having done this, he argued that natural selection could lead to progress as the end result of what today's evolutionists would call a kind of arms race, with ever better features coming from competition with rivals. . . .

Darwin wanted evolution to be what we today might call a professional science. But it was not to be. Darwin was a sick man, so the fate of his ideas had to be entrusted to others—followers like Thomas Henry Huxley in England, Asa Gray in America, and Ernst Haeckel in Germany. Unfortunately, they had aims other than those of Darwin. They all became ardent evolutionists, and all agreed that Darwin's argument in the *Origin*, which showed how many areas of biological inquiry could be tied together under the umbrella of evolution, was definitive. But few, if any, were very keen on natural selection. No one denied it outright, but some thought that the whole issue of final cause was overblown and hence that selection was unneeded, and others thought that final cause still needed explanation and that selection could not really do the job. Either way, people turned to other putative causes, such as Lamarckism, evolution by jumps or what we today might call macromutations (the theory is known as saltationism), or evolution through a kind of internal momentum (orthogenesis). . . .

Evolution seems to have no such immediate function. But, especially combined with the ideology of progress, it could function as the philosophy of the new reformed society—it could have a role as a kind of alternative to Christianity, a kind of secular religion in its own right. It could be a story of origins, a story of humans' exalted place in the process, a vision of where they were going and what they should do to ensure success and triumph. And basically this was how evolution was promoted, especially by the leading gurus of the day, of whom none was more vocal than the English man of science Herbert Spencer. Playing up evolution, his vision was as progressive as anything from the eighteenth century. Spencer saw progress as being a move from the undifferentiated to the differentiated, or what he called a move from the homogeneous to the heterogeneous: "Whether it be in the development of the Earth, in the development of Life upon its surface, in the development of Society, of Government, of Manufactures, of Commerce, of Language, Literature, Science, Art, this same evolution of the simple into the complex, through successive differentiations, holds throughout" (Spencer 1851, 2–3).... Incidentally, it was Herbert Spencer who popularized the word "evolution." Until the middle of the nineteenth century, the term was generally reserved for the development of the individual (*ontogeny*). For evolution in the modern sense (*phylogeny*), most people used words like "transformation" (Richards 1992).

The Synthetic Theory

Things persisted this way into the early decades of the twentieth century and in some respects became even worse when Continental thinkers, notably the embryologist Hans Driesch in Germany and the philosopher Henri Bergson in France, began pushing a kind of neo-Aristotelian theory of life that saw special, nonphysical final forces guiding the path of evolution. Driesch's entelechies and Bergson's élan vital, the foundations of "vitalism," were just not the elements of a forward-looking modern science. Relief finally came with the development of Mendelian genetics and its melding with Darwinian selection, although even here the process took time and was not straightforward (Provine 1971). Gregor Mendel, a Moravian monk who lived in the Austro-Hungarian Empire, discovered the essential principles of heredity in the 1860s. Unlike Darwin, he saw the transmission of characteristics as "particulate," believing that features like color and size can be passed on without the threat of being blended away....

However, it was not until the beginning of the twentieth century that people came to appreciate the importance of Mendel's work and saw that it provided a key element in the story of evolution. You might think that this was a terrible missed opportunity, and that, had Darwin read Mendel's key paper, he would have realized that now he had the answers to all of the issues of heredity that his theory of natural selection demanded. Hence, the theory would have moved forward more quickly, much earlier. This is probably not so. Even when the work of Mendel was discovered, it took time to assimilate it. Many people at first thought it an alternative to Darwin rather than a complement. . . . So even if Darwin had read Mendel—and the monk's work was published in a journal well enough known that, if he searched, Darwin would have found it—there is no reason to think that there would have been a "eureka" moment. Mendel incidentally did read the *Origin*. Interestingly, he never thought of his own work as pertinent to Darwin's problems. Judging from the annotations that Mendel made in the margins of the German translation of the *Origin*, he was far more interested in the theological implications of evolution than in the troubles of heredity that Darwin faced. There is really no surprise here since, after all, Mendel was first and foremost a man of God and only secondarily a plant scientist. . . .

There were fairly straightforward reasons why at first, when Mendel was rediscovered at the beginning of the twentieth century, no one thought that he was speaking to Darwinian issues. Naturally, early geneticists focused on big variations and so tended to favor a kind of saltatory theory of overall change, that is evolution by large jumps. Slowly, however, thanks particularly to work in the second decade of the century by Thomas Hunt Morgan and his students at Columbia University in New York, the nature of the gene was revealed, and it could be seen as the complement to natural selection. Finally, around 1930, a number of highly gifted mathematical biologists, notably R. A. Fisher (1930) and J. B. S. Haldane (1932) in England and Sewall Wright (1931, 1932) in America, showed how Mendelian genes sort themselves and are transmitted in groups (an essential finding, given that selection works only in the group situation), and then it was possible to bring Darwin's work to the completion that it needed. Along with mutation (the coming of new variation, caused by spontaneous changes in genes), natural selection can truly be a significant force for change.

Since by the 1930s the question of the age of the Earth was no longer pressing (the discovery of the warming effects of radioactive decay showed that the Earth is quite old enough for the slow workings of selection), biologists moved rapidly forward with new ideas (known as *population genetics*) that would put empirical flesh on the mathematical skeletons. In Britain a highly vocal supporter of the theory was Thomas Henry Huxley's grandson Julian Huxley (the older brother of novelist Aldous Huxley), who produced a major work that pulled ideas together:

Evolution: The Modern Synthesis (1942). Scientifically, after Fisher the most important figure was the Oxford biologist E. B. Ford (1964), who did groundbreaking studies of selection in populations of butterflies and who gathered around himself in a school of "ecological genetics" a number of younger researchers likewise interested in selection and its effects in nature. Noteworthy were Arthur Cain (1954) and Philip Sheppard (1958), who worked on banding patterns in snails, and Bernard Kettlewell (1973), whose interests were in industrial melanism—the ways in which butterflies change adaptive color patterns as their habitats are changed by the effects of smoke and pollution. In America the influential figure was the Russian-born geneticist Theodosius Dobzhansky, whose *Genetics and the Origin of Species* (1937) was an inspiration to a whole generation of evolutionists. Working alongside him were others, notably the ornithologist/taxonomist Ernst Mayr (author of *Systematics and the Origin of Species* [1942]), the paleontologist George Gaylord Simpson (*Tempo and Mode in Evolution* [1944]), and the botanist G. Ledyard Stebbins (*Variation and Evolution in Plants* [1950]).

By the 1950s Darwin's dream of a mature, professional science of evolutionary biology was realized. Moreover, it was genuinely Darwinian, for although there had been pretenders to the causal throne, notably Sewall Wright's process of genetic drift (random changes in gene frequency in small populations due to the vagaries of breeding), it was recognized that the key factor in organic nature is its adaptiveness, its manifestation of final cause, and that natural selection is a full and satisfying way of explaining this phenomenon. At the same time, progress—and all the moralizing and philosophizing that went along with it—had been expelled. No one was going to use this kind of professional biology as an excuse for quasi-religious speculations about the status of humankind and the obligations that nature lays upon humans.

Yet, for all this, there is one more important factor to the story, and this partly explains why to this day evolutionary ideas remain so controversial to so many. Although evolutionary biology was upgraded from the level of a popular science to the level of a professional science, this did not occur simply out of a disinterested quest for the truth by men who had no aims but the finding of the workings of nature. Virtually every one of the new professional biologists, from Fisher to Stebbins, became an evolutionist because he was attracted to the subject by thinking that it was more than just a scientific theory.

For some, like Dobzhansky, the attraction was explicitly religious. He brought progress—progress to humans, that is—into a kind of overall world picture that saw God working his way through the forces of nature. Ignoring divisions that were crucial to earlier thinkers, Dobzhansky thought that God's grace and an unfurling creation could be combined. Unsurprisingly, he was attracted to the ideas of the Jesuit paleontologist Teilhard de Chardin (1955), who promoted similar ideas. Others, like Julian Huxley in England and Mayr and Simpson in America, were secular in their thinking, but they too liked the idea of progress and thought that it showed that evolution had some kind of direction and meaning.

Because all these seminal evolutionists were people who grew up in the early twentieth century, when much of evolution was only a popular science dripping with metaphysical and moral implications, it would have been a surprise if this had not been the main motivating factor. But, like Darwin himself, this new breed of evolutionists saw that if they were to have professional status for their activities—something they ardently desired as full-time scientists—then they would need to purify their work of its extrascientific aspects. They needed to deal with doctrines of progress and with extrapolated moral exhortations. So they did. They took the extraneous thinking about progress, morality, and the meaning of life out of the science that they produced as professional researchers.

The Past Half Century

That was all fifty or more years ago. The neo-Darwinians in Britain and the synthetic theorists in America built good foundations for their science. Considered just as a science, which it has every right to be, modern evolutionary theory deservedly is one of the most forward-looking and exciting areas of empirical inquiry. Every one of the areas treated by Darwin in the *Origin* flourishes as never before. Selection studies themselves, both theoretical and empirical, have reached a very high degree of sophistication. The work of Rosemary and Peter Grant on the beak size of Galápagos finches is a paradigmatic example of excellent science (B. R. Grant and Grant 1989). The advent of molecular biology has been very important here for offering both new insights (for instance, the significance of drift at the molecular level) and new techniques (the use of genetic fingerprinting for determining heredity).

Social behavior and instinct, detailed in Edward O. Wilson's magnificent survey *Sociobiology: The New Synthesis* (1975), have powerful new models of explanation, like kin selection and reciprocal altruism, confirming Darwin's hunch that the right way to explain the intricacies of social interactions, including group behavior, is from an individualistic perspective. Backing the theoretical works are major empirical studies—for instance, that of Tim Clutton-Brock and his associates on the red deer of Scotland (1982)—that show that evolution has indeed been a powerful factor in the evolution of social behavior. Somewhat controversially, the field (often here known as *evolutionary psychology*) also in a very Darwinian fashion looks at humankind, trying to explain such things as infanticide and sexual behavior and mating patterns in terms of natural selection. Paleontology has been revolutionized.

Not only are there continued major fossil finds—recently the discovery in the Canadian Arctic of a link between fish and amphibians (Daeschler et al. 2006), not to mention the finding in Indonesia of a little humanlike figure (*Homo floresiensis*, naturally known as the "hobbit" [P. Brown et al. 2004])—but molecular techniques have enabled people to explore the past course of evolution with ever greater precision. At the more conceptual level, new theories have been proposed (for instance, the theory of punctuated equilibria of Niles Eldredge and the late Stephen Jay Gould [1972]), and ideas drawn from other areas of evolutionary studies have been applied to understand the patterns of the past (for instance, the way in which the late John J. Sepkoski Jr. [1976] used ecological theories about island biogeography to throw light on past patterns of animal diversity, as well as on the Great American Interchange, an event that occurred about ten million years ago when South and North America joined and animals moved north to south and south to north [Marshall et al. 1982]).

Theories of geographic distribution were revolutionized by the coming of plate tectonics. Until this point, in a tradition embraced by Darwin in the *Origin*, evolutionists had spent many happy hours throwing up hypothetical land bridges and finding ways in which seeds and small animals could cross large bodies of water. Now the moving of the continents did all the work for them. Lystrosaurus, a mammal-like herbivorous reptile found rather more than two hundred million years ago, is fat, short, and squat. It is certainly not an animal that would have roamed far and wide. Today it is found in the same fossil deposits (lower Triassic) on the continents of Africa, (Southeast) Asia, and Antarctica.

This would be inexplicable were it not for the fact that more than two hundred million years ago all those continents touched when they were part of Pangaea. They have since drifted apart. . . . Systematics has been transformed by new cladistic techniques, much aided by the coming of computers that have allowed massive amounts of data to be absorbed, quantified, calculated, and understood (Hull 1988).

Morphology likewise is open to new understandings. Above all, embryology, a subject that the synthetic theorists tended to ignore, has now, especially under the new name *evolutionary development* (evo-devo), been converted from a nonconceptual backwater to the most exciting area of research today in the field of evolutionary studies (S. Carroll 2005). Fantastic new findings have emerged, for instance, about the ways in which organisms as diverse as fruit flies and humans share the same underlying genetic mechanisms for development, and the ways in which organisms are built and variations are produced are among the hottest areas of research. . . . There are, of course, controversies and differences. No one wants to deny the importance—the very great importance—of natural selection. But some evolutionists, particularly in areas like paleontology and embryology, where sometimes questions of adaptive significance are (as was earlier the case for people like Thomas Henry Huxley) not overwhelmingly pressing, think that perhaps other causal factors were significant. This was the underlying theme of the theory of punctuated equilibria, with Gould in particular arguing that many features are not particularly adaptive but are more "spandrel like"—that is, nonfunctional by-products of the evolutionary process (Gould and Lewontin 1979). He accused Darwinian evolutionists of relying too often on "just so" stories, the fantabulous-like tales told by Rudyard Kipling. Today many evo-devo enthusiasts feel much the same way. "The homologies of process within morphogenetic fields provide some of the best evidence for evolution—just as skeletal and organ homologies did earlier. Thus, the evidence for evolution is better than ever. The role of natural selection in evolution, however, is seen to play less an important role. It is merely a filter for unsuccessful morphologies generated by development." Population genetics is destined to change if it is not to become as irrelevant to evolution as Newtonian mechanics is to contemporary physics (Gilbert et al. 1996, 368). No doubt posterity will tell us whether sentiments like these are wise and prescient or merely the unjustified effluvia of enthusiasts for a new discipline that is trying to establish itself as important. Either way, the vigor and excitement of contemporary evolutionary theory are confirmed.

Does this then mean that the nonscientific, more ideological side to evolutionary thinking has now vanished? Does evolution no longer function for many people as a kind of secular religion? Not at all. On the one hand, huge numbers of articles and books (not to mention radio and television programs) show or justify ideological implications ascribed to or drawn from evolution. Edward O. Wilson is a paradigm, for he has poured out a stream of books designed to show that evolution is progressive and that from this flows not just moral exhortations—for Wilson, the preservation of the rain forests—but an alternative view of creation to that of traditional religion:

> If this interpretation is correct, the final decisive edge enjoyed by scientific naturalism will come from its capacity to explain traditional religion, its chief competition, as a wholly material phenomenon. Theology is not likely to survive as an independent intellectual discipline (E. Wilson 1978, 192).

Another scientist much given to this kind of speculation—although he disliked traditional notions of progress and wanted to substitute his own ideas about the randomness of change—was Stephen Jay Gould, particularly in his popular essays, *This View of Life*, published monthly in *Natural History*. Very rarely did Gould fail to draw some kind of moral message from his writings, whether about the racism endemic in our society or the need for conservation. Not that Gould gave traditional religion much more scope than did Wilson.

Any ideas about our being the favored children of God, that we humans might be the reason for the creation, are simply hubris. "Since dinosaurs were not moving toward markedly larger brains, and since such a prospect may lie outside the capabilities of reptilian design ... we must assume that consciousness would not have evolved on our planet if a cosmic catastrophe had not claimed the dinosaurs as victims. In an entirely literal sense, we owe our existence, as large and reasoning mammals, to our lucky stars" (Gould 1989b, 318). Here Gould anticipated the so-called new atheists, like his great British counterpart in the realm of popular science writing, Richard Dawkins. In *The God Delusion* (2006), Dawkins argues that Christianity and other religions are the major sources of humankind's ills, and that we need a secular philosophy as a substitute—a secular philosophy informed by Darwinian thinking, which itself makes appeals to traditional religions not merely otiose but absolutely false. Philosopher Daniel Dennett's *Darwin's Dangerous Idea* (1995) is another exemplar of this kind of literature.

Heading the other way are the many critics of evolutionary thinking. In America today, survey after survey confirms that most Americans do not believe in evolution. Indeed, the great majority believe that the Earth is less than ten thousand years old and that all organisms were created in a burst of divine energy in just six days, followed some time later by their destruction in a universal flood, save only for those lucky pairs that floated away in Noah's ark (Ruse 2005a, Numbers 2006). American evangelical Christianity is not our story here, but the constant attacks on modern Darwinism show clearly that enthusiasts for this kind of religion do not regard evolution as just a scientific theory but as something more, a materialistic, secular alternative to the traditional belief systems that carries within it opposition to all decent and long-held moral norms. In the words of the leading "creation scientist," the late Henry M. Morris: "It is rather obvious that the modern opposition to capital punishment for murder and the general tendency toward leniency in punishment for other serious crimes are directly related to the strong emphasis on evolutionary determinism that has characterized much of this century" (Morris 1989, 148). Apparently, the "notorious Darwinian philosopher Michael Ruse," a well-known "atheistic humanist," has made a major contribution to the moral rot.

More recently, the rather crude literalism of creation science has morphed into the more sophisticated anti-Darwinism of so-called intelligent design theory, which sees aspects of organic life as so complex as to require special creative interventions by the designer. But still the debate today is more than just a conflict of science versus religion and one between competing ideologies. The founder of the intelligent design movement, former Berkeley law professor Phillip Johnson, is explicit in seeing evolution in the old terms: "The Christian philosophy that was overthrown in the 1960s was an easy target because it had become identified with American culture and with worldly ideas like human perfectibility and the inevitability of progress, which are actually profoundly un-Christian" (Johnson 1997, 106). Unsurprisingly, this humanistic metaphysics is linked to moral failings—pornography, gay marriage, abortion, and like transgressions such as socialism.

The history of evolutionary theory falls into three stages. From about 1700 until 1859, it was little more than a pseudoscience riding on the back of the ideology of progress, the notion that humans can make change for the better. It was opposed to traditional religion not so much because it went against the literal truth of the Bible but because it left no place for providence, the belief that change comes only through God's undeserved grace. Charles Darwin's *On the Origin of Species* changed all this. Darwin established the reasonable truth of evolution once and for all. However, his mechanism of natural selection was ignored or downplayed, and the leading evolutionists who followed Darwin, notably Thomas Henry Huxley, were much more interested in using evolutionary ideas as a kind of popular science, almost a secular religion, as an alternative ideology to the Christian religion they saw blocking their way to social reform.

The third stage started around 1930 with the incorporation of Mendelian genetics. Now there was the possibility of building a new professional science of evolutionary change. This occurred, and it is this theory that flourishes today. Yet evolutionists should never forget the past and its influence on the present. Along with the science there is still an ongoing debate about ideology and the extent to which evolution in some way provides a secular alternative to older, overtly religious ways of viewing the world and humankind's status within it.

Chapter 5, Reading 2

The Origin of the *Origin*[1]

CHARLES ROBERT DARWIN WAS born in 1809. His great book, *The Origin of Species*, was published in 1859, when he was fifty. He was to live another twenty-plus years, dying in 1882, by which time the *Origin* had gone through six editions and been extensively revised and rewritten. It used to be the case that it was the sixth edition of 1872 that was most frequently reproduced, but more recently scholars have insisted that the first edition is the really important one—we not only see Darwin's thinking in its original form, but the revisions today are often judged to have been made for less than worthy reasons (in the sense that the criticisms now no longer seem so forceful). It is therefore the first edition that will be the focus of this piece, and my question opening this volume is about its genesis, and the implications that this had for the actual book that Darwin produced. While I do not think that the *Origin* is a particularly mysterious book, I believe that there are aspects to it that are not quite as obvious as we today often assume.

Undistinguished at school, Darwin went first to the University of Edinburgh to study medicine and then (after that proved not to be to his liking) to the University of Cambridge to prepare for the life of an Anglican clergyman. (Janet Browne's biography is definitive [1995, 2002].) We know now that, although Darwin had no formal training as a biologist, by the time he graduated (in 1831) he not only was showing an aptitude

1. Ruse 2009e. Reprinted with permission of Cambridge University Press.

for science but also was long versed in the ways of empirical study and research.... At the end of 1831, Darwin joined *HMS Beagle*, about to start what proved to be a five-year trip mapping the coast of South America and then going on around the world before returning home. Darwin started as a kind of gentleman companion to the captain, Robert Fitzroy, but soon became the de facto ship's naturalist, in which role his earlier scientific activities and training served him very well. The notebooks that he kept show that he was serious and competent right from the start.

The time on the *Beagle* was important for many reasons, not the least of which was that, being away from his Cambridge mentors, Darwin was forced to think independently. This was shown particularly in geology, the science that was most important to him in these early years. Darwin became enthused with the uniformitarian thinking of Charles Lyell in his *Principles of Geology* (1830–33) and broke with the catastrophism of people like Adam Sedgwick (1831), a professor of geology at Cambridge and the man who had taken Darwin on a crash course in Wales in the summer of 1831. In religion, the trip was important because Darwin's rather literalistic Christianity started to fade and he became something of a deist, believing in God as unmoved mover and that the greatest signs of his powers are the workings of unbroken law rather than signs of miraculous intervention.

Most significantly, perhaps because he was now thinking of God as someone whose greatness is evidenced by unbroken law rather than by miracle, Darwin started on the path to evolution. It is generally agreed that Darwin (who knew about evolutionary ideas from reading *Zoonomia*, an evolution-favoring book by his grandfather Erasmus Darwin, as well as from encounters at Edinburgh with the future London professor of anatomy Robert Grant, and from Lyell's discussion of the thinking of Jean-Baptiste de Lamarck) did not actually become an evolutionist on the voyage. But his encounter with the different reptiles and birds on the Galápagos Archipelago shocked him. How could one have different-but-similar forms on islands only a few miles apart? When, on his return to England, Darwin learned that the birds were undoubtedly of different species, this was enough to tip the balance. In the spring of 1837, Charles Darwin slipped over to transmutationism.

For eighteen months, until the end of September 1838, Darwin worked hard looking for a cause of evolution. One suspects that it was the ideal of Newton—much praised by the day's scientific methodological gurus, especially John Herschel (1830) and William Whewell

(1837)—that spurred Darwin here. He wanted to find a force for evolution akin to Newton's force of gravitational attraction. For all that we have Darwin's detailed notebooks—perhaps because the notebooks are so detailed—there has been debate about the exact course of Darwin's thinking. Darwin himself always claimed that he started with artificial selection, realizing that this was the way in which breeders change their animals and plants. Then he started to look for a natural equivalent, and this he found at the end of September 1838 after he had read Thomas Robert Malthus's (1826) treatise on population. More organisms are born than can survive and reproduce. Those that get through will, on average, be different from those that do not.

Through a careful reading of the notebooks that Darwin kept while he was searching for his mechanism—a mechanism that, when discovered, he clearly did think was akin to a Newtonian force—some scholars have concluded that, although in his various sketches and published versions of his theory Darwin does use artificial selection to lead into natural selection, it is unlikely that he really did have the analogy in mind on his way to natural selection (C. Darwin 1987, Herbert 1971, Limoges 1970). He never really thought that artificial selection could do the job, or at least that a natural equivalent would be sufficiently powerful to get full-blooded change. Whether this interpretation is correct is something that has been argued for some time now. My own feeling, looking at some of the material that Darwin read during the crucial discovery months—some material, incidentally, that not only drew attention to artificial selection but also showed that one might expect a natural equivalent, some material that Darwin highlighted particularly—is that he probably did have the analogy in mind. But I would agree that he was more hesitant at the time than his confident later recollections suggest (Ruse 1979).

Darwin did not at once write things up in any formal way. Indeed, we have to work rather carefully through the notebooks to see that he did appreciate the full worth of natural selection. (He did. Jottings later in 1838 about human mental evolution put this fact beyond doubt.) Moreover, it was to be another four years before he actually wrote out what was a 35-page penciled sketch (as we now call it) of his ideas (C. Darwin and Wallace 1958). This was then extended in 1844 to a 230-page essay, which Darwin had fair-copied by the local schoolmaster. It should be added that in his *Autobiography* and elsewhere Darwin referred to 1838 as the point at which he first thought up his species theory, and this may well be true, although there seems to be no written record (nor indeed should there necessarily be).

The Long Delay

Darwin then put things on hold, and having written a letter to his wife asking that in the event of his death she arrange that some competent biologist bring the essay to publication, he turned to a massive eight-yearlong study of barnacles (C. Darwin 1851a, 1851b, 1854a, 1854b). It was not until around 1854 that he turned back to his evolutionary theory.... His friends urged him to get back to the job and to go public, lest he be scooped. Darwin therefore started to write a massive book about his theory. This was interrupted by the arrival, in the early summer of 1858, of the essay by Alfred Russel Wallace, a naturalist and collector in the Malay Archipelago—the essay in which Wallace captured almost exactly the ideas that Darwin had discovered twenty years before. Extracts of Darwin's writings along with Wallace's essay were at once read at the next meeting of the Linnean Society in 1858 and published. Despite stories about the ideas being disregarded, there was immediate interest. Later in the summer, in his presidential address to the British Association for the Advancement of Science, quite favorable notice was made of the papers by Richard Owen. By now, Darwin had launched frenetically into the writing of what he wanted to call an "abstract" of his thinking—a qualification that his publisher, John Murray, wisely declined to accept for a work that in print extended to 490 pages—and so finally *On the Origin of Species by Means of Natural Selection, or the Preservation of the Favored Races in the Struggle for Life*, by Charles Darwin, MA, appeared in November 1859.

There has been and still is considerable controversy over the reasons why Darwin took so long to bring his theory into print.... For the record, I have been marked as one who thinks there was a genuine delay, and I continue to think so (Ruse 1999b). I am not too bothered by the jump between 1839 and 1842 or between 1842 and 1844. Darwin was working flat out on other projects, the geology in particular. Most notably, making him into a household name, Darwin wrote up his account of the trip around the world on the *Beagle*, and what started as a formal report for the Admiralty turned into one of the most popular of travel books at a time when society just loved stories of exploration in distant and strange lands (C. Darwin 1839). Darwin was also newly married, moving to the house in Kent (and having it extended), starting a family, and feeling sick. He had more than enough on his plate at that time.

It is the gap between 1844 and 1858 that fascinates me. I am happy to accept the bits and pieces of new information that come into Darwin's thinking between 1844 and 1859. I have always been impressed by the way that the barnacle work so convinced Darwin of the variation that exists in all natural populations, something that was crucial for a mechanism like natural selection. And let us not forget the "principle of divergence," tied to the tree of life metaphor, where Darwin saw that divergence is the way in which selection maximizes the use that organisms can make of resources. Although I think that in fact there are hints of it even in his notebooks of the late 1830s, I accept fully that Darwin did not really realize the problem and the solution until much later.

However, I have to say that none of this alone or in conjunction really convinces me that this yields the solution. Two things always strike me. First, Charles Darwin was always so ambitious. Never let the friendly, warm, almost casual man and his style deceive you. At the beginning of her biography, Janet Browne speaks of the sliver of ice in the heart of Charles Darwin. I have always thought that this is so. He was not a nasty man in any way, but he did want to make his mark as a scientist, and nothing was going to stand in his way. The sickness was genuine, but he used it to advantage to avoid boring jobs and people. His massive letter writing was sincere, but again and again it was a medium through which Darwin could get others to do jobs for him. My second point is that truly I cannot find all that much difference between the essay of 1844 and the *Origin* of 1859. I have long argued—and continue to argue—that Darwin's theory is a very skillful piece of work. It is, as he truly said, one long argument, not simply one damn thing after another.... The point I make here is that this structure is in the sketch, the essay, and the *Origin*—identical in form and presentation—and much of the evidence is just the same. Even the sub-bits, like the introduction of sexual selection along with natural selection, are the same.

Answering the Question

So I still have the question of the delay. Why did Darwin not publish the essay back in 1844? My answer is twofold. First, he was scared. Not of his wife or anything like that; and I doubt that being labeled a materialist much bothered him. He came from a family and a set (particularly connected with his brother Erasmus) where that was not much of a taunt. In any case, Darwin was not a materialist. He was a deist, and the various writings up to and including the *Origin* make that very clear. (He even added additional references to the creator in later editions.) It was precisely the leaders of his scientific set—those very men who had nurtured him and made his early career possible—whom Darwin feared offending. Eighteen-forty-four was the year in which the notorious evolutionary work the *Vestiges of the Natural History of Creation* was published (Chambers 1844), and the set went after the work with a vengeance. Adam Sedgwick raged against it in the *Edinburgh Review*—it was so vile it must have been written by a woman, but surely no woman could pen such filthy muck. David Brewster (physicist, biographer of Newton, and the inspiration for the flowery passage with which Darwin ends the *Origin*) declaimed against it in the *North British Review*. And Whewell thought it so disgusting that he did not write against it but merely collected selected passages from earlier writings for a little book—*Indications of the Creator* (1845). The first edition did not even mention the *Vestiges* by name. I realize that the reception of *Vestiges* was by no means uniformly negative—Tennyson, for instance, was to use its ideas to finish *In Memoriam*—but for Darwin's group it was anathema. So he knew that he had better stay silent.

The second reason is simply, as many have noted, that Darwin just did not expect the delay to be so long. He set out on his barnacle work thinking that it would take but a year, and it kept stretching on and on as he worked obsessively on the project. One year stretched to eight. The species book—which in the light of the reactions would need very careful documentation—did not get written. I should say that I see here, balancing the ambition, the other side of Darwin's character. He was selfish—call it self-centered if you like—because, as a rich man who had been favored in his youth, he was accustomed to doing what he liked. He became obsessed with a project, and nothing was going to stop him. To put the matter in modern terms, he did not have to write research grants to show that his work would cure cancer. He could just amuse himself, although perhaps "amuse" is not the right word for someone who did work so hard. I see this pattern again after the *Origin*. Why did Darwin not set up a selection lab at Down? He had the money, and there were those who wanted to join him in doing just that. He could see that selection was being downgraded by people like Huxley, but he did not really fight back. Although Darwin did write the *Variation* as an extension of the *Origin* and was sufficiently threatened by Wallace's apostasy (arguing that human evolution demanded divine intervention) that he felt compelled to write the *Descent*, scientifically Darwin went on doing what he had always done, namely, working away on projects—orchids (1878), climbing plants (1880), earthworms (1881)—that caught his fancy.

I see Darwin's sharing his evolutionary ideas with others as part and parcel of this picture. He was not about to share them with Sedgwick and Whewell—still the people who really controlled science—but the younger members of the set had long been discussing origins in a potentially naturalistic way. As soon as he came back from the *Beagle* voyage, Darwin and Owen began chewing the fat over such things. (Pertinently, Owen, the best scientist of them all at this time, was probably well on the way to some kind of Germanic evolutionism, but dared not publish his work because he was so dependent on the established powers. He did not dare accept a knighthood lest he appear too uppity.) Darwin knew full well that when he did publish he would need supporters. So it was quite natural to talk about these things with those who were potential supporters and who, although they may have been cowed by people like Whewell, certainly did not necessarily agree with them.

If the *Origin* is more a product of the late 1830s and early 1840s, then we should judge it on those terms. Let me make five points showing that such an approach pays explanatory dividends. First, take the book's topic. Of course, Charles Darwin was not the first to ask about organic origins. His grandfather Erasmus had done so, for one. And in the 1840s and 1850s people went on asking about the topic—Chambers in the first decade (1844) and Herbert Spencer in the second (1852). However, I wonder if this was something on the front burner of the top professional biologists. Huxley was happy to get on board when the time came, although it took through the 1860s for him to accept that the fossil record showed evolution, and he never taught the topic in his classes (Ruse 1996b, 208). For him, it was indeed the materialism and like elements that were attractive. In the 1830s, however, Darwin's set did rather obsess about the topic—usually very negatively! It was described as the "mystery of mysteries" in a letter from Herschel to Lyell—a letter that became very public thanks to its being reprinted in Charles Babbage's *Ninth Bridgewater Treatise* (C. Darwin 1993, 124). My sense is that Darwin brought the issues back into discussion—incidentally, just at the time when Pasteur was showing the impossibility of spontaneous generation, and so in a way making the whole question of origins a bit iffy.

Second, consider the style of the *Origin*. From the beginning, everyone recognized that it was a remarkably easy read, especially for a work that was doing so much and claiming to be scientific. Richard Owen in his review in the *Quarterly* was quite nasty about this, congratulating Darwin for writing in a way that we have come to expect from the author of travel books and the like—the implication being that, written as it was, this could not be a serious work (R. Owen 1860). Darwin was certainly capable of writing stuff that could be read only by the expert, if at all. Look at the barnacle monographs, for example. But we must think of Darwin's patrons. He may not have had to work, but there were those whose approbation he sought, namely his father and his Uncle Josh (later his father-in-law). Darwin had a rather rocky start—second-rate at school, and dropping out of medicine—and his father was rightly skeptical of his abilities and his willingness to get down to things.

Third, there is the book's structure. People like Huxley, dominating biology by the time the *Origin* was published, could not have cared less about that kind of methodology. Do your anatomy, draw your analogies, trace your phylogenies. It was really all a bit noncausal in its way—certainly noncausal in the sense of forces. Darwin of the 1830s thought otherwise. This was just the time when men like Whewell were trying to define what it is to be a scientist, and that is a major reason why they wrote their books showing the nature of good science. It had to be causal, in a Newtonian sort of way, and hence the debate over Newton's rather mysterious notion of a *vera causa*. This was no abstract debate, nor was the difference insignificant between Herschel and his call for experience and Whewell and his call for consilience. There was debate over geology, with Herschel thinking Lyell did the right kind of stuff (Herschel 1830) and Whewell thinking that the catastrophists did the right kind of stuff (Whewell 1837). Darwin, trained by catastrophists, converted to Lyell, was very sensitive to these issues.

Fourth, take the obvious matter of function and final cause or teleology (Ruse 2003b). People like Huxley were just not interested in these sorts of issues. For them, it was structure and form all the way. In Darwin's language (which he took from others), they were interested in unity of type, not conditions of existence. Huxley declared openly that he was no naturalist. What turned him on were the wonderful structures that were revealed through his work at the dissecting table.

The ends that features served were at best irrelevancies and at worst hindrances to finding true relationships. After the *Origin* also, Huxley and his various scientific friends showed much less interest in adaptation and more in formal issues. It was really not until the twentieth century, after the work of the population geneticists, that function really began to ride again. This was because it was only then that natural selection began to take its place as the dominant mechanism of evolutionary change. And why was there this connection between function and selection? Precisely because, as we have seen, natural selection is a mechanism expressly intended to speak to function, and the reason for this is that Darwin thought that (what John Maynard Smith called) "organized complexity" is the dominant feature of the living world. He thought this because it was precisely what his teachers and senior friends, like Sedgwick and Whewell, were telling him. In a way, by the time the *Origin* appeared, it was already old-fashioned.

Fifth and finally, there is the matter of progress. Pretty much every Darwin scholar now agrees that Darwin accepted some form of biological progress. The question is: What form of progress? I would argue that it is a form that makes selection central, even though many think (with good reason) that the relativity of selection drives a stake through the heart of progress (Ruse 1996b). Darwin saw organisms engaged in what today we call "arms races," with adaptations getting better because of the competition with other lines of organisms. Ultimately, this all leads to intelligence and the evolution of upper-class Englishmen. What makes Darwin's treatment of progress so difficult to discern is that he often seems opposed to the very notion. But what he is against is not the very notion, but a kind of Germanic notion of inevitable upward change for the better—a kind of progress through momentum, which many morphologists saw as analogous to the development of the embryo from blob to fully complete organism.

Darwin was dead set against this. And why? Once again, because such transcendentalism—such *Naturphilosoph* thinking—was hated by Darwin's group in the 1830s. So he had to stay away from it. It is this that makes the *Origin* very different from any of those evolutionary (or quasi-evolutionary) tracts written in the Germanic mode, whether it be Richard Owen's *On the Nature of Limbs* (1849), published ten years before the *Origin*, or Ernst Haeckel's *Generelle Morphologie* (1866), published nearly ten years after the *Origin*.[2]

The Origin of Species is a very great book and a very important book. As I have tried to show, in respects it is also a very puzzling book. I shall be disappointed if the contributors coming after me do not challenge just about every substantive claim that I have made. That is what makes the *Origin*, to this day, a very exciting book.

---

2. Robert J. Richards could not disagree more with this understanding of Darwin and progress. For now, I will simply direct you to his writings where he makes his case—Richards 1992, Richards 2002, Richards 2004.

Chapter 5, Reading 3

Charles Darwin and the *Origin of Species*[1]

WE MUST NOW TURN to the private Darwin: the man who, unbeknownst to the scientific community, was successfully cracking the conundrum Herschel had so aptly named "the mystery of mysteries"—the question of organic origins. Very soon after returning to Britain, probably in the early spring of 1837, Darwin became an evolutionist, and in the fall of 1838, he hit on the mechanism of natural selection brought on by the struggle for existence. In 1842 he wrote a 35-page sketch of his theory (hereafter referred to as the sketch), and in 1844 this was expanded to a 230-page essay (both are included in C. Darwin and Wallace 1958). None of this was made public, though he showed the essay to Hooker. Darwin spent practically the whole of the next ten years concealing his evolutionism, as he labored to produce tomes on barnacle systematics. Only when this was done did he return full-time to his evolutionary work, and in the mid-1850s he began a massive work on natural selection and evolution. He was interrupted by the arrival of Wallace's essay, and after this and short evolutionary extracts from Darwin's earlier writings had been published by the Linnean Society (C. Darwin and Wallace 1958), Darwin rapidly wrote an "abstract" of his ideas. This abstract, *On the Origin of Species by Means of Natural Selection; or, The Preservation of*

1. Ruse 1999a. Reprinted with permission of University of Chicago Press.

Favored Races in the Struggle for Life, was published in November 1859. The private Darwin had joined the public Darwin.

As a teenager, Charles Darwin read Erasmus Darwin's *Zoonomia* (C. Darwin 1958, 49). Moreover, at Edinburgh he became friendly with just about the only Lamarckian in Britain, Robert Grant, who later became a professor in London. Grant expounded Lamarckian evolutionism to him with enthusiasm. So, coupled with his own admission that he greatly admired his grandfather's *Zoonomia*, young Darwin probably had a more sympathetic introduction to organic evolutionism than any other person in Britain at the time (on the other hand, he attended lectures on zoology and geology by Robert Jameson, Cuvier's British editor). However, there was certainly no immediate conversion. Darwin himself, no doubt with truth, later wrote that "it is probable that the hearing rather early in life such [evolutionary] views maintained and praised may have favored my upholding them under a different form in my *Origin of Species*" (C. Darwin 1958, 49). But when he went to Cambridge, to become a clergyman, he did "not then in the least doubt the strict and literal truth of every word in the Bible" (C. Darwin 1958, 57).

When he returned to England he set about writing up his diary into a travel book, published as the *Journal of Researches* (C. Darwin 1839). This set him thinking again about the things he had seen, particularly in the Galápagos Islands. And so in March or April of 1837 Darwin asked himself the question Lyell had refused to pose and, answering it in the obvious manner, moved across the divide and became an evolutionist. The Galápagos finches had to be explained as the natural law-bound product of one parent stock (Herbert 1974). Significantly, it was not until after the voyage, early in 1837, that the ornithologist John Gould convinced Darwin that the finches formed real species, not just varieties (Grinnell 1974, 262).

Breaking with his mentor on this question of evolution did not mean Darwin ceased to be a Lyellian. Though Darwin's evolutionism went against Lyell's steady-state system, Darwin was pushed toward organic evolutionism precisely because he was so committed to the system in the inorganic world. And in another respect Darwin perhaps became more Lyellian than ever before. Lyell had argued that gradual change would be impossible because the transitional forms would be at too great a disadvantage in the struggle for existence: given his dynamic concept of adaptation, Lyell felt that a changing species was bound to succumb. Darwin's first written evolutionary ideas mirrored this concern and led him to speculate about one-step, saltatory species transformations. About this time, he wrote in his notebook: "Not gradual change or degeneration from circumstances: if one species does change into another it must be per saltum—or species may perish." Apparently Darwin first had in mind that species, like organisms, might have definite lifespans and then would (or at least could) change into new species: "Tempted to believe animals created for definite time:—not extinguished by change of circumstances" (Herbert 1977, 10:247n).

It is plain that Darwin was thinking frantically and that nothing was very stable in his mind. In the summer of 1837, about three months after he became an evolutionist, he decided he could think more systematically if he kept notebooks devoted to the organic origins question. He kept these "species notebooks" for two years, right through the time when he discovered natural selection, and they are invaluable guides to tracing the minutiae of his thought. Let us examine these notebooks and see how, about eighteen months after his first rudimentary speculations, Darwin came upon the mechanism for which he is so famous.

From the first notebook it seems clear that Darwin's earliest speculations on the nature and causes of evolution did not last long. By midsummer 1837 the Lyellian worries about the possibility of gradually changing species had diminished. The Galápagos experience not only turned Darwin toward evolutionism, it influenced his thinking about the causes of evolution (Grinnell 1974). In particular, Darwin had in mind the model of a group of organisms, isolated from all others, evolving into a new species. "Let a pair be introduced and increase slowly, from many enemies, so as often to intermarry—who will dare say what result. According to this view animals on separate islands, ought to become different if kept long enough apart, with slightly different] circumstances.—Now Galapagos tortoises, mocking birds, Falkland fox, Chiloe fox.—English and Irish Hare" (Darwin 1980, 10). In such a model, as this passage shows Darwin recognized, the threat from external competition is eliminated; thus Lyellian fears for the adaptedness of changing species vanish. Darwin therefore felt free to posit gradual organic evolution based on small changes rather than sudden leaps. And he did switch to such minute changes, though we shall see remnants of saltatory changes in his thought for some time.

In the first species notebook, Darwin had tended to let variation look after itself. It occurred, and Darwin thought it was in some way a function of environmental conditions. But he had been concentrating on organisms evolving in isolation and thinking of the threat to an organism as coming primarily from outside its own species; though, as a passage quoted above shows, he was very conscious that adaptive failure spells extinction. Thus he had been able to shelve the question of adaptation, at least for newly evolving organisms. His first concern was to permit change—any kind of change. But about the beginning of 1838 Darwin began to worry in earnest. After all, isolation simply sets aside the problem of adaptation; it does not eliminate it. Even if they are protected from outside threats, isolated organisms like those on the Galápagos Islands are building up adaptations, and as a good Lyellian—as well as the protege of those rabid natural theologians Sedgwick and Whewell—Darwin knew this problem had to be solved. "With belief of transmutation and geographical grouping we are led to endeavour to discover causes of changes—the manner of adaptation (wish of parents??)" (De Beer 1960, 2:227). Just how does one link new variation with the fact of organic adaptation? "Can the wishing of the Parent produce any character in offspring? Does the mind produce any change in offspring? If so, adaptation of species by generation explained?" (De Beer 1960, 2:219). But this is just wild guessing, and so Darwin turned to the only possible source of information about new variations, the way they are passed on, and their relation to adaptation. He turned to the domestic world, the world of the animal and plant breeder.

At this point we must tread carefully, for from here on the thought processes we find in the notebooks and the account Darwin later gave of his road to natural selection do not quite coincide (Limoges 1970, Herbert 1971). In one of many similar accounts of his discovery, Darwin wrote: "I came to the conclusion that selection was the principle of change from the study of domesticated productions; and then, reading Malthus, I saw at once how to apply this principle" (C. Darwin 1903, 1:118). This seems fairly clear. In the domestic world, Darwin saw that the selection of minute useful variations could lead to overall adaptive advantages; and though he could not at first understand how it happened in the wild, he could see by analogy that such a process was possible. But in the notebooks we do not get this picture. Darwin quickly realized (if he had not realized it before) that the art of breeding consists in picking out favorable variations, and that the variations one normally needs will be small ones. "All these facts clearly point out two kinds of varieties.—One approaching to nature of monster, hereditary, other adaptation" (De Beer 1960, 2:4).

Although he already thought of adaptation in a dynamic way, never before had Darwin realized what pressure there is on organisms or how critical and dynamic is the concept of adaptation: "Population is increase at geometrical ration in far shorter time than 25 years—yet until the one sentence of Malthus no one clearly perceived the great check amongst men" (De Beer 1960, 4:135). Consequently, "One may say there is a force like a hundred thousand wedges trying [to] force every kind of adapted structure into the gaps in the economy of nature, or rather forming gaps by thrusting out weaker ones" (De Beer 1960, 4:135).

In some manner (to be discussed shortly), this reading of Malthus was enough. Darwin had all he needed. Malthus showed that the survival and reproductive differentials come from the desperate struggle for existence. To survive, an organism must have an adaptive edge not only over members of other species, but over members of its own species (Vorzimmer 1969). Darwin was able to connect this with what he knew about selection, which demands the very differential reproduction the struggle provides; and he saw that an organism's adaptive edge over other organisms could not be defined independent of the struggle. Adaptation was a function of the peculiarities of organisms that won out. Winning, and only winning, was what counted. Thus was born the idea of natural selection as an evolutionary mechanism. Some organisms win, because they have helpful characteristics the losers lack; and in the long run these characteristics add up to full-blown adaptations and significant evolutionary change. As Darwin later wrote, "Here, then, I had at last got a theory by which to work" (C. Darwin 1958, 120). And it is not long (November 27, 1838) before we find him referring to his mechanism to explain change: "An habitual action must some way affect the brain in a manner which can be transmitted.—this is analogous to a blacksmith having children with strong arms.—The other principle of those children which chance produced with strong arms, outliving the weaker ones, may be applicable to the formation of instincts, independently of habits" (N, in H. E. Gruber and Barrett 1974, 42).

It seems likely that sometime between 1839 and 1842, when he wrote the sketch, he was led to one of his subsidiary evolutionary mechanisms, sexual selection. Man selects not only for qualities that will aid his livelihood—heavier cows, shaggier sheep, bigger vegetables—but on occasion also for qualities that give him pleasure. These qualities tend to be of two kinds—combative strength, as when one breeds a fiercer bulldog or cock, and beauty, as when one breeds a fancier pigeon. Darwin mirrored these qualities in his analysis of selection. Natural selection corresponds to selecting for things that help man survive. Sexual selection corresponds to selecting for things that give man pleasure: Darwin in turn divides sexual selection into selection through male combat, where the stronger male gets the female(s), and selection through female choice, where the more attractive male gets the female(s) (C. Darwin and Wallace 1958, 48–49, 120–21).

In 1842 and 1844, Darwin wrote out the preliminary versions of his theory, which in essence were very little changed in the *Origin*. We have a discussion of domestic selection. We get the struggle, then the analogical counterpart of domestic selection, natural selection. Then, along with discussions of the nature of sterility and the like, we get the mechanism of selection applied to all the problem areas mentioned above. In short, we have something Darwin himself could accept as a properly structured theory. Before we can continue chronologically, however, we must pause and go back. For Darwin's contemporaries, his teachers and his seniors in the scientific network, religious questions were a most important barrier to acceptance of an organic evolutionary theory. Why were they not a barrier for Darwin?

Darwin and Religion

At the center of revealed religion, based on faith and revelation, is the Bible, and we know that when he went up to Cambridge Darwin took the Bible literally. When he left Cambridge to join the *Beagle*, his faith was fairly orthodox, but during the voyage it started to crumble. Undoubtedly the major reason for this was Darwin's growing conviction that the Bible, particularly the Old Testament, was incompatible with science, particularly uniformitarian geology (C. Darwin 1958, 185). As Darwin became committed to science, he became more and more committed to the rule of law, which in turn excluded miracles. But for Darwin, Christianity without miracles was nothing (at least as a divinely inspired religion), and so his adherence to Christianity faded away (C. Darwin 1958, 86–87). That he rested the case for Christianity so thoroughly on miracles was perfectly natural, for he was brought up on Paley's *Evidences of Christianity* (C. Darwin 1958, 59). As we have seen, in the traditional English empiricist manner, Paley made the whole truth of the Christian revelation entirely dependent on the genuineness of the biblical miracles. So when miracles went for Darwin, Christianity went too.

After the *Origin* was published, Darwin became something of an agnostic about the existence of God (C. Darwin 1958, 94; Mandelbaum 1958). Until that point it is probable that, although no Christian, he was neither atheist nor even agnostic. He was a deist of a kind—believing in an unmoved creator who worked entirely through unbroken, unchanging law. He therefore accepted a natural religion based on reason and sense. This is the language Darwin used in the *Origin* (1859, 488), and it seems improbable that he was hypocritically using such a vocabulary for tactical reasons—he was going to ruffle the Christian feathers anyway. Certainly Darwin was a deist while he was discovering his theory, for he constantly used such language in his notebooks, when he was talking only to himself (H. E. Gruber and Barrett 1974, 154). But one suspects there is more to the story. Lyell was a deist in the sense ascribed to Darwin, but religion continued to be a major antievolutionary stumbling block for Lyell. Why did Darwin remain unawed by the question of man, feeling none of the emotions Lyell felt about man's peculiarity and special status? In the first species notebook (late 1837), man was firmly put in his place: "People often talk of the wonderful event of intellectual man appearing. The appearance of insects with other senses is more wonderful" (De Beer 1960, 2:207). And no sooner had Darwin grasped natural selection than he was speculating on how it might apply to man. Man was just not the barrier for Darwin that he was for so many others.

Darwin took this attitude for a number of reasons, feeling sure that man and his origin, like everything else, must come beneath the rule of unbroken law. First, unlike others of his group, Darwin had firsthand experience of savages in the wild—the natives of Tierra del Fuego. Their primitive lifestyle clearly impressed on Darwin how very non-peculiar and close to the brutes man is. . . . Second was Darwin's family background. Although Darwin himself at one point believed the Bible literally and intended to become a clergyman, his grandfather Erasmus was at best a weak deist, quite able to believe in evolution, whom Coleridge thought an atheist (Coleridge 1895, 1:152); his father Robert, who had an overwhelming influence on Darwin, was an unbeliever (C. Darwin 1958, 87); his uncle Josiah Wedgwood was a Unitarian (he supported Coleridge for a number of years [Meteyard 1871]); and, most important of all, Charles's older brother Erasmus had become an unbeliever by the time Charles returned from the *Beagle* voyage. One suspects that all of these things must have influenced Charles, making him fairly unimpressed on the subject of man. There is no doubt that Erasmus directed or confirmed Charles's thought, for Emma wrote to Charles just after their marriage, tactfully bemoaning Erasmus's bad influence (C. Darwin 1958, 236).

Third was the intellectual community with which Darwin was mixing in London while he was formulating his ideas. Not only was Darwin exposed to believing scientists like Lyell and Whewell, he was also intimate with men whose religious ideas were—or could be taken to be—much more tolerant of evolutionism. Lyell was good friends with Babbage (C. Darwin 1958, 108) and read the *Ninth Bridgewater Treatise* (De Beer 1960, 5:59), which ascribed everything to the rule of law. Also, Darwin was much in the company of Carlyle, a close friend of his brother Erasmus, and was indeed greatly taken with him, for he wrote to Emma just before they were married (January 1839) that "to my mind Carlyle is the best worth listening to of any man I know" (Emma Darwin 1915, 2:21). There apparently was little scientific sympathy between Darwin and Carlyle, but Carlyle's unorthodox religious beliefs may have fallen on fertile ground. In particular, his natural supernaturalism, with its deliberate refusal to see any part of the creation as more or less marvelous or miraculous than any other, may well have influenced Darwin, as it undoubtedly influenced Huxley. Be this as it may, Darwin was certainly mixing in company that was not as ready to exalt man's status as were Whewell and Lyell.

Finally, it seems that Darwin simply cared less about religion than many other men. He admitted this himself (C. Darwin 1958, 91). Darwin thought about the subject as much as most men, if not more—his studies forced him to do this. But essentially he just wanted to get on with his science, whatever the consequences. From Darwin one never gets the burning religious zeal to be found in, say, Sedgwick or, in a different sense, Huxley (Herbert 1977).

Nevertheless, though he was putting a strain on adaptation, particularly as a bond between God and his creation (a fact he fully recognized [1838, 70]), this does not deny the great importance of adaptation for Darwin. He could not accept such adaptation as the product of divine intervention; but he knew from his scientific/religious background that any adequate biological theory must meet the problem head on, and he tried to do this with his mechanism of natural selection. Indeed, Darwin's sensitivity about adaptation had a somewhat paradoxical result. Darwin was educated and was doing his great creative work in the 1820s and 1830s, when the concept of design through adaptation was at its peak. By the time he published his work, as we have seen, this strand of natural theology had been much buffeted. Hence, though his nontheological theory is often portrayed as taking the teleology out of biology, if anything Darwin was bringing it back in! Adaptation, with its orientation toward ends, was a more significant facet of the organic world for Darwin than it was for Huxley. Because the creative Darwin was a man of the 1830s rather than the 1850s, after the *Origin* biology was in a sense brought back to a teleology based on adaptation, from which it had started to stray.

The Long Wait

We come now to the major puzzle in the Darwinian story. By the middle of 1844 Darwin had completed a version of his theory in the 230-page essay. Although he made some changes when he wrote the *Origin*, they were comparatively minor. Why then did Darwin not publish his work at once, as soon as he had got something down on paper, rather than finish some geological work and then plunge into a massive, eight-year systematic study of barnacles? The usual reason given for the delay is some version of that offered by Huxley. The only person to whom Darwin showed his essay (when it was just finished) was Hooker, with whom a friendship was rapidly ripening (C. Darwin and Wallace 1958, 257). Unconverted, Hooker suggested that, before publishing, Darwin might well deepen his understanding of biology. Darwin saw the value of this and turned to barnacles with just such an aim. Thus, as Huxley wrote, "Like the rest of us, he had no proper training in biological science, and it has always struck me as a remarkable instance of his scientific insight, that he saw the necessity of giving himself such training, and of his courage, that he did not shirk the labor of obtaining it" (F. Darwin 1887, 1:347).

There are two reasons why this sanitized version of the cause of the delay is unconvincing. First, there is no great difference between the essay and the *Origin*. Darwin became more convinced that the smallest of new variations ("individual differences") are the building blocks of evolution, and this was probably influenced by his barnacle work.... Second, he was not the man to let a little ignorance stand in the way of the publication of a bold and sweeping hypothesis. I say this not sarcastically but literally—Darwin could not wait to get into print on the subjects of coral reefs and Glen Roy. He was ambitious and wanted to make his mark in the scientific community. A few blank spots were not going to stop him from publishing his solution to what his group took as the major scientific problem of the day. He was realistic enough to know how good his solution was. The true answer has to be sought in Darwin's professionalism, just as his success at finding and developing his theory must be sought there. Darwin was not an amateur outsider like Chambers. He was part of the scientific network, a product of Cambridge and a close friend of Lyell, and he knew well the dread and the hatred most of the network had for evolutionism. . . . Darwin knew his theory was much better than Chambers's—"better" in that it more adequately answered the problems as then understood—but it was evolutionary and materialistic nonetheless, and it was certainly not going to make its author very popular. When telling Hooker of his evolutionism, Darwin confessed that it was like admitting to a murder (F. Darwin 1887, 2:23). It was a murder—the purported murder of Christianity, and Darwin was not keen to be cast in this role. Hence the essay went unpublished.

Also, Darwin had no idea the delay would be so long. Stimulated by a strange barnacle he discovered while on the *Beagle*, Darwin decided to do a little work on barnacles (C. Darwin 1958, 117-18). This project exploded into a full-length study, taking a great deal of time, and was further dragged out by the constant, severe illness that crippled him. Days, weeks, and months were lost when he was unable to work. From the pushy, vibrant young man of the 1830s, Darwin was reduced to an invalid. And so year after year the essay on species lay untouched—with strict instructions that it be published in the event of his death (C. Darwin and Wallace 1958, 35-36). Darwin had no desire to be ignored by posterity.

On the Origin of Species

Darwin's first chapter, "Variation under Domestication," deals with animal and plant breeding. He argues that organisms with many different forms, like pigeons, have common ancestors and that the main reason for these diverse forms, besides use and disuse, is man's power of selection. The second chapter, "Variation under Nature" establishes the widespread variation in the wild. Darwin was concerned with very small variations, "individual differences," as opposed to larger changes, "single variations" (C. Darwin 1859, 44-45). He showed more confidence in their ubiquitous existence in the *Origin* than in the essay, perhaps in part because of his work on barnacles. In the essay he explicitly allowed that natural evolution might occasionally occur through a saltation from one form to another (certainly not, however, across a species [C. Darwin and Wallace 1958, 150; Vorzimmer 1963]). In the *Origin* all natural changes are smooth, fueled by individual differences. The next two chapters are the crucial ones. First, we get the derivation of the "Struggle for Existence." ... Then in the fourth chapter, "Natural Selection," Darwin goes on to derive his key mechanism.

This fourth chapter of the *Origin* contains two important subsidiary discussions. First, Darwin shows that he has definitely relinquished geographical isolation as an essential element in evolutionary speciation. Relying entirely on the smallest of variations, Darwin reveals his fear that isolated populations, tending by their very nature to be small, might not be injected with enough new variation to cause significant change. He therefore gives up exclusive reliance on isolation in return for being able to argue that (splitting) speciation might occur between subgroups of large populations, where there definitely would be ample new variation. Second is the "principle of divergence." Arguing analogically from pigeons, Darwin suggests that "the more diversified the descendants from any one species become in structure, constitution, and habits, by so much will they be better enabled to seize on many and widely diversified places in the polity of nature, and so be enabled to increase in numbers" (1859, 112). In other words, the reason there are so many kinds of species and so much splitting is that it confers a selective advantage.

Next we get the fifth chapter, "Laws of Variation," a subject about which Darwin candidly admitted "our ignorance... is profound" (1859, 167). He seemed to think there were two basic kinds of variation. One type was more or less random (with respect to needs), and Darwin suspected these variations might be due to conditions impinging in some way on the reproductive system. Those of the other type do seem to involve a direct response to the environment: "I believe that the nearly wingless condition of several birds, which now inhabit or have lately inhabited several oceanic islands, tenanted by no beast of prey, has been caused by disuse" (1859, 134).

The sixth chapter deals with "Difficulties on Theory." One problem Darwin tackles in this chapter is the absence of intermediate forms between species. Why is it that we do not "everywhere see innumerable transitional forms?" (1859, 171). Darwin posited that, excluding cases where speciation forms through isolation and where one would thus expect no intermediates, a group intermediate between two diverging varieties will tend to be smaller than the main groups because the intermediate groups will be in the rather narrow zones between the larger areas in which the main groups are adaptively diverging.

A seventh chapter titled "Instinct" follows, in which Darwin explains that instincts, like structures, vary; that they can be of great adaptive value; and that it is reasonable to think of them as subject to and produced by natural selection. Following this comes the chapter "Hybridism." Here he argues, as one might expect given the Darwinian position, that there is a very gradual gradation between perfect fertility and perfect sterility (1859, 248), thus blurring the distinction between variety and species (1859, 276). Darwin also suggests that sterility is a by-product of the laws of growth and heredity—it is "incidental on other differences, and not a specially endowed quality" (1859, 261). These reproductive barriers between species are not deliberately fashioned by selection. This discussion concludes just about every element of Darwin's solution to the problem of the origin of species—the problem of how and why organisms split into distinct groups as opposed to the problem of organic origins simpliciter. Darwin presented his solution piecemeal through the first half of the *Origin* rather than offering it all in one unit. Speciation occurs because organisms change in response to changing environments and because the more diversified and specialized they become, the more efficiently they can exploit their environments.

We come next to the geological chapters, numbered 9 and 10, respectively: "On the Imperfection of the Geological Record" and "On the Geological Succession of Organic Beings." The first of these chapters is very Lyellian—explicitly so. Darwin admits that, prima facie, the geological record raises major problems for an evolutionary theory like his own: he recognizes the objection that there was inadequate time for the slow process of natural selection and that the abrupt transition from one species to another (as revealed by the record) disproves evolution, as does the first appearance of life in all its sophistication. In reply to the first objection, Darwin counters that there was indeed enough time.... In reply to the second objection, Darwin argues that the gaps in the fossil record can be explained by the inadequacy of the record—fossils were not deposited, we have failed to find the fossils, and so on. In any case, speciation occurs most frequently when the ground is being elevated, thus causing new stations (as on islands). But since this is not a time of deposition, no transitional fossils are being laid down. Third, why did life appear full-blown at the beginning of the Silurian? This problem, Darwin admits, worries him. It is not because the older the rock, the more metamorphosed it automatically becomes, and that all pre-Silurian rocks are so metamorphosed that they can show no fossils. The Silurian is too rich in fossils to allow that aging necessarily metamorphoses. Ever ready with a hypothesis, Darwin suggested that pre-Silurian organisms might have flourished on continents where oceans are now. And he guarded his hypothesis against contrary evidence like deep-sea borings by suggesting that the fossils of these organisms may now be metamorphosed by the great weight of the ocean above them (1859, 306–10)!

The second geological chapter was much more positive. For instance, it is necessary to Darwin's theory, unlike Lamarck's (and in certain respects Chambers's too), that there be no escalator-like evolution, where the loss of a particular species is compensated by the evolution of the same species at a later time. For Darwin, a species has only one chance, and he was pleased that the fossil record confirms this (1859, 313). . . . Divergence was fundamental for Darwin, and since man is not ontologically different from the rest of creation, there could be no unilinear man-directed progression. We have divergent evolution leading to ever greater adaptive specialization. But within this framework Darwin was prepared to concede some vague subjective sense of progression: he too, as Lyell had always feared, linked progression and evolution.

The two chapters following the geological discussion, numbers 11 and 12, on "Geographical Distribution," contain some of Darwin's strongest cards, as our knowledge of his path to discovery might lead us to suspect. . . . Penultimately, we get a grab-bag chapter: "Mutual Affinities of Organic Beings: Morphology: Embryology: Rudimentary Organs." The natural system is explained as simply a function of common descent. Morphological problems dissolve in the same way. Take the classic homology between the hand of man, paw of the mole, leg of the horse, flipper of the porpoise, and wing of the bat. Naming Owen to support his position, Darwin argues that this problem is too much for the doctrine of final causes (1859, 435; Owen had hardly adopted Darwin's solution!). But it can all be easily explained by a theory of descent with modification owing to natural selection. . . . Finally comes a chapter titled "Recapitulation and Conclusion," noteworthy primarily for containing what must be the understatement of the nineteenth century: "Light will be thrown on the origin of man and his history" (1859, 488).

Darwin and the *Origin* therefore were not the natural culmination of a long line of evolutionists and their writings. Yet Darwin and his work did not spring from nowhere. For direct links, as well as intellectual sympathies, we must look primarily to the scientific group from which he came. Above all, we see in Darwin the influence of Lyell. We know that Darwin's geology was Lyellian through and through, and the same is largely true of his attack on the organic origins problem, even though he was doing just what Lyell could not do. Darwin as geologist satisfied all three of our divisions of Lyellianism, and as might be expected we find that the evolutionist Darwin was a complete actualist. He wanted to explain the origin of organisms by causes of a kind we see about us at present, in both the domestic and the natural worlds. Similarly, Darwin was a uniformitarian. He wanted no causes of an unknown intensity. Thus, after some initial speculating (itself a function of Lyellian worries), he wanted no super saltations creating new species at a leap. His link to steady statism was more tenuous. In the inorganic world he accepted it entirely and believed it ultimately fueled evolution. In the organic world Darwin was in one sense going against Lyellian steady statism, but in other senses he was not. He was not a normal progressionist: for him, progression was almost incidental and evolution could (and sometimes did) go against it. Also, Darwin, like Lyell, saw new species appearing and old disappearing on a fairly regular, continuous basis. Even here he did not break completely with his mentor.

Chapter 5, Reading 4

Darwinian Evolution[1]

Human Prehistory

WHAT IS THE BEGINNING of a scientific story about human origins? Cover our options and play it safe. The big bang, which started everything, occurred about 13.8 billion years ago (Morison 2014). Whether there was anything before it or what caused it is a matter of speculation. The universe as we know it—the Sun and the planets—is about 4.5 billion years old. Nothing lasts forever, and it is thought that the Sun is about halfway through its life time.

The causes of the origin of life are still in dispute (Bada and Lazcana 2009). In the sense of working according to established, unbroken laws, no one in the scientific world has any doubt that these causes are natural. The origins are findable, and one day perhaps soon will be found. Nobel Prizes for the winners! What we do know is that life seems to have appeared about as soon as it could have appeared, meaning as soon as the Earth and—especially—the water on its surface had cooled enough to allow life to flourish. You are not going to get much life if everything, everywhere, is 1,000°C. No one can be precisely accurate about these matters—Archbishop Ussher (1581–1656) pinned down the first day of creation as Sunday, October 23, 4004 BC—but general opinion is that life started around 3.8 billion years ago (bya).

1. Ruse 2021b. Reprinted with permission of Cambridge University Press.

For the first half of life's history, the cells were simple—prokaryotes—without complex nuclei and other cell parts. Then, about 2 bya, some prokaryotes fused and more complex cells—eukaryotes—were formed. Complex life was off and running (Benton 2009). The big event is generally thought to be the Cambrian explosion—about 550 million years ago (mya). This was when most of the major groups (technically known as "phyla") appeared—arthropods (insects), chordates (animals with a notochord, a kind of skeletal rod), mollusks (snails), and more. Because we humans have backbones, it is the chordates that matter to us. Evolution took us through our own particular subgroup (the vertebrates) from fish, to amphibia, to reptiles, to mammals and birds (Harari 2015, Reich 2018). We humans are mammals, our ancestors appearing about 225 mya, ratlike, nocturnal, and keeping well out of the way of those lumbering reptile brutes, the dinosaurs.

Mammals gave rise to the primates, about 50 mya or rather older, and now (from our perspective) the story starts to get really interesting. First you get monkeys—although we are their descendants, the actual groups of animals (species) from which we come are now extinct—then the great apes, and finally the line that is going to lead to humans. Members of this line are known as "hominins." It is now thought, much to the surprise of many people about fifty years ago, we split off from the other great apes about 6 or 7 mya, and, even more to the surprise of many people about fifty years ago, our line split first from the gorillas and then from the chimpanzees. In other words, our closest great-ape relatives are the chimpanzees, and they are more closely related to us than they are to gorillas.

As the late Stephen Jay Gould used to emphasize, the history of life is not like a poplar tree, pushing straight to the top (Gould 1988). Leaving aside the complexifying fact that genes can be transferred across lines, in respects making a network a more appropriate metaphor, the history is much more like a bush than a tree. The line leading to us kept splitting and splitting. . . .

Darwinian Theory

Back to human prehistory. What brought all of this about? In one sentence: Charles Darwin's theory of evolution through natural selection. Separate two things. First, the fact of evolution, that all organisms, past and present, including humans, came by a natural developmental process from, as Darwin said, "one or a few forms." Second, the causes or mechanism(s) that brought this all about. Darwin first became convinced of the fact of evolution (Browne 1995). This happened in March 1837, and then he spent eighteen months looking for an explanation, which he found at the end of September 1838. This search for a cause was not idiosyncratic. A graduate of the University of Cambridge, where unsurprisingly the work of the earlier Isaac Newton was taken to be the model of good scientific practice, Darwin searched for the biological equivalent of the causal force of gravitational attraction. At the end of September 1838, he found it. Natural selection! After sitting on his ideas for some fifteen to twenty years—in 1842 he wrote a 35-page "sketch" and in 1844 he wrote a 230-page essay—spurred by the arrival of an essay with nigh-identical ideas by the naturalist Alfred Russel Wallace—he finally published his theory. *On the Origin of Species by Means of Natural Selection, or the Preservation of Favoured Races in the Struggle for Life* appeared toward the end of 1859.

Although it was accepting the fact of evolution that led Darwin to look for causes, in the *Origin* he started with the causes—following Newton, who gave the laws of motion and of gravitational attraction—and only then moved on to the fact, the tree of life—following Newton, who explained the heliocentric view of the universe. First, he introduced artificial selection, the picking and choosing that farmers and breeders do with animals and plants to get ever better specimens. He stressed how powerful this is, at the same time emphasizing that the changes are in the direction of properties—fatter cattle, shaggier sheep, more beautiful songbirds—the selectors want. There is an underlying program of design. To this end, one breeder said that selection "is the magician's wand, by means of which he may summon into life whatever form and mould he pleases." Another, speaking of the breeders of sheep, concluded: "It would seem as if they had chalked out upon a wall a form perfect in itself, and then had given it existence" (C. Darwin 1859, 31).

Now Darwin was off and running, showing how there is a natural equivalent of breeders' selection and that this too produces design-like organisms. Preparing the way, he argued that in all populations one finds variation—something needed for the building blocks of evolution. Although he had assumed it before, a decade-long study of barnacles had given empirical evidence of the ubiquity of this variation (C. Darwin 1851b). Then come the key arguments. First, invoking the ideas of the late-eighteenth-century political scientist (and Anglican clergyman) Thomas Robert Malthus (1826), Darwin argued that in natural populations we are going to have a "struggle for existence." The potential rate of reproduction is geometric, whereas the potential rate of food increase is arithmetic, and the former—one, two, four, eight . . .—outstrips the latter—one, two, three, four. "Hence, as more individuals are produced than can possibly survive, there must in every case be a struggle for existence, either one individual with another of the same species, or with the individuals of distinct species, or with the physical conditions of life." Adding: "Although some species may be now increasing, more or less rapidly, in numbers, all cannot do so, for the world would not hold them" (C. Darwin 1859, 31).

Seizing on this conclusion, and invoking that variation for which he had argued, Darwin asked: "Can the principle of selection, which we have seen is so potent in the hands of man, apply in nature? I think we shall see that it can act most effectually." He further asked whether it can "be thought improbable, seeing that variations useful to man have undoubtedly occurred, that other variations useful in some way to each being in the great and complex battle of life, should sometimes occur in the course of thousands of generations?" He answered himself:

> If such do occur, can we doubt (remembering that many more individuals are born than can possibly survive) that individuals having any advantage, however slight, over others, would have the best chance of surviving and of procreating their kind? On the other hand, we may feel sure that any variation in the least degree injurious would be rigidly destroyed. This preservation of favourable variations and the rejection of injurious variations, I call Natural Selection. (C. Darwin 1859, 80–81)

The action of natural selection over generations leads to wholesale change, evolution. What is crucial about the wholesale change brought on by natural selection is that, as in the domestic case, the change is not random. It is in the direction of design-like features—the hand and the eye—that will help their possessors in the struggle for existence (55). Adaptations! Thanks to natural selection, "we see beautiful adaptations everywhere and in every part of the organic world" (C. Darwin 1859, 61).

Finally, Darwin was ready to introduce the fact of evolution, using the metaphor of a tree. "The affinities of all the beings of the same class have sometimes been represented by a great tree. I believe this simile largely speaks the truth" (C. Darwin 1859, 129). Life grew upward through time, with new branches representing new kinds of organism: "The limbs divided into great branches, and these into lesser and lesser branches, were themselves once, when the tree was small, budding twigs; and this connexion of the former and present buds by ramifying branches may well represent the classification of all extinct and living species in groups subordinate to groups" (C. Darwin 1859, 129). We arrive at the present. "As buds give rise by growth to fresh buds, and these, if vigorous, branch out and overtop on all sides many a feebler branch, so by generation I believe it has been with the great Tree of Life, which fills with its dead and broken branches the crust of the earth, and covers the surface with its ever branching and beautiful ramifications" (C. Darwin 1859, 129–30).

The rest of the *Origin*, a good three-fifths, is devoted to showing how selection explains and is in turn confirmed by what we know of the nature and behavior of organisms in nature. Starting this detailed overview—what Darwin's mentor at Cambridge, the philosopher William Whewell, called a "consilience of inductions" (Whewell 1840)—Darwin turned to social behavior as is exemplified by the hymenoptera—the ants, the bees, and the wasps. This was a subject of much interest in the nineteenth century, not least because so many people kept hives of bees. Why and how, for instance, do we find sterile workers in a nest, helping others but apparently going against the rule of adaptations, always helping in the struggle for existence and reproduction? Darwin had no good grasp of genetics, but he offered a proto-version of what is now known as "kin selection" (Maynard Smith 1964). If you can help close relatives to reproduce—like the queen and your fertile sisters—then you are passing on your own features vicariously.

Turning to the artificial selection analogy, Darwin pointed out how we can pass on the desired features of the slaughtered steer by returning to the breeding stock from which it came. If its fertile parents have fertile offspring, that will do the trick. Note that here, as always, Darwin thinks in terms of benefit to the individual, not the group. That goes right back to the beginning. Remember: "There must in every case be a struggle for existence, either one individual with another of the same species, or with the individuals of distinct species, or with the physical conditions of life" (C. Darwin 1859, 63). In modern terms, Darwin was an ardent "individual selectionist" as opposed to "group selectionist."

Corresponding to the question about the fossil record, there are the shared underlying forms between different species—underlying forms that Darwin's contemporary, the anatomist Richard Owen, called "archetypes," likening them to the Forms of Plato (R. Owen 1849). Darwin explained the similarities—homologies—in terms of evolution through selection. Darwin was no fanatical pan-selectionist, insisting that every facet of the organic world had to be shown to be adaptive. Indeed, he had stressed this very point, while insisting that selection always comes first, and indirectly may explain the nonadaptive. "It is generally acknowledged that all organic beings have been formed on two great laws—Unity of Type, and the Conditions of Existence. By unity of type is meant that fundamental agreement in structure, which we see in organic beings of the same class, and which is quite independent of their habits of life. On my theory, unity of type is explained by unity of descent." Adding, however, that this means that "the law of the Conditions of Existence is the higher law; as it includes, through the inheritance of former adaptations, that of Unity of Type" (C. Darwin 1859, 206).

We come to embryology, by confession Darwin's own favorite. Why the similarity of embryos of organisms that are so very different as adults? Simply because, in the womb, conditions are similar for all and so there is no pressure on selection to separate them. Selection kicks in only as they move to adulthood. And so finally to the most famous passage in the whole history of science. Evolution, evolution through natural selection brought on by the struggle for existence, explains all.

> Thus, from the war of nature, from famine and death, the most exalted object which we are capable of conceiving, namely, the production of the higher animals, directly follows. There is grandeur in this view of life, with its several powers, having been originally breathed into a few forms or into one; and that, whilst this planet has gone cycling on according to the fixed law of gravity, from so simple a beginning endless forms most beautiful and most wonderful have been, and are being, evolved. (C. Darwin 1859, 490)

Humans

What about that most interesting of organisms, Homo sapiens? From the first, Darwin had always been stone-cold certain that we humans are part of the picture. In one of his private notebooks, written just after he had read Malthus and seized on what it all meant, we get the first explicit reference to natural selection—applied to humans and not just to humans but to our mental abilities. "An habitual action must some way affect the brain in a manner which can be transmitted.—This is analogous to a blacksmith having children with strong arms.—The other principle of those children, which chance? produced with strong arms, outliving the weaker ones, may be applicable to the formation of instincts, independently of habits" (C. Darwin 1985, N 42). (Note that, like Spencer, Darwin was—and ever was—a believer in Lamarckian heredity, the inheritance of acquired characteristics.) In the *Origin*, however, fearing that too much discussion of humans would swamp everything, Darwin stayed away from the topic, dropping in only one provocative comment so people would not think he was being cowardly. "In the distant future I see open fields for far more important researches. Psychology will be based on a new foundation, that of the necessary acquirement of each mental power and capacity by gradation. Light will be thrown on the origin of man and his history" (C. Darwin 1859, 488).

After the *Origin*

For our purposes, the story of evolution from the *Origin* to the present can be told quickly. The big gap in Darwin's theorizing was an adequate theory of heredity. This did not arrive on the scene until the beginning of the twentieth century, when people discovered the experiments on pea plants and their interpretation by the monk Gregor Mendel, working in the Austro-Hungarian empire, back in the 1860s (Bowler 1989). Mendel's breakthrough was to see that the units of heredity, what we now call genes, are particulate—apart from occasional random changes (mutations) they are transmitted entire from generation to generation. At first, many thought Mendelism a challenge to Darwinism—take one or the other—but gradually they were seen as complementary, Darwinian selection working on Mendelian genes (Provine 1971). This was all brought together by the theoreticians around 1930—in England, Ronald A. Fisher (1930) and J. B. S. Haldane (1932), and, in America, Sewall Wright (1931, 1932). A few years later the experimentalists and naturalists got into the act. In America, the Russian-born fruit-fly geneticist Theodosius Dobzhansky (1937), the German-born ornithologist and systematist Ernst Mayr (1942), the paleontologist George Gaylord Simpson (1944), and then, bringing up the rear, the botanist G. Ledyard Stebbins (1950). In Britain, the overall systematizer was Julian Huxley

(1942)—he was the grandson of Thomas Henry and older brother of Aldous—and the chief naturalist was E. B. Ford (1964) and his school of "ecological genetics." The new theory or revitalized theory—known as "Neo-Darwinism" in Britain and as the "synthetic theory" in America—was off and running. . . .

A couple of points and then we can move to matters more philosophical. First, the modern theory is completely and utterly Darwinian, in the sense that it is his causal mechanism of natural selection—sexual selection too, which today is often subsumed beneath the generic natural selection—that is absolutely central. This said, in the modern theory selection does have a rather different place from that given to it in the *Origin*. There, it is up front and center. We go from the population pressures to the struggle for existence and from there to natural selection. Nothing is rigidly formal, certainly not mathematical, but it seems that (as always) Darwin was trying to be Newtonian in offering laws (all populations have a tendency to increase in numbers geometrically) from which one can deduce conclusions, like the struggle and selection.

The modern theory is likewise Newtonian in this respect—in the trade it is called being "hypothetico-deductive"—but much tighter and more formally mathematical. What is crucial is that selection is deposed from its central position (not role or importance) and the modern theory starts with genetics, Mendelian—more recently, as we shall see, molecular—thinking extended out from individuals to populations (Ruse 2006). The key hypothesis is the Hardy-Weinberg law. Genes are found on the chromosomes, generally in the heart of the cell, the nucleus, and because chromosomes come in pairs there are always two genes in an organism occupying the same paired position (locus). . . .

Molecular Biology—Friend or Foe?

The second point to be mentioned is the coming of molecular biology, the double helix of 1953. At first, the fear, particularly held by eminent evolutionists, was that Darwin's theory would be pushed to one side by the molecules. Fifteen years after Watson and Crick, we find Ernst Mayr writing that "enthusiastic but poorly informed physical scientists have lately tried very hard to squeeze all of biology into the strait jacket of a reductionist physical-chemical explanation" (Mayr 1969, 128). With the rhetorical flourish that characterized so much of his writing (and his whole personality), he concluded a critique of molecular intrusions into the evolutionary field by saying: "It is futile to argue whether reductionism is wrong or right. But this one can say, that it is heuristically a very poor approach. Contrary to the claims of its devotees, it rarely leads to new insights at higher levels of integration and is just about the worst conceivable approach to an understanding of complex systems. It is a vacuous method of explanation" (Mayr 1969, 128). Mayr's worries were already dated. As an eminent Anglican theologian said at the end of the nineteenth century about the coming of Darwinism in an age of Christianity, it came disguised as a threat and proved to be a friend (A. Moore 1890). The almost comical—certainly ironic—thing is that molecular biology proved to be the handmaiden of evolutionary biology, not its master (or mistress). Evolutionary biology set the problems, and molecular biology set about giving answers.

Chapter 5, Reading 5

Darwinism[1]

CHARLES DARWIN PUBLISHED *On the Origin of Species* in 1859 and *The Descent of Man* in 1871. He changed the world. Although there were those who continued to stand firm against evolution—indeed, as is well known, there are still those who continue to stand firm against evolution—generally, even the religious accepted that organisms, including humans, are the end point of a long, slow process of natural development. As in the Hans Christian Andersen tale about the lad who said openly that the king has no clothes, so when Darwin said "evolution," nigh everyone said that they had known it all along! Natural selection had more mixed success. Everyone accepted it to some extent. Thomas H. Huxley, for instance, always had some doubts about its universal power and applicability, but when it came to humans physically, he was fully convinced of its overwhelming importance. This said, the scientific community was slower in coming to full acceptance, and it was more in the popular domain that natural selection—and even more sexual selection—was a huge success. Poets, novelists, politicians, and many others harped on and on about its importance.

---

1. Ruse 2018a. Reprinted with permission of Princeton University Press.

Modern Evolutionary Biology

Modern evolutionary biology slowly moved from pseudoscience to popular science. Evolutionary theory became a professional science, in the sense of something studied in university departments and with senior researchers and graduate students, grants, journals, and so forth, starting around 1930 and picked up—particularly in England (where it became known as neo-Darwinism) and in America (where it became known as the synthetic theory of evolution)—over the next decades (Ruse 1996b). By 1959, somewhat arbitrarily choosing the hundredth anniversary of the *Origin*, one had (to use a somewhat hackneyed term) a fully functioning paradigm.

This was a Darwinian theory, in the sense that natural selection played (and continues to play) the central causal role, a status brought about by the melding of selection with the newly found and developed theory of heredity, Mendelian (and then later molecular) genetics. At the beginning of the twentieth century, the work of the somewhat obscure Moravian monk Gregor Mendel was rediscovered, and with this, the big hole in Darwin's theorizing could be filled. Thanks particularly to the work in the second decade of the century by Thomas Hunt Morgan and his associates at Columbia University, it was seen that the crucial unit of heredity—the gene—is a physical thing (now known to be long threads of nucleic acid) on the chromosomes in the nuclei of cells. These genes maintain their integrity from generation to generation, thus giving selection something stable and heritable on which to act. However, every now and then the genes spontaneously change (*mutate*)—much is now known about the causes but the important thing is that Darwin's insight was correct, the changes are random both in not appearing to order and in not necessarily bringing on new features of any use to the possessor.

Adaptations—characteristics with ends, with purposes—are as vital to modern evolutionary biology as they were to Darwin. Final cause talk, thinking of organisms in terms of design, is all important. One thing realized by today's evolutionists is that Darwin was unduly pessimistic in thinking (as he did) that we will never see natural selection in action. In the right circumstances, it is readily observable. A justly celebrated demonstration of selection in action producing features that are directed toward ends is that of the couple Peter and Rosemary Grant and their long-term study of *Geospiza*—better known as "Darwin's finches"—on an islet in the Galápagos Archipelago (P. Grant 1986, P. Grant and Grant

2007). The Grants demonstrated unambiguously that in times of plenty, the beaks of the species they were studying were relatively fine and all purpose—for cracking seeds, eating insects, and whatever—but that during times of drought, when the only available foodstuffs seemed to be hard-shelled nuts, the beaks evolved in a direction of stubbiness and strength. To put the matter teleologically: Why did the successful finches have stubby beaks? For the purpose of breaking up nuts with hard shells. And that was a very good thing from the viewpoint of the finches. . . . Paleontologists are into this game too. They think in terms of design, as if someone had sat down and built an organism to achieve a certain end. They are looking for purposes, for functions, for ends.

Refinements

It is worth noting two more points about modern thinking—extensions on Darwin's day. First, many worry (understandably) about how selection can possibly be effective if new variations are random in not arriving in time to order, as it were. A major advance in our thinking in this respect is due to the Russian-born American geneticist Theodosius Dobzhansky (1900–1975). He made much of what is known as "balanced superior heterozygote fitness" (Ruse 1982, 137). Genes are paired with mates on corresponding chromosomes. Sometimes these genes are identical (homozygotes) and sometimes different (heterozygotes). An interesting situation ensues when heterozygotes do better in the struggle (are fitter) than either homozygote. A famous case in point concerns the awful genetic disease sickle cell anemia. A person born a homozygote for a certain gene is going to die (without drastic medical intervention) from anemia, at the age of four. However, a person who is a heterozygote with one sickle cell gene and one normal gene is going to be fitter than a homozygote for the normal gene. The reason is simple: namely, that being a heterozygote gives you a natural immunity to malaria, one not possessed by those with two normal genes. It is for this reason that the sickle cell gene is found only in parts of Africa where malaria is endemic, or in populations from such areas, particularly North Americans of African descent. The important point to note is that no matter how bad the sickle cell gene, it will persist in the population—in the "gene pool"—because of its virtues for heterozygotes. . . .

Dobzhansky seized on this fact and generalized, assuming that such a phenomenon is widespread and that consequently any population is going to carry a large variety of genes—those from the same chromosome position are known as *alleles*—on which selection can act immediately without waiting for favorable mutations (Dobzhansky et al. 1977). Superior heterozygote fitness is not the only putative way of getting in-group variation. Selection for rareness would also do the trick. Suppose a predator has to learn something about its prey before it can strike—color markings, for instance. A rare form would be at a selective advantage and thus start to spread, until it was so common that the predator would more quickly learn to seek it out, so positive selection would ease off and you might expect a balance between different forms. What is exciting is that at this point molecular biology (barely ten years after the discovery of the DNA model) came to the aid of organismic (evolutionary) biology by showing through new techniques (gel electrophoresis) that natural populations do harbor huge amounts of genetic variation (Lewontin 1974). All in all, selection is highly plausible as an important creative force.

The second point about modern evolutionary thinking is that no one (starting with Darwin, as we have seen) thinks that every last thing about organisms, living or dead, has to be (or had to be) adaptive. The late Stephen Jay Gould, paleontologist and popular science writer, made much of this. Greatly influenced by German biology, he focused on the homologies between organisms, stressing their importance for establishing the fact of evolution but their irrelevance for proving the force of natural selection (Gould 1977, 2002). More broadly, in a well-known article, coauthored by Richard Lewontin, "The Spandrels of San Marco," Gould argued strongly that Darwinian evolutionists assume far too readily that living nature is adaptive, that it is full of purpose (Gould and Lewontin 1979). He felt that evolutionists slide into some kind of panadaptationism, thinking that every last organic feature has to be functional, the product of natural selection. Referring to the Leibnizian philosopher in Voltaire's *Candide*, he accused evolutionists of Panglossianism, thinking that these must be the best of all possible features in the best of all possible worlds. And to make the case complete, supposedly, evolutionists invent "just so" stories—thus named from Rudyard Kipling's fantasy stories—with natural selection scenarios leading to adaptation.

As a counter, Gould (and Lewontin) drew attention to the triangular decorative aspects of the tops of pillars in medieval churches, arguing that although such "spandrels" seem adaptive, they are in fact by-products of the builders' methods of keeping the roof in place. "The design is so elaborate, harmonious, and purposeful that we are tempted to view it as the starting point of any analysis, as the cause in some sense of the surrounding architecture." This, however, is to get things precisely backward. "The system begins with an architectural constraint: the necessary four spandrels and their tapering triangular form. They provide a space in which the mosaicist worked; they set the quadripartite symmetry of the dome above" (Gould and Lewontin 1979, 582). Who knows but that we have a similar situation in the living world? Much that we think adaptive is merely a spandrel, and such things as constraints on development prevent anything like an optimally designed world. Perhaps things are much more random and haphazard—nonfunctional—than the Darwinian thinks possible.

A deal of ink—a very great deal of ink—was spilled over these claims. General reaction by Darwinian evolutionists—who make up perhaps 95 percent of this population—was that much that Gould said was true but well known already (Maynard Smith 1981). One phenomenon bringing on the nonadaptive is so-called genetic drift. This is where the vagaries of breeding—the chance encounters between organisms—can be sufficiently powerful to counter the effects of selection. This was the basis of a theory put forward in the early 1930s by the American population geneticist Sewall Wright—the "shifting balance theory of evolution" (1931, 1932). Drift is most likely to occur in small populations and, based on an extensive study of shorthorn cattle, Wright argued that evolution proceeds by drift, creating innovative new features when large populations are fragmented, with these features then spreading through the whole group when the fragmentation comes to an end. As it happens, there has been much criticism of Wright's overall theory (Coyne et al. 1997). Yet no one denies that drift probably does have some role—and it is agreed that it probably has a major role at the molecular level below the winnowing effects of the struggle for existence. . . . More generally, no one denies that many features are going to be nonadaptive, or perhaps were once adaptive and no more. Why do vertebrates have four limbs rather than six like the insects? John Maynard Smith argued that this may be a relic of when vertebrates were aquatic and two limbs fore and two limbs aft were very effective for raising or lowering the body immersed in water (Maynard Smith 1981). There is nothing sacrosanct about numbers. There are some fossil vertebrates with eight or nine digits rather than five.

So where are we today in evolutionary thinking? In the world of organisms, adaptation is the norm—the hugely well-justified null hypothesis—and it is your task to make the contrary case if you so wish. Purpose thinking rules, and it is cherished (Ruse 2003b). In a good Kantian sense, today's biologists use end-directed thinking and language when they are dealing with organisms.... There is no implication that there is a designer any more than there is an implication that you are really thin and slimy when I accuse you of having wormed your way into my affections and trust for your own nefarious ends. The metaphor remains, and Kant was right—you simply cannot do this kind of biology without it.

Conflict over Progress

Turn now to purpose in history. You might think that this is going to be a very short discussion. Natural selection is opportunistic. What works in one situation does not necessarily work in another. There is no reason to expect a forward direction to evolution, even one interrupted by reversals and sidestepping. Moreover, Mendelian/molecular genetics is adamant. There is no direction to the new variations—the building blocks—of evolution. No "higher" or "lower." That is an absolute. There is no value in the course of evolution. Stephen Jay Gould was eloquent. There is no direction and so evolution apparently can go whichever way. Progress to humans is just not on. "A noxious, culturally embedded, untestable, nonoperational, intractable idea that must be replaced if we wish to understand the patterns of history" (Gould 1988, 319). Making facetious reference to that celestial body that hit the Earth sixty-six million years ago, wiping out the dinosaurs and making possible the age of mammals, Gould wrote: "Since dinosaurs were not moving toward markedly larger brains, and since such a prospect may lie outside the capabilities of reptilian design . . . we must assume that consciousness would not have evolved on our planet if a cosmic catastrophe had not claimed the dinosaurs as victims. In an entirely literal sense, we owe our existence, as large and reasoning mammals, to our lucky stars" (Gould 1989b, 318).

Interestingly, Gould notwithstanding—and we shall see that he is more complex and convoluted than one might expect on first sight—many of today's leading evolutionists are quite open about their beliefs in biological progress. The most distinguished member of the fraternity,

Edward O. Wilson, Harvard professor and world-leading specialist on ants and on sociobiology (the evolution of social behavior) is unequivocal. "The overall average across the history of life has moved from the simple and few to the more complex and numerous. During the past billion years, animals as a whole evolved upward in body size, feeding and defensive techniques, brain and behavioral complexity, social organization, and precision of environmental control—in each case farther from the nonliving state than their simpler antecedents did" (E. Wilson 1992, 187). Adding: "Progress, then, is a property of the evolution of life as a whole by almost any conceivable intuitive standard, including the acquisition of goals and intentions in the behavior of animals." Elsewhere he writes of the "pinnacles" of social evolution, judging that we humans have won that competition outright (E. Wilson 1975). Part of the ongoing problem is that of defining biological progress in terms that are not flagrantly circular. If you define progress in terms of being humanlike, which is basically the move of Wilson, it is hardly surprising to find that we have won.

In the *Origin*, Darwin admitted candidly that paleontologists have a sense of progress. "The inhabitants of each successive period in the world's history have beaten their predecessors in the race for life, and are, in so far, higher in the scale of nature; and this may account for that vague yet ill-defined sentiment, felt by many paleontologists, that organization on the whole has progressed" (C. Darwin 1859, 345).

What is striking is how most evolutionists more or less take biological progress for granted—for all that they are prone to deny it when in public and totally sober—and go on to argue from there. There are a number of reasons for this. One is that since we are asking questions about progress, necessarily we are at the end of the evolutionary process and so there is an inclination to think that we must have won. There tends not to be much fellow feeling with warthogs at a time like this. Second, bound up with this first point is the sense that if warthogs feel that they are so very important, why don't they speak up and say so? Because we can ask questions about progress, we tend to judge progress in these terms. Perhaps warthogs judge progress in terms of wallowing in mud and letting the world pass by except when out feeding or copulating. Would a warthoggian Aristotle judge that inferior? Third, in a tradition going back to Diderot, there is a tendency to read hopes of cultural progress into the biological world and come up with confirmatory biological progress. Uniquely, scientists live in a world where—the social constructivists and other relativists notwithstanding—there is real progress. Newton was better than Aristotle, and Einstein was better than Newton. Creationism is wrong, and evolution is right. Why wouldn't one expect to find biological progress, especially since we are the ones responsible for scientific progress?

Arms Races

Yearning sentiments aside, is there any reason, any Darwinian reason, to think that progress will occur? Specifically, is there any Darwinian reason to think that biological processes will lead to human beings? Or, at least, is there any Darwinian reason to think that biological processes will lead to what have been called "humanoids," that is, humanlike beings with intelligence and so forth? . . . Humans have evolved, so obviously they could evolve. If you were to allow the hypothesis of multiverses—an infinite number of other universes, some (presumably an infinite number) like ours—then presumably (updating the argument of the atomists) somewhere, sometime, humans were going to evolve. To say otherwise is to say that they couldn't. However, even if this is so—and there are serious critics of multiverses—we hardly have anything one would be inclined to call "progress." It starts to sound like huge arrogance to say that, in a situation where presumably one has billions of life-forms everything in some sense bows down to us—or even to the billions of us in various galaxies. We can put the values in, if we want, but we are not reading them out of nature. There is nothing in biology itself to say we are better. Although, this said, there is equally nothing to stop us looking for Darwinian reasons for thinking that the nature of the evolutionary process is such that humanlike beings are (best scenario) necessarily going to emerge or (less attractive scenario) at least very likely to emerge. There are two popular proposals.

The first builds on the insight of Darwin about competition leading to improvement, particularly the competition between evolving lines leading to improvement. This idea about "arms races" was elaborated in most detail by Julian Huxley in a little book at the beginning of the last century. In nature, "if one species happens to vary in the direction of greater independence, the inter-related equilibrium is upset, and cannot be restored until a number of competing species have either given way to the increased pressure and become extinct, or else have answered pressure with pressure, and kept the first species in its place by themselves too discovering means of adding to their independence." Eventually: "it comes to pass that the continuous change which is passing that through the organic world appears as a succession of phases of equilibrium, each one on a higher average plane of independence than the one before, and each inevitably calling up and giving place to one still higher" (J. Huxley 1912, 115–16).

One who has (without acknowledgment) picked up enthusiastically on this kind of thinking is Richard Dawkins. "Directionalist common sense surely wins on the very long time scale: once there was only blue-green slime and now there are sharp-eyed metazoan" (Dawkins and Krebs 1979, 508). He too finds the key in arms races. As one who embraced computer technology early and enthusiastically, perhaps expectedly Dawkins notes that, more and more, today's arms races rely on computer technology rather than brute power, and—in the animal world—he finds this translated into ever bigger and more efficient brains. No need to hold your breath about who has won. Dawkins invokes a notion known as an animal's EQ, standing for "encephalization quotient" (Jerison 1973). This is a kind of cross-species measure of IQ that takes into account the amount of brainpower needed simply to get an organism to function (whales require much bigger brains than shrews because they need more computing power to get their bigger bodies to function), and that then scales according to the surplus left over. Dawkins writes, "The fact that humans have an EQ of 7 and hippos an EQ of 0.3 may not literally mean that humans are 23 times as clever as hippos! But the EQ as measured is probably telling us something about how much 'computing power' an animal probably has in its head, over and above the irreducible amount of computing power needed for the routine running of its large or small body" (Dawkins 1986, 189).

Others endorse similar lines of thinking. For instance, there is reason to think that shellfish are in arms races with predators, putting ever greater resources into thicker, tougher shells, with the predators developing ever more efficient methods of boring into shells and extracting the contents. However, not every Darwinian biologist is that enthused by arms races. The fossil evidence, for instance, does not show unambiguously that prey and predators have become ever faster. And even if arms races are ubiquitous, it does not follow that intelligence will always emerge. Having high intelligence means having large brains, and having large brains means having ready access to large chunks of protein, the bodies of other animals. There were no vegans in the Pleistocene. Sometimes—as cows and horses demonstrate—it is just easier to get your food in other ways, especially if you are living on grassy savannahs. Despite his enthusiasm for progress, Jack Sepkoski put matters colorfully and definitively: "I see intelligence as just one of a variety of adaptations among tetrapods for survival. Running fast in a herd while being as dumb as shit, I think, is a very good adaptation for survival" (Ruse 1996b, 486). So the overall answer seems to be that although arms races may well lead to intelligence, there is no guarantee that this will happen, and given that we have only the one instance (admittedly successful) to go on, it would be rash to argue with too much confidence that progress up to humans is the norm.

Channeling

The other favored approach to getting progress out of the Darwinian system works on the theme of ecological niches and organisms finding them and occupying them. We often think of the broad niches occupied by organisms—animals particularly—water, earth, and air. Why not just add on another—culture—and suppose that it was waiting to be occupied and finally protohumans found it and moved in? Gould of all people floated some idea like this. He thought that, if not on our Earth, then somewhere in the universe this might have happened. He quoted Theodosius Dobzhansky: "Granting that the possibility of obtaining a manlike creature is vanishingly small even given an astronomical number of attempts . . . there is still some small possibility that another intelligent species has arisen, one that is capable of achieving a technological civilization." Gould commented, "I am not convinced that the possibility is so small." He gave an argument that evolutionary convergence (where two different lines evolve essentially similar adaptations to survive and reproduce) suggests that even though major intelligence has arisen but once on this Earth, it is quite possible that elsewhere in the universe it has arisen quite independently. "But does intelligence lie within the class of phenomena too complex and historically conditioned for repetition? I do not think that its uniqueness on earth specifies such a conclusion. Perhaps, in another form on another world, intelligence would be as easy to evolve as flight on ours" (Gould 1985a, 412).

Using much the same argument as Gould, the paleontologist Simon Conway Morris (as a Christian) is very keen to argue for the inevitability of the appearance of humans. He argues that only certain areas of what we might call "morphological space" are welcoming to lifeforms (the center of the Sun would not be, for instance), and that this constrains the course of evolution (Conway Morris 2003). Again and again, as Gould argues, organisms take the same route into a preexisting niche. The saber-toothed, tigerlike organisms are a nice example, where the North American placental mammals (real cats) were matched right down the line by South American marsupials (thylacosmilus). There existed a niche for organisms that were predators, with catlike abilities and shearing/stabbing-like weapons. Darwinian selection found more than one way to enter it—from the placental side and from the marsupial side. It was not a question of beating out others but of finding pathways that others had not found.

Conway Morris argues that, given the ubiquity of convergence, we must allow that the historical course of nature is not random but strongly selection constrained along certain pathways and to certain destinations. Most particularly, some kind of intelligent being was bound to emerge. After all, our very own existence shows that a kind of cultural adaptive niche exists—a niche that prizes intelligence and social abilities. "If brains can get big independently and provide a neural machine capable of handling a highly complex environment, then perhaps there are other parallels, other convergences that drive some groups towards complexity." Continuing: "We may be unique, but paradoxically those properties that define our uniqueness can still be inherent in the evolutionary process. In other words, if we humans had not evolved then something more-or-less identical would have emerged sooner or later" (Conway Morris 2003, 196).

Does this do the trick and is this progress? One might question positive answers to both questions. Even if it exists, why should we or anyone else necessarily or even probably enter the culture niche? Life is full of missed opportunities. Maybe Gould is right and most times evolution would have gone other ways and avoided culture entirely. Warthogs rule supreme. Huxley always argued that now that humans occupy the culture niche, no other animal is going to be able to enter (J. Huxley 1942). Perhaps other animals (dinosaurs) would have prevented our animals (mammals and then primates) from making their way to the door. In any case, many wonder if it is right to think that niches are just waiting out there, ready to be conquered and entered. Do not organisms create niches as much as find them? There was hardly a niche for head lice, for instance, until vertebrates like us humans came along. Should we expect that there was a niche for culture, just waiting there, like dry land or the open air? Perhaps there are other niches not yet invented. We cannot imagine something other than consciousness; but take heed of the wise warning of J. B. S. Haldane: "Now my own suspicion is that the Universe is not only queerer than we suppose, but queerer than we can suppose" (Haldane 1927, 286). For all their talk about analogy, Christians tend to think that their God can get up to some pretty clever tricks, way beyond their ken. Perhaps these are not all supernatural abilities, but simply abilities that were omitted from our evolution. Perhaps, far from being the best, we are a short side path and very limited in the true scheme of things. No more than in the case of arms races do we get much guarantee of either human emergence or a sense that we are in some way superior and for this reason we won. Purpose is there in Darwinian biology, through and through. Thanks to Darwin, many enthusiasts think we have come a long, long way. We have purpose in the individual feature—the eye exists in order to see. Equally, although there are some (including myself) who are not so enthusiastic on this score, many think we have purpose in history. Humans are the destined end point, thus far. Have we arrived at the bright Elysian shore? Many Darwinians think we have. Others—who have greater or less degrees of enthusiasm for natural selection—are not so certain.

6

Progress and Directionality in Evolution

Chapter 6, Reading 1

Progress[1]

It is a huge job to understand humans from an evolutionary perspective, if indeed that is in any way possible. Especially since, even if you wanted to, for practical as well as ethical reasons, you cannot really experiment on us. But there is more to the matter than just the science, or rather just the straight science. There is the fact that we are humans. The conclusions matter in a very personal way. For Christians and indeed for members of other major faiths (especially Judaism and Islam) it is virtually a priori obvious that what evolution has to say about humankind is going to be interesting, and perhaps threatening. It is central to Christianity that we humans are special, we are particularly favored by God because we alone are "made in his image." What this means precisely is a matter of some debate. Most Christians are not biblical literalists, thinking that God created humans some six thousand years ago, on the sixth day, out of mud or some such thing. They may not even think that we have to be exactly as we are. Perhaps for instance humans might have six fingers or green skin. I am not sure about sex, but perhaps even this is an option. It has always been fundamental to the Judeo-Christian perspective that God is not a sexual being, even though traditionally we refer to God as "Father" and as "he." But, as Saint Augustine pointed out around the end of the fourth century, we do have to be intelligent beings and creatures with a moral capacity. Without that, nothing makes sense.

1. Ruse 2012c. Reprinted with permission of Cambridge University Press.

More than this, our existence has to be necessary in some sense. It cannot just be a matter of chance that humans or humanlike creatures exist. I don't see any reason why we shouldn't exist on the moon or somewhere else if that is a viable option. (Obviously as things stand the moon is not a viable option.) But exist we must (Plantinga 2009). . . .

What does evolutionary biology, Darwinian evolutionary biology in particular, have to say about human uniqueness and our inevitable appearance? Start with history, for it tells us much about the case (Ruse 1996b). Above all, it tells us that the concern with the special status of humans goes back to the beginnings of evolutionary hypothesizing. We have seen that it was not really until the beginning of the eighteenth century that people first started to speculate on natural developmental origins for life. What we should understand now is that these speculations did not occur in a vacuum, but rather were the direct outcome of a metaphysical or theological battle about the nature and status of humankind (Ruse 2005b). I have explained earlier that although there were indeed those who took the Bible more or less literally, it has always been part of Christian tradition (articulated by Saint Augustine) that if need be one can understand the text metaphorically or allegorically in order to harmonize it with other beliefs, from science and philosophy and so forth.

More secular thinkers embraced the ideology or metaphysic of progress, the belief that through our intelligence and labors we humans can make things better, in science, in education, in healthcare, and more. Evolution was part and parcel of this ideology. People read progress into the organic world, seeing it as an upward ladder from the primitive to the complex, from what they called the "monad" to what they called the "man." This may have been a bit of a stretch, but it was not such an odd thing to do, because, even though there was no proper fossil record or anything like that, there was a tradition going back to the Greeks of seeing the living world as connected in a series from least to most, the so-called "great chain of being" (Lovejoy 1936). Having done this, they promptly turned around and saw the progress of the living world as justification for their beliefs in social progress! Charles Darwin's grandfather Erasmus has been mentioned, but he was far from alone. The French encyclopedist Denis Diderot is a good example. "Just as in the animal and vegetable kingdoms, an individual begins, so to speak, grows, subsists, decays and passes away, could it not be the same with the whole species?" (Diderot 1943, 48, quoting *On the Interpretation of Nature*, 1754). He made no bones about being a progressionist and seeing a link between his social views and his scientific speculations. "The Tahitian is at a primary stage in the development of the world, the European is at its old age. The interval separating us is greater than that between the new-born child and the decrepit old man" (Diderot 1943, 48, quoting *Supplement to Bougainville's Voyage*, 1772). Progress-infused thoughts continued unabated in evolutionary speculations right down to the middle of the nineteenth century. In fact, it was often the very possibility of promoting social and cultural progress through talk about the fossils and today's organisms that motivated people to get involved in such speculations.

Darwin and Progress

We come now to Charles Darwin and the *Origin of Species*. With some good reason, many people think that Darwin rang the death knell on thoughts of biological progress, or if he did not do the job fully he started the process that finished with the coming of modern genetics. Natural selection may not be a tautology, but it does relativize change. . . . The coming of Mendelian genetics is often thought to have finished what Darwin started. Darwin himself was always insistent that the new variations that appear in populations, the raw building blocks of evolution, are random—not in the sense of uncaused—although he did not know the causes (he speculated that injury to the generative organs might play a role) he was sure there were causes—but in the sense of not appearing according to need. To suppose that they did appear as needed seemed, to Darwin, to put in a directedness, a teleological impulse, that is contrary to modern science. He was very critical of his American supporter Asa Gray when the latter did suppose that the variations of evolution have some kind of direction. For Darwin, this was to take the whole matter outside of science (Ruse 1979). And the arrival of modern genetics underscored this very point. The whole point about the new variations, what we now call *mutations*, is that they do not occur according to need. Today, we know a lot about the causes and we know that the rate of mutation can usually be quantified, but individual mutations are random both in their not appearing just when needed and even more in being unrelated to need. An organism may be able to use them, but it is a crapshoot all of the way. So once again it seems that thoughts of overall direction to the course of Darwinian (or neo-Darwinian) evolution are stymied.

Consequently, after the *Origin*, people—including Charles Darwin himself—went right on being enthusiastic progressionists, in biology that is. "The inhabitants of each successive period in the world's history have beaten their predecessors in the race for life, and are, in so far, higher in the scale of nature; and this may account for that vague yet ill-defined sentiment, felt by many paleontologists, that organization on the whole has progressed" (C. Darwin 1859, 345). By the end of the book, it is clear that in the mind of Charles Darwin, the sentiment is not so very vague. "And as natural selection works solely by and for the good of each being, all corporeal and mental endowments will tend to progress towards perfection" (C. Darwin 1859, 489). Or consider the closing lines of the *Origin*.

> From the war of nature, from famine and death, the most exalted object which we are capable of conceiving, namely, the production of the higher animals, directly follows. There is a grandeur in this view of life, with its several powers, having been originally breathed into a few forms or into one; and that, whilst this planet has gone cycling on according to the fixed law of gravity, from so simple a beginning endless forms most beautiful and most wonderful have been, and are being, evolved. (1859, 490)

In many respects, *The Descent of Man* is a more popular book than the *Origin*—it is packed with anecdotes and is well larded with Darwin's own opinions and value judgments. Among which are the beliefs that we humans are at the top of the heap and among humans the inhabitants (or former inhabitants) of Europe are the very best. A neat combination of an endorsement of capitalism and the virtues of progress in culture gets the desired result.

> In all civilized countries man accumulates property and bequeaths it to his children. So that the children in the same country do not by any means start fair in the race for success. But this is far from an unmixed evil; for without the accumulation of capital the arts could not progress; and it is chiefly through their power that the civilized races have extended, and are now everywhere extending, their range, so as to take the place of the lower races. (C. Darwin 1871, 1:169)

Had nothing changed with the coming of the theory of evolution through natural selection? Actually, a little did change. Although Darwin may have been as much of a biological progressionist as earlier evolutionists, he showed his greater degree of scientific sophistication in realizing that you cannot just talk of progress without at some level trying to define it, or rather the things that make for progress. Everyone knows that what you are after is evolution up to beings that have properties that coincide with "humanlike," but to avoid circularity or triviality you have got to have some kind of independent criteria. If you define biological progress in terms of "humanlike" then obviously you are going to get progress in some sense. We humans are still here, and we are the final product of evolution (or one of the final products). But this is not much of a conclusion. You have got to find criteria separate from simply "humanlike"—you have got to show that "humanlike" emerges independently. Relying on the best recent German biology, Darwin (writing in the third edition of the *Origin* in 1861) opted for some kind of organization and differentiation and specialization.

> Natural selection acts, as we have seen, exclusively by the preservation and accumulation of variations, which are beneficial under the organic and inorganic conditions of life to which each creature is at each successive period exposed. The ultimate result will be that each creature will tend to become more and more improved in relation to its conditions of life. This improvement will, I think, inevitably lead to the gradual advancement of the organization of the greater number of living beings throughout the world. But here we enter on a very intricate subject, for naturalists have not defined to each other's satisfaction what is meant by an advance in organization. Amongst the vertebrata the degree of intellect and an approach in structure to man clearly come into play. It might be thought that the amount of change which the various parts and organs undergo in their development from the embryo to maturity would suffice as a standard of comparison; but there are cases, as with certain parasitic crustaceans, in which several parts of the structure become less perfect, so that the mature animal cannot be called higher than its larva. Von Baer's standard seems the most widely applicable and the best, namely, the amount of differentiation of the different parts (in the adult state, as I should be inclined to add) and their specialization for different functions; or, as Milne Edwards would express it, the completeness of the division of physiological labor. (C. Darwin 1861, 133)

How does this tie in with evolution, or more precisely, how does this tie in with natural selection and human nature? Darwin was acutely sensitive to the fact that there was an issue here. He could not simply expect progress to emerge. Even more importantly, although he was happy to accept German thinking to define progress, he did not want to get identified with the kind of Germanic upward Progressivism one finds in German Romantic philosophers (like Fichte and Schelling) at the beginning of the nineteenth century—a sort of world force that was supposed by Hegel to be pushing the whole of life ever upward (Richards 2003). It was for this reason that on the flyleaf of his own copy of *Vestiges* Darwin wrote, "Never use higher and lower." It was not the progressivist sentiment that he repudiated, but the belief (clearly endorsed by Chambers) that life has a kind of upward momentum, all of its own. Natural selection was the mechanism for Darwin, and he recognized that sometimes it stands still or goes backward, and we cannot expect instant or constant biological progress. Often, indeed, even using specialization as your measure, it is hard to know what to think:

> We shall see how obscure a subject this is if we look, for instance, to fish, amongst which some naturalists rank those as highest which, like the sharks, approach nearest to reptiles; whilst other naturalists rank the common bony or teleostean fishes as the highest, inasmuch as they are most strictly fish-like, and differ most from the other vertebrate classes. Still more plainly we see the obscurity of the subject by turning to plants, with which the standard of intellect is of course quite excluded; and here some botanists rank those plants as highest which have every organ, as sepals, petals, stamens, and pistils, fully developed in each flower; whereas other botanists, probably with more truth, look at the plants which have their several organs much modified and somewhat reduced in number as being of the highest rank. (C. Darwin 1861, 133–34)

Nevertheless, as intimated in a passage given above, ultimately Darwin thought that natural selection really does make for upward change, because the winners overall will be better than the losers. And this ties in with the definition of progress.

> If we look at the differentiation and specialization of the several organs of each being when adult (and this will include the advancement of the brain for intellectual purposes) as the best standard of highness of organization, natural selection clearly leads towards highness; for all physiologists admit that the specialization of organs, inasmuch as they perform in this state

their functions better, is an advantage to each being; and hence the accumulation of variations tending towards specialization is within the scope of natural selection. (C. Darwin 1861, 134)

After Darwin

Leave critical comment for the moment, for we shall be coming back to the modern-day equivalent of Darwin's own thinking. Accept simply that even if Darwin worried about progress, few others did. Almost to a person, people went on thinking of evolution as progressive. Extreme but highly influential was Darwin's contemporary Herbert Spencer. Here he is in a passage written two years before the *Origin*, but to be repeated by him nonstop for the rest of his very long life (into the twentieth century). Like Darwin in being influenced by Germanic thinking, Spencer also adopted a criterion of progress that involved division and specialization, or as he called it a move from the homogeneous to the heterogeneous:

> Now we propose in the first place to show, that this law of organic progress is the law of all progress. Whether it be in the development of the Earth, in the development of Life upon its surface, in the development of Society, of Government, of Manufactures, of Commerce, of Language, Literature, Science, Art, this same evolution of the simple into the complex, through successive differentiations, holds throughout. From the earliest traceable cosmical changes down to the latest results of civilization, we shall find that the transformation of the homogeneous into the heterogeneous is that in which Progress essentially consists. (Spencer 1857, 246)

In Spencer's eyes, everything obeys this law. Compared with other animals, humans are more complex or heterogeneous; compared with savages, Europeans more complex or heterogeneous; and compared with the tongues of other peoples, the English language more complex or heterogeneous. One should say that there was a certain inevitability about all of this that was alien to Darwinism. Nor was there any striving to fit this thinking to more selection-based notions like the division of labor. Although he had thought of natural selection before the *Origin* was published, for Spencer it was never a major part of the evolutionary story. A kind of Germanic progressivist determinism was there for all to see. Causes just keep multiplying effects and this leads to ever greater complexity, and somehow value or worth gets pulled in along the way.

This all seems to occur in waves. Something disturbs the natural balance, forces work to regain the balance, but when this occurs we have moved a stage higher. An overall process of "dynamic equilibrium."

What about the coming of Mendelism? Take two of the key figures bringing together Darwinian selection and Mendelian genetics: Ronald A. Fisher in Britain and Sewall Wright in America. Both were ardent progressionists! Fisher made much of something that he called the "fundamental theorem of evolution." There is a kind of perpetual upward momentum to the evolutionary process, as the result of selection at work. "The rate of increase in fitness of any organism at any time is equal to its genetic variance in fitness at that time" (Fisher 1930, 35). In other words things are getting better all of the time, proportional to the amount of variation in a population.

Sewall Wright took a very different path from Fisher. We saw there also the elements that make very easy the case for Wright's enthusiasm for all kinds of progress, in culture and in biology (Ruse 2004). Much influenced by his work at the United States Department of Agriculture on the genetics of cattle breeding, and even more influenced by an enthusiasm for the ideas of Herbert Spencer, Wright saw the key to innovative evolutionary change in the breaking up of large populations into small groups in which genetic drift would be effective, and then recombination into the larger mass once again. He thought that in the small groups new innovative adaptations would appear thanks to drift and that these would then spread in the reunited whole. Combining this with the metaphor of an adaptive landscape, Wright saw organisms as moving from one peak to another thanks to the new adaptations. One could of course regard the overall landscape as akin to a waterbed, with a new peak in one area being balanced by a new valley in another, and with no overall upward gain in height. However, this was decidedly not Wright's own vision, as he saw organisms inching step by step toward greater heights—"dynamic equilibrium"—ultimately perhaps into one universal world soul.

Biological progress was alive and well in the 1930s. The theoreticians gave the lead and the empiricists were happy to follow. There is good evidence that this was a major attraction for the leading researchers in the field—Dobzhansky, Mayr, and Simpson in America and overwhelmingly Julian Huxley (the grandson of T. H. Huxley) in Britain. For the last named, his whole intellectual life was a mission to replace conventional Christianity with what (in the title of one popular book) he was to call "religion without revelation" (1927). Deeply influenced by the thinking of the French vitalist philosopher Henri Bergson (1907), Huxley argued that the key to understanding lies in control and independence—basically the ability to manipulate one's environment and to break loose from its constraints. Thus judged, humans are clearly the winners in the race to the top. Curiously for one who for all of his very long life carried high the banner of Darwinism, or more precisely who inherited the role of top Darwinian, Julian Huxley was always a bit ambivalent about the workings and power of natural selection. Perhaps given his grandfather's like ambivalence, his stance was not so curious after all. On the one hand, he did think that selection had a hand in biological progress. "It should be clear that if natural selection can account for adaptation and for long-range trends of specialization, it can account for progress too. Progressive changes have obviously given their owners advantages which have enabled them to become dominant" (J. Huxley 1942, 568). On the other hand, like Bergson (who was in turn influenced by Spencer), he saw an inevitability to the upward rise of humankind. "One somewhat curious fact emerges from a survey of biological progress as culminating for the evolutionary moment in the dominance of Homo sapiens. It could apparently have pursued no other general course than that which it has historically followed" (J. Huxley 1942, 569).

Biological Progress Today

Move on now from history to the present. My suspicion is that, with good reason, many if not most evolutionary biologists today would say that all of this talk about biological progress really is past history. They would say that today no reputable evolutionary biologist, certainly no Darwinian biologist, believes in biological progress. Certainly we do not find discussions of it in the literature, in the major journals like *Evolution* and *American Naturalist*. This last point is certainly true, but the story is a little more complex than that, and history tells us why. (It is curious how few scientists think that history can tell them anything. This includes evolutionary biologists, who, as a matter of professional commitment, believe that the biological past can tell us virtually everything.) The reason why biological progress fell out of favor in professional evolutionary biological circles was not that people suddenly stopped believing in it. It was because increasingly they realized that parading so obviously a value-laden notion as progress right up front in their science was antithetical to their aims as successful professional scientists. From the Scientific Revolution on there was an ideology, particularly in the physical sciences, that science should in some sense be purely objective knowledge. In the felicitous phrase of Karl Popper, it should be "knowledge without a knower" (1972, 109)—not, obviously, in the sense that there is no knower involved, but that the nature of the knower (male, female, Christian, Jew, British, American, gay, straight) should be entirely irrelevant.

For various reasons—not the least the rise of Nazi Germany and its condemnation of "Jewish science"—this ideology burned fiercely in the 1930s and 1940s. Yet here was evolutionary biology, just now rising from many years of (at best) museum-level displays and wild hypothesizing, still parading front and center beliefs about human superiority—often still parading views about white human superiority. If this was not enough, there were other branches of the biological sciences, most notably the growing molecular branches, that were competing for students and grants and university places and so forth. So a deliberate effort was made to downplay and remove talk of biological progress from the professional publications. By the 1950s, as you search the literature, it is harder and harder to find discussions of upward rise anywhere. The one exception was Julian Huxley, who went on promoting biological progress to the end. But it was long since he had held a university position, and increasingly he was professionally isolated from mainstream science. When, in the late 1950s, Huxley (1959) endorsed the ideas of the French paleontologist Pierre Teilhard de Chardin, SJ (1955), who was promoting a progressionist view of life leading up to humans (and beyond to Christ), the reactions were scathing (Medawar 1967). It was not so much that Julian Huxley was wrong, but that he was openly doing things that his community saw as flagrantly opposed to the status building in which they were engaged.

What does this mean for today? The progress-expelling campaign was obviously successful, and one suspects that a major effect has been that today's professional evolutionist has moved on. He or she is not drawn to the field because they have progressivist yearnings, and for them keeping biological progress out of the discussion is not something that has to be done deliberately. The wonder would be that one might want to bring it in. However, having said this, since the main motivation for expelling biological progress was more social than intellectual, one might expect to find that there would still be evolutionists today who have sympathies of one sort or another with progress in some sense, and that these would come out, if not in the fully professional writings then in the more popular (or philosophical) writings about evolution and its history. And this is indeed so. It does not take much searching to find thoughts of biological progress and claims that link it to modern evolutionary biology. Today's most distinguished living evolutionist, Edward O. Wilson of Harvard University, is open in his fervent belief in biological progress.

> The overall average across the history of life has moved from the simple and few to the more complex and numerous. During the past billion years, animals as a whole evolved upward in body size, feeding and defensive techniques, brain and behavioral complexity, social organization, and precision of environmental control—in each case farther from the nonliving state than their simpler antecedents did. (E. Wilson 1992, 187)

Adding: "Progress, then, is a property of the evolution of life as a whole by almost any conceivable intuitive standard, including the acquisition of goals and intentions in the behavior of animals."

With views like this coming down from on high (a.k.a. Harvard University), it is little surprise that in the general domain also biological progress is alive and well. In the popular mind, in the way that evolution is presented in popular print or on radio or television, the course of biological history is almost inevitably represented as one leading in a fairly straight line up to Homo sapiens. To the person in the street, it would come as a significant surprise to discover that this is not the universal commitment of all practicing evolutionary biologists.

Reasons for Biological Progress

Recognizing then that biological progress will be something that yields beings humanlike (although at the risk of circularity it cannot make direct reference to being humanlike), and reserving for a moment discussion of how this might be done, let us look at some recent suggestions for achieving such progress. Some, I think, are perhaps of broader interest but can be ruled out quickly as suggestions meriting scientific status or consideration. Going back briefly to Christianity (and other faith systems) the simple fact is that it cannot allow the possibility of failure. For the Christian, humans or humanlike beings had to emerge. That is not an option that can fail (Ruse 2000a, 2010). The obvious solution therefore is to put God himself up front doing the job. That was what Asa Gray and other so-called "theistic evolutionists" always did (Gray 1876). They made God responsible for the new variations, the mutations in today's language, and he directed enough of them that humans were bound to emerge. Today's chief enthusiast for this kind of thinking is the American physicist-theologian Robert John Russell, who thinks that God works down at the quantum level, directing changes as needed (Russell 2008). As far as we are concerned, his actions are hidden because we can see only the average effects, but he makes sure that at the individual level of change the right variation appears at the right time.

Hidden or not, this is not science, and there are those who think it is not very good Christianity either. If God does work in this sort of way, why (if he is an all-powerful, loving Father) does he not also work to prevent some of the horrendous mutations that cause untold human suffering? The trouble with getting God involved in the day-to-day workings of evolution is that, although he may get credit for the good things, he makes himself directly liable for the bad things also. So let us move on to suggestions that, be they true or false, are at least putatively scientific. There are at least three suggestions, although as we shall see, the third is perhaps better considered an alternative to progressivist readings of the evolutionary record. The first, endorsed most enthusiastically today by Richard Dawkins (1986), stands in direct line back to Darwin. It picks up on his belief that selection leads to competition, with one group or line outcompeting another group or line, and that this leads bit by bit to overall improvement (Dawkins and Krebs 1979). Humans emerge as the end of the line.

Richard Dawkins may not share Julian Huxley's enthusiasm for Spencer and Bergson, but he is very much in favor of progress. "Directionalist common sense surely wins on the very long time scale: once there was only blue-green slime and now there are sharp-eyed metazoa" (Dawkins and Krebs 1979, 508). And arms races are the key. In particular, using today's refinements, he points out that, more and more, arms races rely on computer technology rather than brute power, and—in the animal world—Dawkins translates this into bigger and bigger brains. No prizes are given for guessing who has won. He refers to a notion known as an animal's EQ, standing for "encephalization quotient" (Jerison 1973). This notion is a kind of cross-species measure of IQ that factors out the amount of brainpower needed simply to get an organism to function (whales require much bigger brains than shrews because they need more computing power to get their bigger bodies to function), and that then scales according to the surplus left over. Dawkins writes:

> The fact that humans have an EQ of 7 and hippos an EQ of 0.3 may not literally mean that humans are 23 times as clever as hippos! But the EQ as measured is probably telling us something about how much "computing power" an animal probably has in its head, over and above the irreducible amount of computing power needed for the routine running of its large or small body. (Dawkins 1986, 189)

The second major attempt to articulate and justify a Darwinian basis for a progressivist rise in life, from the blob to the human, comes from the work of the British paleontologist Simon Conway Morris (2003). He starts his case from the fact that only certain areas of potential morphological space will be able to support functional life. As a Darwinian he adds to this the assumption that selection is forever pressing organisms to look for such potential, functional spaces. From this he draws the conclusion that, if such spaces exist, in the full course of time they will be occupied. For better or for worse, Conway Morris seems to think that the occupation will probably occur sooner rather than later, and probably many times. Now comes the key step. He highlights the way in which life's history shows an incredible number of instances of convergence—meaning by this cases where the same adaptive morphological space has been occupied again and again....

The overall conclusion that Conway Morris draws is that, because convergence is almost the norm rather than the exception, we must allow that the historical course of nature is not random but strongly selection constrained along certain pathways and to certain destinations. Most particularly, movement up the order of nature, the chain of being, is no accident. For all of the contingency in the Darwinian evolutionary process, such a progress was predestined. Sooner or later therefore some kind of intelligent being (often called a *humanoid*) was bound to emerge.... In the end came humankind, less by chance and more by Darwinian destiny.

The third proposal is, as I have said, less a proposal for progress per se than an alternative solution. Stephen Jay Gould was always very hostile to notions of biological progress. He was forever fulminating against it, because he thought it was racist and sexist and much more. Particularly (and at an obvious personal level) he disliked the way that progressivist ideas had been used to belittle the status and talents of Jews (Gould 1981). He spoke of biological progress as "a noxious, culturally embedded, untestable, nonoperational, intractable idea that must be replaced if we wish to understand the patterns of history" (Gould 1988, 319). He argued that there is nothing inevitable about the emergence of humans. Making joking reference to the asteroid that hit the Earth sixty-five million years ago and wiped out the dinosaurs, making possible the age of mammals, he wrote:

> Since dinosaurs were not moving toward markedly larger brains, and since such a prospect may lie outside the capabilities of reptilian design (Jerison, 1973; Hopson, 1977), we must assume that consciousness would not have evolved on our planet if a cosmic catastrophe had not claimed the dinosaurs as victims. In an entirely literal sense, we owe our existence, as large and reasoning mammals, to our lucky stars. (Gould 1989b, 318)

And yet Gould did not want to deny that there may be something to the idea of an increase in life's complexity. Although unacknowledged, he used an idea anticipated (at the very least) in an early jotting by Charles Darwin himself. Somehow complexity just emerges because in a way you are always going to be building on what you have already.

> The enormous number of animals in the world depends of their varied structure & complexity—hence as the forms became complicated, they opened fresh means of adding to their complexity—but yet there is no necessary tendency in the simple animals to become complicated although all perhaps will have done so from the new relations caused by the advancing complexity of others It may be said, why should there not be at any time as many species tending to dis-development (some probably always have done so, as the simplest fish), my answer is because, if we begin with the simplest forms & suppose them to have changed, their very changes ton tend to give rise to others. (C. Darwin 1987, E 95–97)

Darwin certainly thought of the complexity as ultimately adaptive. Immediately after the passage just quoted, he added: "It is quite clear that a large part of the complexity of structure is adaptation." I am not sure that Gould felt this way, but he did see complexity as growing during the course of life's history (Gould 1989a). Life is a bit asymmetrical. Necessarily, it started simple. Necessarily, it cannot get less simple. However, it can get more complex. Not through any guiding power, but because this is the way that things are. It is a rather like the old tale of the drunkard and the sidewalk. On one side, the sidewalk is bounded by a wall, and on the other side lies the gutter. It may take a long time, but in the end the drunkard will end in the gutter. This is not through any conscious choice, but because the drunkard can fall off the sidewalk and he cannot walk through the wall. His random staggering will eventually lead to the gutter. So it is with evolution. There is no progress in nature, but there is direction. Will this direction eventually end up with humans or humanlike beings? Interestingly, although he was not optimistic about this one-off world of ours, he rather thought that if you extend your gaze out to the rest of the universe and to the many life-forms that surely exist out there, one might reasonably expect to find the evolution of intelligent beings of some sort (Dick 1996).

Similar ideas have been promoted recently by paleontologist Daniel McShea and philosopher Robert Brandon. They see a kind of non-Darwinian upward momentum to life's history. Introducing what they call the "zero-force evolutionary law" (ZFEL) they write: "In any evolutionary system in which there is variation and heredity, in the absence of natural selection, other forces, and constraints acting on diversity or complexity, diversity and complexity will increase on average" (McShea and Brandon 2010, 3). Interestingly they circle back to Spencer, acknowledging that they write in his spirit. It seems that for them (as was the case for Spencer) things just naturally keep complexifying—one cause leads to several effects and these in turn multiply. They are not committed to the kind of surging view that characterizes dynamic equilibrium—although Gould's non-Darwinian punctuated equilibrium surely echoes—but they do see an upward drive as part of the ontology of the universe.

Is There Biological Progress?

In concluding this chapter, let us turn now to some critical thoughts. First, what about the very notion of progress itself? What about higher and lower and what about features that do increase or improve through life's history and that characterize human beings above all others? As soon as you start to look at it carefully, complexity begins to be an awfully slippery concept. Dawkins has some suggestions (1992, 2003). Drawing on ideas from information theory, he argues that more complex organisms would require physically longer descriptions than less complex organisms.

> Notwithstanding Gould's just skepticism over the tendency to label each era by its newest arrivals, there really is a good possibility that major innovations in embryological technique open up new vistas of evolutionary possibility and that these constitute genuinely progressive improvements (Dawkins 1989; Maynard Smith and Szathmáry 1995). The origin of the chromosome, of the bounded cell, of organized meiosis, diploidy and sex, of the eucaryotic cell, of multicellularity, of gastrulation, of molluscan torsion, of segmentation—each of these may have constituted a watershed event in the history of life. Not just in the normal Darwinian sense of assisting individuals to survive and reproduce, but watershed in the sense of boosting evolution itself in ways that seem entitled to the label progressive. It may well be that after, say, the invention of multicellularity, or the invention of metamerism, evolution was never the same again. In this sense, there may be a one-way ratchet of progressive innovation in evolution. (Dawkins 1997a, 1019–20)

None of this is to say that we humans should value fellow humans less than or even equal to other animals (or plants). I think all of us would think it morally wrong were someone to save a dog at the expense of a child. But it is to say that it is not readily apparent that Darwinian evolutionary theory sits comfortably with most proposed notions of biological progress. Perhaps an ecumenical conclusion is warranted. The eminent evolutionary biologist Francisco J. Ayala has said: "Well, I would say that by many definitions, including very biologically meaningful definitions, humans are more progressive than any other organism," adding "'progressive' is an evaluative term that demands a subjective commitment to a particular standard of value" (Ayala 1988, 95). Leave it at that.

Let me sum up. There is no clear-cut answer to the question about the relationship between Darwinian evolutionary theory and hopes of progress, especially progress up to human beings. There are arguments suggesting that, despite the contingency of the Darwinian process, some kind of advance is not only possible but likely. I think few on the scientific side would argue for some sort of necessity. Perhaps, all other things being equal, complexity will inevitably increase over time. But are all other things always equal and, even if they are, does complexity in itself yield the kind of result (humanlike qualities) that we want? Ultimately the only people who are really going to be upset by these conclusions are probably the religious, for whom the arrival of human beings is a no-compromise demand. This is not a scientific problem but a theological problem, and my suspicion is that believers should therefore search for some kind of theological solution. Supposedly scientific solutions, like that of Robert J. Russell, that put God into the natural creative process generally (always?) turn out to be unsatisfactory for both science and theology. For what it is worth, my own solution (at least for the Christian) is to play on the fact that God is outside time and space and that he can create as many universes as he wants (Ruse 2010b). Since we have evolved, it was possible for us to evolve by natural causes. At some point, we were (or would be) bound to appear. God is not waiting around for this to happen, as would be the case for us. You might think that this is all an awful lot of hard work and waste—all of those nonproductive universes. But really you have that already in our universe. All of those galaxies doing nothing productive, at least in the sense of producing humanlike beings. In the nineteenth century there was a lot of angst about this apparent waste, leading those with a taste for natural theology like the Scottish physicist and biographer of Newton David Brewster and the historian and philosopher of science William Whewell to argue at length about possible solutions. One cannot truly say that much light was thrown on the issues, so perhaps these are theological worries we can leave to those who care. (See Whewell 2001, especially my introduction.)

In a way, those of us who take seriously David Hume's point about the impossibility of deriving values from facts (Hume 1940) would have realized from the outside that the attempt to get progress out of the evolutionary process was in some sense doomed to failure. At most, one could get something that coincides with human nature but no guarantee that this something is of any worth. My own take is that we will probably never get quite what we want, but that no failure will quell the feeling that somehow there really is progress up to humankind (Ruse 1996). I think we are caught in a kind of Cartesian situation. If you doubt Descartes's *cogito, ergo sum*, then you reaffirm its truth. To doubt is to think is to be. The same is true if you ask whether there is biological progress leading up to humans. On the one hand, you are drawing attention to the fact that we are still around and are among the last products of evolution. This in itself seems to give us a special place at the top of the tree. On the other hand, as Simpson pointed out, you are showing that you have the ability to ask about whether there is progress. This seems to give us special abilities. Either way, we won—and that is surely what progress is all about.

Chapter 6, Reading 2

The Problem of Progress[1]

THE IDEA OF PROGRESS at the cultural level is that of humans making things better—education, health, material comfort, safety, and so forth (Bury 1920). It is important to stress that it is humans making things better. The very essence of progress is that we do it ourselves. It is debated as to whether the Greeks had much idea of progress. They certainly don't seem to have had the commitment that we find later, and often the picture is more of a cyclical nature—like a sine curve—than that of something getting ever higher. In the *Republic*, for instance, Plato envisions the decline and fall of the ideal state. With the coming of Christianity, progress fell out of favor. There was a rival ideology, providence. This is the idea, in its usual Augustinian form, that everything is due to God—his grace—and everything good comes through the blood of the Lamb. We humans are helpless, mired in original sin, and save for the sacrifice on the cross we are doomed to everlasting misery. Whereas progress stressed it is all up to us, providence insisted that we unaided could do nothing. . . .

Things really started to change by the eighteenth century, the Age of the Enlightenment. Thanks to science, to increasing prosperity, to ever more secular philosophical analysis, increasingly there was the conviction that we don't need God so much. We can do things ourselves. Progress started to edge out providence. And it is here that evolution starts to come into the story (Ruse 1996b). As the Romantics asked,

1. Ruse 2021e. Reprinted with permission of Cambridge University Press.

could it be that as individual organisms develop, so whole groups might develop analogously? What would this mean? Most obviously, from the simple to the complex, from the undesirable to the desired, from the blob to the human? Even before Romanticism started to take form, several started to speculate in this way. This is Denis Diderot, the French philosopher and creator of the huge compendium of knowledge, the *Encyclopédie*, writing in the middle of the eighteenth century: "Just as in the animal and vegetable kingdoms, an individual begins, so to speak, grows, subsists, decays and passes away, could it not be the same with the whole species" (Diderot 1943, 48). Here he uses the organic analogy, but, in this pre-Romantic era, it was truly the cultural progress/biological progress that was fundamental. . . .

Keeping it in the family, let us turn to Erasmus Darwin, late-eighteenth-century physician, grandfather of Charles Darwin, who was an enthusiast for taking Progress in the cultural world—from now on I shall follow convention and speak of the cultural notion as Progress, with the *p* capitalized—and applying it analogically to the organic world, progress—the biological notion, without a capital *p*. Breaking into verse.

> Organic Life beneath the shoreless waves
> Was born and nurs'd in Ocean's pearly caves;
> First forms minute, unseen by spheric glass,
> Move on the mud, or pierce the watery mass;
> These, as successive generations bloom,
> New powers acquire, and larger limbs assume;
> Whence countless groups of vegetation spring,
> And breathing realms of fin, and feet, and wing.
> Thus the tall Oak, the giant of the wood,
> Which bears Britannia's thunders on the flood;
> The Whale, unmeasured monster of the main,
> The lordly Lion, monarch of the plain,
> The Eagle soaring in the realms of air,
> Whose eye undazzled drinks the solar glare,
> Imperious man, who rules the bestial crowd,
> Of language, reason, and reflection proud,
> With brow erect who scorns this earthy sod,
> And styles himself the image of his God;
> Arose from rudiments of form and sense,

An *embryon* point, or microscopic *ens*!
(Erasmus Darwin 1803, 1:11)

Biological progress! And to underline the point, explicitly Darwin tied biological progress in with cultural Progress. The idea of organic progressive evolution "is analogous to the improving excellence observable in every part of the creation; such as the progressive increase of the wisdom and happiness of its inhabitants" (Erasmus Darwin 1801, 2:247–48)....

Progress, cultural, continued to underpin progress, biological. As the nineteenth century got underway, this melded in with the organic analogy. Herbert Spencer saw Progress everywhere, complementing his beliefs in progress everywhere. Moving rapidly forward, just before the middle of the twentieth century, Julian Huxley, waxed strong on the subject. He was always a fanatical progressionist and for the life of him could not keep progress out of his biology writings. In his major overview of the field—*Evolution: The New Synthesis* (1942)—there was progress, front, back, and throughout. A no-nonsense, brutally strong progress. "One somewhat curious fact emerges from a survey of biological progress culminating in for the evolutionary moment in the dominance of *Homo sapiens*. It could apparently have pursued no other course than that which it has historically followed" (J. Huxley 1942, 569).

To bring the story up to the present, we have Edward O. Wilson and his fervent belief in biological progress. "The overall average across the history of life has moved from the simple and few to the more complex and numerous" (E. Wilson 1992, 187). Modern writers stress the interconnections of culture and biology, where cultural Progress is not just an analogy for biological progress, but where culture and biology are intertwined in one overall progressive picture. We learn from evolutionary biologist Daniel E. Lieberman that four or so million years ago there were climate changes, with less forest and more savannah (Lieberman 2013). There was therefore ecological space to branch out—which our ancestors did, unlike those of the other great apes. Whether we moved out from choice or because of necessity given that others were more successful in the forests is a good question.

We know that a lot of characteristically human features, particularly bipedalism and the freeing up of the forelimbs and hands, came because of these moves. Bipedalism is not that good for fast running—literally, my four-legged cairn terriers run circles around me—but being up on two legs does make possible ongoing running for distances without breaks. You may not be able to outrun your prey, but you can keep going until literally they drop dead from exhaustion. Helping is the fact that being upright cuts down on the amount of the body exposed to solar rays. Unlike the prey, you are not overheating from the sun. Whether as cause or effect, or a combination of both, the first protohumans were hunter-gatherers—out after game (mainly male) or picking fruits and berries and roots (a big job for females). They were notable for their cooperation, working together and sharing. Other great apes are not much into this. Most probably the growth in brain size was part and parcel of the causal process—bigger brains mean more efficient hunter-gathering. More efficient hunter-gathering means ever more success biologically.

All of this at first was confined entirely to Africa, and then two mya bands started to venture forth, finally covering the whole globe. Modern humans appeared about two hundred thousand years ago and moved out of Africa into the Middle East about eighty thousand years ago. Things start to pick up about fifty thousand years ago, which is when modern Homo sapiens started to spread around the globe. There was development of language, more sophisticated tools, and so forth. With these moves and development, we start to see the evolution of distinguishing features, notably skin color. Near the equator you need dark skin to counter the ill effects of ultraviolet radiation; in more temperate climes, you need light skin to produce vitamin D. The big breakthrough to agriculture came about thirteen thousand years ago. An Ice Age struck, the hunter-gatherer lifestyle was no longer that sustainable, populations that had started doing a bit of cultivation and the like proved to be at a great selective advantage, and the rest as they say is history. . . .

In ten thousand years, the number of humans increased one hundred-fold—from about five or six million when agriculture got underway to about six hundred million at the time of Jesus. And obviously this was just a start. We keep going, adding almost as many again to the beginning of the nineteenth century. (Then things really took off, with around seven or eight billion today.) There is really no need to labor the point. Humans have been a huge success. And if you keep thinking at the purely biological level, as Al Jolson told us truthfully, you ain't seen nothing yet. The human population is growing at eighty-three million a year. Incidentally, don't think that truly all of this P/progress is really just at the cultural level. There is ongoing feedback. The move to agriculture meant the availability of milk products. Normally, adult humans are lactose intolerant and cannot digest such products. The right genes came along and made all possible. As needed. The Irish are lactose tolerant; in Asia, where agriculture did not yield these products, people are still lactose intolerant. Progress leading to progress, leading back to Progress.

All in all, there is no surprise that, in the general domain also, biological progress is alive and well. Not just in the laboratory of Edward O. Wilson, in the popular mind, in the way that evolution is presented in popular print or on the radio or television, the course of biological history is almost inevitably represented as one leading directly to Homo sapiens.

The Naysayers

What about Darwinian evolutionary theory and the warthoggian issue? Does Darwinian evolutionary theory tell you that humans are superior to warthogs? Stephen Jay Gould was rhetorically flamboyant on the topic. No surprise. By the 1980s, biological progress was "a noxious, culturally embedded, untestable, nonoperational, intractable idea that must be replaced if we wish to understand the patterns of history" (Gould 1988, 319). Big surprise. Gould saw no inevitability to the emergence of humans.

The sociobiology controversy convinced Gould that thoughts of biological progress leads to racism. Humans are not only top, but certain groups of humans are even toppier. The American immigration laws of the 1920s, essentially ending immigration by European Jews, was proof in itself. As a child of such Jews, Gould wanted nothing of it (Gould 1981). So, he wrote often and passionately against biological progress. And, to be fair, his thinking against biological progress was based on the very essence of Darwinism. The building blocks of change, variations or, in terms of genetics, mutations, are random in the sense of not appearing as needed. Pollution changes the background from white to black and so you need a new color for your camouflage. Black instead of white. You might as likely get bright red or psychedelic green. And, what might be available and what might work is a veritable crapshoot. Not much direction there.

Then there is natural selection itself. It is relativistic. Being large is not inherently good in itself. Elephants may be free from predators, but they need a lot of roughage as feed. The hobbit on the island of

Flores makes the point. Food was scarce so there was a premium on being small. If the island were overflowing with fruit and vegetables, but its denizens threatened by indigenous predators, the premium would be on being large. There is no inevitable change in one predetermined desirable way. Darwin's theory seems to have built-in opposition to directed change, to inevitable progress, even if you define it in some nonnormative way like larger brains and bipedalism....

Complexity

And yet! The conviction persists that we are superior. As G. G. Simpson used to say, if other organisms disagree with this judgment, let them speak up and make their cases (Simpson 1964). Perhaps, Gould and other skeptics notwithstanding, thanks to evolution, there is some mark of progress that does keep increasing through time. Go back to Clifford and his argument for panpsychism.

> The only thing that we can come to, if we accept the doctrine of evolution at all, is that even in the very lowest organism, even in the Amoeba which swims about in our own blood, there is something or other, inconceivably simple to us, which is of the same nature with our own consciousness, although not of the same complexity. (Clifford 1901, 2:38–39)

Dawkins is right about this. "Once there was only blue-green slime and now there are sharp-eyed metazoa"—and sharp-eyed metazoa are more complex than blue-green slime.... Unfortunately, it is not yet time to declare victory. Indeed, McShea cautions that there may be no victory.

> Is a human more complex than a trilobite overall? The question seems unanswerable in principle because the types of complexity are conceptually in dependent. The aspects of other measures, such as size, have this same independence: a balloon can be larger than a cannonball in volume but smaller in mass. Likewise, a trilobite might have fewer parts but more interactions among parts. Thus, it is hard to imagine how a useful notion of overall complexity could be devised. (McShea 1996, 480)

Adding to the woes, even if there is a measure of complexity, this doesn't mean that things are necessarily more desirable—which is a mark of genuine progress—or that Darwinism promotes such complexity increase.... Is this a surprise to any Darwinian? In the Darwinian world, it is not complexity as such that counts, but complexity as superior adaptation. As McShea points out, this just isn't necessarily true (1991).

Why Progress? Arms Races

Complexity, I am afraid, raises more problems than it solves. Go at the problem another way. Can we show in some sense that evolution does promote organisms with ever better adaptations? There are at least three proposed strategies. The first takes us right back to the heart of the cultural Progress/biological progress analogy. It centers on the notion of arms races, where lines of organisms compete against each other thereby improving their adaptations. The prey gets faster and in tandem the predator gets faster.... This military analogy idea was developed in detail by Julian Huxley in his first book, *The Individual in the Animal Kingdom* (1912). First, the cultural side, using naval warfare as the example. "Halfway through the century, when guns had doubled and trebled their projectile capacity, up sprang the *Merrimac* and the *Monitor*, secure in their iron breast-plates; and so the duel has gone on" (J. Huxley 1912, 115). Concluding: "Each advance in attack has brought forth, as if by magic, a corresponding advance in defense." Then, the biological: "With life it has been the same: if one species happens to vary in the direction of greater independence, the inter-related equilibrium is upset, and cannot be restored until a number of competing species have either given way to the increased pressure and become extinct, or else have answered pressure with pressure" (115). Adding: "So it comes to pass that the continuous change which is passing through the organic world appears as a succession of phases of equilibrium, each one on a higher average plane of independence than the one before, and each inevitably calling up and giving place to one still higher" (116).

Jumping to the present, Richard Dawkins goes right along with this kind of thinking. He brings up the increasing employment by competing nations of ever more sophisticated computer technology. In the animal world, Dawkins sees the evolution of bigger and bigger brains. As with Huxley's independence, so with Dawkins's brainpower. We won!

Why Progress? Convergence

The second progress-producing process centers on convergence. The key starting point is that of the niche, a space that organisms occupy. Fish, for instance, have colonized the ocean niche. Birds, the air-above-us niche. British paleontologist Conway Morris argues that only certain areas of potential morphological space will be able to support functional life (2003). This puts constraints on the direction of evolution. Not all pathways are open. But those that are open tend to be used again and again—*convergence*. The same adaptive morphological space is shared by different occupants. The most dramatic and well-known case is that of saber-toothed-tiger-like organisms, where the North American placental mammals (real cats) were matched right down the line by South American marsupials.

From examples like this, Conway Morris concludes that the historical course of nature is not random but strongly selection constrained along certain pathways and to certain destinations. In other words, life's history was far from haphazard as someone like Sepkoski rather implies. And this opens the way for progress. It happened and had to happen. Come what may, sooner or later some kind of intelligent being (a *humanoid*) was bound to emerge. After all, our own very existence shows that a kind of cultural adaptive niche exists—a niche that prizes intelligence and social abilities. It was waiting for us and we grabbed it! Humans came into their own. It was their Darwinian destiny! "We may be unique, but paradoxically those properties that define our uniqueness can still be inherent in the evolutionary process. In other words, if we humans had not evolved then something more-or-less identical would have emerged sooner or later" (Conway Morris 2003, 196).

Why Progress? Complexity

Turn to a third and final attempt to get progress from, or despite, the Darwinian process. This takes us back to the notion of complexity. Somehow—as Darwin rather implies in notebook musings—it just emerges because, in a way, you are always going to be building on what you have already. Earlier, we saw Darwin toying with this strategy: "The enormous number of animals in the world depends of their varied structure & complexity" (C. Darwin 1987, E 95). We saw also that natural selection was not left entirely out of the mix, or though perhaps more accurately, we might say that Archdeacon Paley was not left entirely out of the mix: "It is quite clear that a large part of the complexity of structure is adaptation" (C. Darwin 1987, E 97).

Others, more recently, have taken up the case. Somewhat surprisingly given his earlier thinking, the already encountered paleontologist Daniel McShea, backed by his colleague at Duke University, philosopher Robert Brandon, sees in life's history the potential for just that kind of upward momentum to life's history that Darwin toyed with. Favoring a somewhat inclusive notion of complexity, in terms of number of part types, they introduce what they call the "zero-force evolutionary law" (ZFEL). They write: "In any evolutionary system in which there is variation and heredity, in the absence of natural selection, other forces, and constraints acting on diversity or complexity, diversity and complexity will increase on average" (McShea and Brandon 2010, 3). It seems that for them things just naturally keep complexifying—one cause leads to several effects and these in turn multiply. Thus, all falls into place.

Why the Enthusiasm?

What do we say to all of this? Gould is right, Sepkoski even more so. Darwinism is not friendly to biological progress. There must be something fishy going on. Darwin was a great revolutionary. He was no rebel. Cultural Progress was the foundation of Darwin's personal, especially family, life. He was going to find progress in his theory, a move eased by the fact that, as a teenager, he read his grandfather's progress-impregnated speculations on evolution. True, Darwin knew the problems, as reflected by his notebook comment—"There is no necessary tendency in the simple animals to become complicated" (C. Darwin 1987, E 95). But, he had to get it somehow, so one is really not surprised by the progressivist sentiments of the *Origin* or the proto–arms race speculations that came shortly after that book was published.

Julian Huxley was rather differently motivated. More than one person has remarked on the paradox that, although Huxley was the great Darwinian synthesizer, his underlying commitment to natural selection was less than enthusiastic, at best. Huxley did not get his progressivist thinking from Darwin. He was open that he got it from the French philosopher Henri Bergson, whose *L'évolution créatrice*, published in 1907, greatly influenced Huxley (he was fluent in French) and was completely the underlying philosophy of *The Individual in the Animal Kingdom*. Bergson's enthusiasm for Spencer proved infectious. Remember: "The organic world appears as a succession of phases of equilibrium, each one on a higher average plane of independence than the one before" (J. Huxley 1912, 116). From Spencer back to Schelling and a philosophy of nature rising teleologically to the top.

With McShea and Brandon, we are right back in the nineteenth century: "We identify the intellectual ancestor of our view as Herbert Spencer." Whatever else, they have the self-regard for their accomplishments as had that Victorian bachelor long ago.

> The scope we claim for the ZFEL is immodestly large. The claim is that the ZFEL tendency is and has been present in the background, pushing diversity and complexity upward, in all populations, in all taxa, in all organisms, in all timescales, over the entire history of life, here on Earth and elsewhere. (McShea and Brandon 2010, 134)

Not a lot of Darwinian selection to be seen here. And, of course, there isn't any. The paradox of McShea is answered. How, at one point, he can be so skeptical about complexity and progress and, at another point, endorse the link. When he is considering the problem from a Darwinian perspective, where adaptation is all important, complexity does not equal progress. When he is considering the problem from a Spencerian perspective, where adaptation is secondary, complexity does equal progress.

Causal Inadequacies

Natural selection simply does not guarantee winners and losers, and so you look with a pinch of salt at the arguments for biological progress in a Darwinian world—a world of the machine. Start with arms races. They seem plausible. The prey gets faster, the predator gets faster. Note though that this is comparative progress, which is not normative. Even if arms races can produce comparative progress, you have still got to show that they occur and that they are reasonably frequent, if not the norm. . . . This said, not every evolutionist thinks that arms races are all that prevalent. Paleontologist Robert Bakker is skeptical about mammalian arms races between predators and prey (1986). And in any case, unless you are convinced that what we know of shellfish evolution translates directly into the necessary evolution of human beings, the peak of biological perfection, the support of the arms race hypothesis seems less than overwhelming. One has a nasty feeling that, although Dawkins trumpets that it is comparative progress he is after, truly with his emphasis on humans and brains, hopes of absolute progress lurk.

The same seems true of Conway Morris's claims about niche occupation. Many biologists would challenge the way in which Conway Morris conceives niches. He thinks of them as objective things, "out there," waiting to be occupied. Throwing cold water on this belief, Richard Lewontin writes: "Organisms not only determine what elements of the outside world are relevant to them by peculiarities of their shapes and metabolism, but they actually construct, in the literal sense of the word, a world around themselves" (Lewontin 2002, 54). This is a biological worry. And now a theological worry: Is Conway Morris making a case too strong for its own good? Remember J. B. S. Haldane. Not only is the world queerer than we think it is, but it is queerer than we can think it is. Who is to say that there are not niches out there waiting for beings to evolve to the point where they can enter and achieve a higher state of being? Beings to whom we humans are but children—perhaps warthogs?

Values

What do we conclude? Simply this. Darwin and, assuming that this is what they believe, McShea and Brandon were right that we do see an increase in some notion of complexity. This is because, if you start at the bottom, there is nowhere to go but up. It is surely the case that natural selection, through arms races and the like, had a hand in this. Darwinian selection played a major role in driving the evolution of humans up from the other great apes. However, this is not to say that, in any overall sense, Darwinian selection promotes progress. . . . Darwin's theory just does not guarantee (absolute) progress. And expectedly and revealingly, the more you seem to get a guarantee of progress from science, as from Julian Huxley or Simon Conway Morris or Edward O. Wilson or Dan McShea and Robert Brandon, the less you seem Darwinian.

I suppose that, even in a Darwinian world, if you believe in an infinite number of multiverses, human evolution was bound to happen, an infinite number of times. Almost paradoxically, Gould believed something of this nature: "I can present a good argument from 'evolutionary theory' against the repetition of anything like a human body elsewhere;

I cannot extend it to the general proposition that intelligence in some form might pervade the universe" (Gould 1985b, 409). Unfortunately, you are also going to have infinite numbers of also-rans—would-be humans with the IQs of turnips and the sporting ability of Michael Ruse. I do not know whether in such a case one would want to speak of progress. So I am not sure you solve a theological problem we have encountered—how, given the non-directionality of the Darwinian process, we can reconcile Darwinism with the underlying Christian assumption that, given God's creative powers, humans were bound to appear. I doubt God wants multiverses full of turnip-IQ, soccer-playing Ruses. I suppose you might argue that God, seeing the possibilities, picked the one he wanted. Like Dobzhansky I worry a bit about this. We don't really have directed evolution, at least not as we normally think of directed evolution. Yet, if quantum processes really are random, I am not sure that beforehand even God could tell which possibility would produce humans. Just that some will do so. . . . Progress is a value-impregnated notion. Darwinism eschews (absolute) values. Never the two shall meet.

Chapter 6, Reading 3
Evolution and Progress[1]

EVOLUTION IS THE CHILD of progress. Enlightenment hopes of ongoing social improvement were translated into beliefs about upward organic development through time since the Earth began (Richards 1992). As so often with family relationships, evolutionists tend to be ambivalent about their parentage. A critic says, "Progress is a noxious, culturally embedded, untestable, nonoperational, intractable idea that must be replaced if we wish to understand the patterns of history" (Gould 1988, 319). A defender responds: "I do think that progress has happened, although I find it hard to define precisely what I mean" (Maynard Smith 1988, 229).

Charles Darwin was torn on the subject, cautioning himself never to speak of "higher" and "lower," yet filling the *Origin* with flowery passages about the upward rise of life. Usefully, we find in his thinking a distinction between what we might label comparative progress, meaning the adaptive advance of one line of organisms over others, and absolute progress, meaning improvement up a scale of fixed value. This latter has been characterized as a directed change toward that which is better (Ayala 1988). Enthusiasts for absolute progress generally think that humans come out on top.

---

1. Ruse 1998b. Reprinted with permission of Oxford University Press.

Comparative Progress

Comparative progress is a Darwinian notion, centering on selection. At the microlevel, all would agree that it occurs, although there is much debate about its precise nature and extent. Much attention has been paid recently to one particular form, the so-called arms race, in which organisms compete and evolve, throwing up methods of attack and defense in a way analogous to human weapon development (Dawkins 1986; N. B. Davies et al. 1989). Controversy arises when one tries to take the hypotheses and findings of microevolution and apply them to the long timescale, the concern of the macroevolutionist. There seem to be two particular points of dispute: namely, that over significant, new adaptations—*innovations*—and that over protracted shifts—*trends*.

Innovations

Innovations supposedly open the way to the occupation of new ecological niches or to the seizing of niches already occupied (Nitecki 1990). One has an "adaptive breakthrough." Defining *innovation* as something that has crossed a functional threshold, it has been claimed that they are the "mainsprings" of macroevolution (Jablonski and Bottjer 1988). Many evolutionists—particularly paleontologists—simply take innovation as given, worrying more about its identification and implications. Endothermal homoeothermy—the very essence of what it is to be a bird or a mammal—has been offered as a paradigmatic example of such a phenomenon (Liem 1988). With such an adaptation, one can do many things barred to reptiles, particularly those involving nocturnal niches where there is no sunlight to warm the body.

The pressing concern of these enthusiasts is to find causal reasons why innovations should prove innovative. It is doubted that the phenomena could come about by pure chance, for some periods of the history of life seem to have been very much more prone to innovation than others. Opinion divides on two hypotheses, which may work more together than against each other. The ecological hypothesis suggests that an innovation is more likely to succeed in an empty niche, and that is why (for example) there was so much innovation in the early Paleozoic (Erwin et al. 1987). The genomic hypothesis suggests that the genomes of earlier organisms were less canalized and had fewer epistatic interactions. Hence the organisms were more open to radical redesign (Valentine and Erwin 1987).

Unfortunately, the fossil record yields few crucial predictions enabling one to make a decision between these hypotheses. Not that this is a matter of any great regret to some, who criticize the whole notion of innovation, arguing that the comparative evidence simply gives no support to the idea that some adaptations gave their possessors a major advantage over their rivals. Cracraft argues that close study shows that avian flight—something that a defender of innovation would see as a direct consequence of endothermal homoeothermy—crumbles into a long series of not very innovative parts (Cracraft 1990). Indeed, he believes that virtually all supposed important innovations are without genuine "ontological status" regarded as real evolutionary phenomena. As is so often the case with evolutionary disputes, one senses that much of the difference between Cracraft and those whom he criticizes rests on different definitions of what one would count as "innovative," as well as the selective use of examples supportive of one's own case.

Trends

Trends likewise find evolutionists divided. Some, particularly students of marine invertebrates, state flatly that the fossil record shows *trends*, meaning paths up to improved adaptation (Jackson and McKinney 1990). Such trends include forms of growth, rates of growth, and potential for greater habitat choice. The favored explanation of these trends— a phenomenon that has been referred to as "escalation" (Vermeij 1987)—rests on some form of extended arms race. It is claimed that as predators get more efficient at gaining their prey—for marine invertebrate predators, an increased ability to break or cut through shells, and the like—so also there was the evolution of yet stronger shells, resisting being broken or otherwise torn apart.

Critics of trends take a number of lines. One is to admit that there are phenomena that seem prima facie as if they are trends, but deny that they are caused by selection, or at least that as trends they are caused by selection (Benton 1987). Perhaps they are a function of other (possibly nonorganic) factors. A popular suggestion here is that mass extinction might have been a key causal factor, a key chance causal factor from the point of selection. Niches may have been opened up and organisms may have moved into such available space, without any significant adaptive pressure. This being so, it is hardly appropriate to talk of "improvement." And in any case, even if we do have selection-driven trends, we should not automatically assume that arms races are the key causal factors. We might have two taxa evolving in different, independent ways (Gould and Calloway 1980).

Another critical line is to deny that the fossil record is all that trendy anyway (Benton 1987). One can do this by contesting the phenomenal claim. Often, there is a confusion between increased variance in a group and a changed mean, with the former being taken as evidence for the latter. Again, one might feel that close examination shows all sorts of fluctuations, incompatible with genuine trends. Or one can argue that the patterns in the fossil record are little more than one would expect from any random process. Raup points out that the fossil record is a Markov-type time series, where any move (in a character) is necessarily but one step from the previous case (Raup 1988). The importance of this point is that most Markovian time series show trends, simply as a matter of course! They are what you expect, and moreover the increases and decreases recorded in such randomly generated trends tend to score as statistically significant. The mathematics of the case gives us "progress."

A Synthesis

The last word on comparative progress has not yet been said. Often science advances by synthesizing the strong points of opposing positions (615). Recently, it has been argued that we can get a viable notion of progress if we recognize, with the proponents, that key adaptations—where natural selection is improving functions, without undue cost to the rest of the organism(s)—can be crucial; but where extinction (and hence the emptying of niches) may be no less necessary (Rosenzweig and McCord 1991). Speaking of "incumbent replacement," these synthesizers agree that if an organism already occupies a niche, then a rival with a key adaptation may be unable to invade. However, after extinction, the old occupier type and the new key adaptation–possessing type compete on equal terms, and then the new type may well win. They illustrate their case with the example of the replacement of straight-necked turtles by those with flexible necks.

Absolute Progress

Absolute progress is no less a matter of debate than is comparative progress, nor is it less a matter of controversy. In part, this is the result of a very public campaign by Gould to expose its supposed shortcomings and to expel all of its traces from professional evolutionary studies (Gould 1988). In his bestseller, *Wonderful Life* (1989b), Gould tells of the finds of weird Cambrian fossils, with preserved soft bodies, in the Burgess Shale in British Columbia, Canada. On the basis of the wide range of types, with many orders no longer in existence, he argues that we have no right to claim that survival success was anything more than happenstance. Most probably, one of the forms in the deposit (*Pikaia*) was a chordate; but there are no grounds for thinking that its persistence was anything but luck. Run life's tape again, and the picture would be quite different. Hence we have no right to think that there was progress, especially not progress to humans.

To which one can only say that although Gould may be right, he has certainly not proved his case. At most he has shown—although this is certainly worth showing—that one cannot say much on the basis of the fossils alone. Perhaps the prevertebrate did have a key adaptation. The record cannot say. Nor, argue friends of progress, can it say what Gould would have it say. Although he is against progress, he does think that evolution has a direction—you can tell which end is up (Gould et al. 1987). Think of a major group of organisms like the mammals, and consider the clades (evolving branches) of which it consists. Is there any difference between the early clades of the group and those appearing later? Gould suggests that the early clades tend to be "bottom heavy," in the sense that they show maximum diversification earlier in their histories rather than later. Clades appearing later tend to be "top heavy." But, counter Gould's critics, this is simply a statistical artifact (Kitchell and Macleod 1988). The pattern observed is no more than one might expect on a random basis. One might add that, if Gould's finding is well taken, it cries out for an explanation in terms of adaptive breakthroughs, hardly a happy prospect for one as critical of pan-selectionism as Gould.

Complexity

Turning to the positive case for absolute progress, the most venerable criterion of improvement—with roots back to Aristotle's *De anima*—is that centering on complexity (Ruse 1988c). It is not that people value complexity in itself, but that they regard it as a flag for other desirable qualities, like intelligence. Of course, everybody recognizes that you do not get a simple inevitable evolution of the more complex from the less complex. At least, every Darwinian recognizes that sometimes simplicity pays. Measuring variation in a single dimension along the vertebral column, it can be shown that in the return to the water, mammalian backbones underwent a significant simplification (McShea 1991). Yet this simplification was not only evolution, but was evolution in a sustained adaptive direction—an "improvement" by any reasonable criterion.

To use complexity as a criterion, one must think of the overall picture of evolution. A recent extended case for this kind of progress has been made by Bonner (1988). Empirically, Bonner shows that there is a general organic trend to increased size of the largest organisms. This he explains simply by the fact that although niches at the bottom may be filled, there is always room at the top. But bigger organisms require a more complex support system; and, defining complexity as a simple function of the number of different cell types an organism has, Bonner concludes that there has indeed been such an evolution in the history of life.

There are many holes in this argument, starting with the fact that Bonner does nothing to show that the supposed evolution of size is other than an epiphenomenon of Markovian randomness. This apart, and ignoring the assumption that all differences of cell type have equal weight with respect to complexity, is it necessarily the case that great size spells complexity? The shrew is many orders of magnitude smaller than the big whales. Is it that much less complex? Or what about the shrew compared to the large dinosaurs? In any case, what about humans? Intuitively, we seem pretty complex. How do we rate vis-à-vis the dinosaurs? Bonner himself seems to recognize that he has a problem here, for by the time he gets to humans he is talking about behavior. Which may make sense, but is neither an obvious synonym for complexity nor for size.

One may think that one could save the argument by suggesting that advance be grounded in terms of coding DNA present in the genome, which in some sense would give a measure of complexity/progress (Maynard Smith 1988). Unfortunately, the correlation is not very exact. Also, one should not forget that the genome only gives instructions for making an organism; it is not the organism itself. Progress could come in getting more from less, getting a more complex organism from less DNA. In which case, a reduction in DNA would be a sign of advance.

Darwinism

Darwin himself tried to get absolute progress from comparative progress (Ospovat 1981). He thought that out of the competitive selective process some features would emerge that would simply be better than their alternatives, on any reasonable value scale. This argument is still extant. Dawkins decries "earlier prejudices" about progress (1986). However, in his great enthusiasm for arms races, he does rather imply that some features, specifically intelligence, are more equal than others. Also he thinks in terms of adaptive breakthroughs, including the biggest breakthrough of them all: "the evolution of evolvability" (Dawkins 1989). Needless to say, humans look pretty good on this picture.

The most overt, living Darwinian enthusiast for absolute progress is E. O. Wilson. He is convinced that there has been such progress and that we won the race (E. Wilson 1978). Recently, in an attempt to clarify his position, he has taken to distinguishing between "success" and "dominance" (E. Wilson 1990). *Success* is to be defined in terms of the longevity of a species and of all of its descendants through geological time. *Dominance*, on the contrary, is to be measured both in terms of the abundance of a group compared to other groups and in terms of overall "ecological and evolutionary impact" on all other organisms. By these measures, it would be too early yet to judge of human success, but we are clearly very dominant.

This suggestion does not in itself speak to causes, although Wilson would argue that the clarification introduced is a first step to a selection-based explanation of instances of success and dominance. One serious problem seems to be that, as presented, success simply goes to the oldest organisms that still have descendants. The reptiles are more successful than the mammals because they appeared first. One surely needs to modify the definition to take into account design potential for success? One might then say that the mammals were/are successful because their special adaptations made them adaptable in the face of change and disaster, and hence they could/did survive.

I suppose also one might say that humans do not promise to be a very successful organism, since they have the ability to produce twenty-first-century technology and yet are caught with Stone Age emotions. This does not seem much like progress. Likewise, although dominance may correspond to a recognizably intuitive concept, it is not necessarily that which we think is maximized when we speak of progress. The AIDS virus bids fair to be dominant, but we would hardly think its continued spread to be a matter of progress.

Yet, let us not end this discussion on an entirely negative note. A sophisticated approach to the fossil record may yield some positive support for some notion of progress, specifically in the Wilsonian sense of "success." If adaptations are getting better in some absolute sort of way, then perhaps taxa (species, genera, etc.) ought to last longer. With respect to genera, although a definitive case cannot yet be claimed, there are significant signs of improvement in survivorship—measured as a function of decrease in extinction rate—during the Phanerozoic (Raup 1988). Indeed, we could get as much of a change in generic half-life from seven mya to about fifteen mya. Obviously, even if well taken, this conclusion does not pick out one species as being superior to all others, nor does it relate success to other supposed marks of progress, such as complexity.

A Biological Anthropic Principle?

Concluding our survey, we see that the notion of progress continues to be of concern to evolutionists—especially those interested in macroevolution—as it continues to be a topic of controversy. Many evolutionists feel distinctly uncomfortable in discussing the very notion, and there is certainly a tendency to push such discussions into the semipopular realm (Maynard Smith 1992, Dawkins 1986, Gould 1989b). This is surely a function of the fact that absolute progress, certainly, is seen to have a blatant value component incompatible with the ideal of modern science. Note, however, that although in this discussion (by our very definition) we have allowed that there is such a component, this does not in itself say what value(s) are being endorsed. In flat opposition to just about every progressionist mentioned above, one prominent evolutionist (G. C. Williams) has argued recently that progress entails going in the face of that which is adaptively advantageous, because he feels (with T. H. Huxley) that that which is morally good is rarely if ever that which is biologically good (Williams 1989).

The question does remain why, for all its problems, progress of some kind remains so seductive a notion for so many evolutionists. No doubt there are many reasons; but my suspicion is that a major causal factor is some sort of biological version of the so-called anthropic principle: our understanding of the world is a function of our abilities to understand the world (Barrow and Tipler 1986). We are organisms, end products of the evolutionary process, with the ability to ask questions about progress. Perhaps this alone is enough to turn us to favorable thoughts of progress (Ruse 1988).

Chapter 6, Reading 4

Charles Darwin and Progress[1]

WE COME NOW TO Charles Darwin, author in 1859 of *On the Origin of Species by Means of Natural Selection, or the Preservation of Favored Races in the Struggle for Life*. I shall discuss his ideas in some detail, for he is the key figure in the history of evolutionary thought. . . . Tradition has it—a tradition founded primarily on his own self-deprecating remarks in his *Autobiography*—that Darwin's childhood and youth were intellectually undistinguished. He was second-rate at school, he was (after two years) a dropout from the medical program in Edinburgh, and he was then an idle undergraduate at the University of Cambridge.

However, although no one would suggest that Darwin was an infant prodigy, it is clear that he showed an early interest in and aptitude for natural science. Even before he went to Cambridge he had considerable training, theoretical and practical. Through the tutelage of his older brother, Erasmus, and through first-rate lectures at Edinburgh, Darwin received a thorough grounding in the chemistry of the day. Additionally, in Scotland, thanks to ardent pursuit of extracurricular activities, he developed considerable knowledge of and skill at marine invertebrate zoology, as well as favorable exposure to evolutionary ideas from that articulate spokesperson for the cause, Robert Grant.

At Cambridge, Darwin fell into the company of the science clique, the members of which (including such professors as Adam Sedgwick,

1. Ruse 1996a. Reprinted with permission of Harvard University Press.

of geology; John Henslow, of botany; and William Whewell, then of mineralogy) saw real talent in the young man. Continuing his informal science education, for the full three years of undergraduate life Darwin attended Henslow's very comprehensive course in the principles of botany. Although by the time of his graduation in 1831 Darwin was hardly a fully trained scientist, there is therefore little surprise that, through his university connections, he was offered and accepted the de facto post of naturalist on *HMS Beagle*, which was to spend five years surveying in the Southern Hemisphere, primarily around South America. It was on the basis of his diaries written during this trip that Darwin was to write a still charming travel book, which established his reputation firmly with the Victorian public (C. Darwin 1839).

On the *Beagle* voyage, Darwin developed rapidly into a more than competent natural philosopher. Trained by Sedgwick, he was however much impressed by Lyellian uniformitarianism, from which stance he made his own contribution to geology when he gave the (still accepted) reasons behind the formation of coral reefs, showing how they are a function of the coral growing upward as the seabed beneath sinks gradually (C. Darwin 1842). Work like this brought the newly returned Darwin right into the heart of the Oxford- and Cambridge-linked, middle-class scientific community. Nevertheless, it is clear that, from the very start, Darwin's horizons were fixed above geology. He wanted to solve what Herschel referred to as "the mystery of mysteries": the origins of organisms, in all of their variety (Cannon 1961). It was this that led to evolution (Oldroyd 1984).

Hindsight often makes it difficult to discern the exact course and importance of events. However, with respect to his conversion, there is little reason to doubt Darwin's own recollection of the stunning impact of a Pacific island group visited by the *Beagle* in 1835 (Sulloway 1982b, 1985). He could not get over the fact of "the South American character of most of the productions of the Galápagos Archipelago, and more especially by the manner in which they differ slightly on each island of the group" (C. Darwin 1958, 118–19). This is not to say that there was an immediate, onboard conversion to evolutionism; rather, full insight came later (spring of 1837), back in Britain, when Darwin learned that the birds of different islands belong to different species (Sulloway 1982a).

Why should this (somewhat trivial) piece of information make so much difference? By this stage of his life Darwin, who had earlier been a practicing Christian, had moved toward his family's deism. It simply did not make sense for God to have put similar but different animals on different islands, within sight of each other, especially since the habitats were essentially identical. Even less did it make sense for God to have ensured carefully that the denizens of the Galápagos were South American–like rather than European-like. Who would care? The explanation had to lie elsewhere and Darwin found it in evolutionism. Ancestors had come to the islands and then evolved as they moved from island to island (C. Darwin 1987).

This takes us to the fact of evolution. What of causes? What of evolution as theory? Initially, Darwin seems to have supposed that there might be limited terms for species, ending with sudden switches to new forms. Then, more confidently, he favored some kind of Lamarckian causal process, in which the inheritance of acquired characters played a major role. He always kept this as a secondary mechanism, but soon queried its adequacy as a general cause. Continuing his search, Darwin next realized that animal and plant breeders effect great changes by picking or selecting those organisms with desired characteristics and breeding from them alone (Ruse 1975a). But, how could such a process occur in nature? The insight came late in September of 1838, when Darwin read the *Essay on the Principle of Population* by Robert Malthus (1826). He realized at once that the central doctrine of that work on human societies applies throughout the living world. There is an ongoing, potentially explosive population pressure, against an invariable background of limited supplies of space and food. Hence, there will necessarily be a struggle for existence. Yet, this struggle need not be entirely negative, for success is not chance. It is rather a function of the possession of various peculiar, useful features—and the nonpossession of such features by the losers. And so there will be a process akin to the breeders' picking, a process Darwin labeled "natural selection."

Given enough time and an ongoing supply of new variation, there will be full-blown evolution. But more than this followed from Darwin's line of thought. As one who had been reared at Cambridge on an almost unrelieved diet of the works of Archdeacon Paley, Darwin felt strongly—with every established middle-class scientist in Britain at that time—that any adequate force must be able to explain that all-pervasive facet of the organic world: its teleology, its adaptedness. Every organism, in just about every way, seems as if it were designed to be what it is. And to this, Darwin thought his new cause could speak.

Natural selection is as much a force directed toward organization and functioning as is Paley's God (Ruse 1999b).

A mechanism is not a full-blown theory. Darwin realized this, and so he worked furiously to embed his central notion in a complete framework. In 1842, that is four years after discovering natural selection, Darwin wrote up a short (private) sketch of his ideas, and two years later he expanded this into a full-length essay (C. Darwin and Wallace 1958). He did not publish. We shall probably never truly know the full reason for this reticence—assuming that there is a full reason. We know that Darwin had fallen sick by this time, from a mysterious illness that was to plague him for the rest of his life. We know his wife, whom he married in 1839, would most probably have been upset by a public airing of his ideas, and Darwin was not about to cause trouble in that direction. We know that *Vestiges*, published in 1844, caused tremendous controversy, and Darwin had no wish to alienate his fellow scientists—especially given that Chambers's greatest critics, like Sedgwick, were not only Darwin's friends but often his sometime teachers.

I myself doubt that Darwin realized that the delay would become as long as it did—but long it became, as Darwin buried himself, physically in the countryside and intellectually in a massive study of barnacle taxonomy (C. Darwin 1851a, 1851b, 1854a, 1854b). Finally, in 1858, when Darwin had started to write an immense tome on evolution—overwhelming with fact and footnote—his hand was tipped. A young naturalist, Alfred Russel Wallace, sent to Darwin (of all people!) a short essay containing Darwin's own ideas in perfect cameo. Thus spurred, Darwin wrote up his theory in a full yet manageable volume, and *On the Origin of Species* toward the end of 1859.

Having discussed the struggle for existence—a well-known notion in the post-Malthusian era of mid-Victorian Britain—he began an intentionally seductive presentation of the central mechanism:

> Can the principle of selection, which we have seen is so potent in the hands of man, apply in nature? I think we shall see that it can act most effectually. . . . Can it . . . be thought improbable, seeing that variations useful to man have undoubtedly occurred, that other variations useful in some way to each being in the great and complex battle of life, should sometimes occur in the course of thousands of generations? If such do occur, can we doubt (remembering that many more individuals are born than can possibly survive) that individuals having

> any advantage, however slight, over others, would have the best chance of surviving and of procreating their kind? On the other hand, we may feel sure that any variation in the least degree injurious would be rigidly destroyed. This preservation of favorable variations and the rejection of injurious variations, I call Natural Selection. (C. Darwin 1859, 80–81)

Now, with the mechanism made explicit, Darwin could turn to evolution as fact, and so the rest of the *Origin* (a full two-thirds) consists of a journey through the various branches of biology, as evolution through selection is applied to major claims and theory problems. Hence, on the one hand the reader is persuaded of the value of such an approach, and on the other hand facts that we know about the living world point us in one shared direction, namely toward evolution through natural selection. . . .

By the end of the 1860s, the codiscoverer of natural selection, Wallace, had fallen into apostasy over human origins, arguing that many aspects of human nature—our hairlessness and our thinking capacity, for instance—just could not have come through natural selection. The savage has the potential for deep thought, but because in the general course of life he or she never uses it, it could not have been produced by natural selection (Wallace 1870b, Kottler 1974, Smith 1972). Agreeing in part with the premises of Wallace's argument, Darwin turned in search of an alternative conclusion. In sexual selection he found his answer. Rivalry for mates centers on human notions of beauty, which bring on the evolution of those human attributes for which Wallace supposed we must have a nonnatural cause. Hairlessness, intelligence, and the rest come from intraspecific struggle for mates. Hence, like all other organisms, humans are an entirely natural product of the evolutionary process.

Darwin's Biological Progressionism: Before the *Origin*

I come now to the question of Darwin and progress, and the related question of Darwin and Progress. Does he, in any sense, show progressionist tendencies? Was he influenced by ideas of progress? Did he have an evidential base? Begin with the question of biological progress, and turn to the first jottings on evolutionism. Right at the beginning of one of the earliest notebooks, when Darwin was edging toward transmutationism but had not yet grasped selection, we see that he was thinking in terms of high and low with mammals at the top. Yet even from the beginning, Darwin was struggling with progress. He believed in it, but he was not quite sure what he believed in. Most definitely, Darwin knew what he did not want to believe in, namely any kind of simple, unilinear, monad-to-man progressionism. The Galápagos factor was crucial here, for it led Darwin to think of change as sparked by geographical isolation (as on islands), with consequent evolution to differing forms. In other words, transformation always has splitting at its heart. For Darwin, the tree of life (or "coral" of life, as he speculated on calling it) was fundamental.

In addition, however progress may be defined, Darwin always appreciated that an organic descent from higher to lower might be a smart life strategy move. That first (preselection) mechanism, a kind of Lamarckian adaptation of structure, sparked by new habits, certainly opened the path for successful degeneration. Moreover, according to the outside influences, such degeneration was indeed a possibility. Others more knowledgeable than he were telling Darwin that simple progressionism was paleontologically false. Owen, at this crucial point in time working on Darwin's *Beagle* fossils, was uncompromising: "The different organized forms which have succeeded each other do not display regularly progressive stages of complication, or perfection of Structure. Plants and animals exhibiting different degrees of Complication of Structure have coexisted at different periods" (R. Owen 1992, 222).

Ladderlike progressionism is impossible, and Darwin was thus led into comments about why we should expect to find insect instinct happily thriving alongside human intelligence. Nevertheless, for all the qualifications, ultimately one branch of the tree of life leaves all others behind. Why? Already, we have some of Darwin's speculations. In a way, progress is an artifact of the very fact of evolution itself. You have to start from the bottom up—necessarily you go from simple to complex, and you cannot have complex organisms (vertebrates, for instance) without simple organisms for them to subsist on. Yet, obviously, these are more necessary than sufficient conditions for progress. Once given plants, why should vertebrates evolve? Why is there more on our planet than thick, green jungle? It is at this point, when he is trying to generate a positive force for progress, that we start to see Darwin running into real difficulties.... I assume that this comment referred not only to Lamarck but also to the spirit forces of the *Naturphilosophen*.... Essentially, however, Darwin was caught on the horns of a Lyellian dilemma. If geological change be directional, hotter to cooler (say), then organic progress might follow. But Lyell (and Darwin following him) interpreted uniformitarianism as implying a nondirectional, steady-state Earth history. Hence, overall directional organic change seems impossible.

The early Darwin rested uneasily on the question of progress. He needed both definition and cause. With respect to the first, the need for an adequate understanding of what one might mean by "progress," not to mention "higher" and "lower," intensified when, with the publication of *Vestiges*, critics lambasted Chambers's suggestions. It is no wonder that we find Darwin at once starting to disavow any such problematic terms. Yet, at the same time, he started to inquire of his friends as to their understanding of the notions! Thus to Hooker: "On what sort of grounds do Botanists make one family of plants higher than another" (C. Darwin 1985–2023, 3:301). Response: "Generally speaking in Botany highness and lowness are synonymous with complexity and simplicity of structure" (C. Darwin 1985–2023, 3:313). Gradually, however—thanks particularly to the work of Germanic scientists—Darwin began to see his way to an adequate (and by the 1850s, fairly conventional) notion of highness and lowness. A rise up the scale required increasing differentiation, specialization, division of labor. Remember Von Baer: "The more homogeneous the whole mass of the body is, so much the lower is the grade of its development. The grade is higher when nerves and muscles, blood and cell-substance, are sharply distinguished" (Von Baer 1828, 1:207). Darwin began to see how ideas like these could be incorporated sympathetically into his own thinking.

Causally speaking with respect to progress, Darwin needed a more directed, a less passive and active, force at the center of his evolutionism. He had to move away from seeing change as just a reaction to outside forces and more as a process with its own internal dynamic. Grasping the significance of selection did not do this per se—the adaptation it produces might occur merely in response to external forces—but selection does open the way to inter-organic tensions. In other words, it might be an activity which in itself alone could fuel change—even progressive change. Organisms themselves, in relation to each other, cause change and splitting (Ospovat 1981). Thus Darwin was led to develop—or, rather, given the early speculations, to elaborate—what he called his "principle of divergence": "The same spot will support more life if occupied by very diverse forms.... This I believe to be the origin of the classification and affinities of organic beings at all times; for organic beings always seem to branch and subbranch like the limbs of a tree from a common trunk, the flourishing and diverging twigs destroying the less vigorous—the dead and lost branches rudely representing extinct genera and families" (from a letter written to Asa Gray, Sept. 5, 1857 [C. Darwin and Wallace 1958, 266–67]). The point is that you get more success by diversifying. It must be emphasized that Darwin always thought of selection as working for the benefit of the individual rather than the group (Ruse 1980). So he was not claiming that selection was working to produce divergence and variety, merely that more organisms (taken as a whole) can be supported on the same resources. And this was the kind of total life history within which Darwin's progressionism was framed.

Darwin's Biological Progressionism: The *Origin*

These various thoughts bring us to the Darwin of the *Origin*. After twenty years of hard work, he was beginning to think that he was getting on top of the problem of progress. He wrote on the topic, in some detail, to Hooker, at the end of 1858. First, on December 24, he promoted the claim that "species inhabiting a very large area, and therefore existing in large numbers, and which have been subjected to the severest competition with many other forms, will have arrived, through natural selection, at a higher stage of perfection than the inhabitants of a small area" (C. Darwin 1985–2023, 7:221). And then, on December 31, he expanded and qualified his thinking:

> Your letter has interested me greatly; but how inextricable are the subjects which we are discussing! I do not think I said that I thought the productions of Asia were higher than those of Australia. I intend carefully to avoid this expression, for I do not think that anyone has a definite idea what is meant by higher, except in classes which can loosely be compared with man. On our theory of Natural Selection, if the organisms of any area belonging to the Eocene or Secondary periods were put into competition with those now existing in the same area (or probably in any part of the world) they (i.e. the old ones) would be beaten hollow and be exterminated; if the theory be true, this must be so. . . . I do not see how this "competitive highness" can be tested in any way by us. And this is a comfort to me when mentally comparing the Silurian and Recent organisms.—Not that I doubt a long course of "competitive highness" will ultimately make the organization higher in every sense of the word; but it seems most difficult to test it. (C. Darwin 1985–2023, 7:228–29)

And yet, for one who had worked for so long on the topic, the tone of these letters was not as confident as one might have expected. All was not entirely well. The problem of progress has been moved sideways in a not entirely satisfactory manner. As Darwin wrestled to preserve—or, rather, strengthen—the notion of advance, while at the same time stressing even more the treelike nature of his evolutionary thought, he now offered us what we might call relative or comparative progress. This was what Darwin called "competitive highness": a process whereby organisms pursue a particular route, trying to better their rivals as they do so. But, in itself, this was not an analysis of an absolute notion of progress, which (culminating as it does in humans) was Darwin's original goal—and which appeared still to be his hope. There is a gap here: Darwin skipped over from relative progress to absolute progress. His implicit argument seems to be that, if you compared specializations between branches, perhaps even thinking of who would beat out whom, you could get some idea of which organisms come out on top on an overall scale. Implicit or not, this was now the official line; notwithstanding the tensions between the two notions of progress, inasmuch as any tack was taken in the *Origin*, it is this. The tree of life is the fundamental metaphor, and—for all of the branching—through it the very picture of evolution is of one stretching ever upward. To support this idea Darwin made only one rather sweeping statement, which presciently, effectively blurred relative and absolute progress: "The inhabitants of each successive period in the world's history have beaten their predecessors in the race for life, and are, in so far, higher in the scale of nature; and this may account for that vague yet ill-defined sentiment, felt by many paleontologists, that organization on the whole has progressed" (C. Darwin 1859, 267). Apart from a repetition of this sentiment in the conclusion, there was no detailed discussion.

By the third edition of 1861 (C. Darwin 1959), when it was clear that he was going to be treated as a responsible scientist—perhaps spurred also by a correspondence he had just had with Lyell on the topic—Darwin felt somewhat more confident, at least on the question of relative progress, arguing that selection

> will, I think, inevitably lead to the gradual advancement of the organization of the greater number of living beings throughout the world.

He made appeal to the authority of Von Baer and referred specifically to the significance of "the completeness of the division of physiological labor."

Then, he drew a connection between efficiency, specialization, and highness:

> If we look at the differentiation and specialization of the several organs of each being when adult (and this will include the advancement of the brain for intellectual purposes) as the best standard of highness of organization, natural selection clearly leads towards highness; for all physiologists admit that the specialization of organs, inasmuch as they perform in this state their functions better, is an advantage to each being; and hence the accumulation of variations tending towards specialization is within the scope of natural selection. (C. Darwin 1959, 222)

But, although there was a final promise about linking relative to absolute progress, as it turns out this was virtually an end to the matter. The geological section simply reprinted the original passage about the "vague yet ill-defined sentiment" with a comment about specialization thrown in. Somehow, relative progress leads to absolute progress—and if you are not now convinced you will have to take it on trust.

Darwin's Biological Progressionism: Post-*Origin*

As with evolution in general, there was one species about which Darwin showed no ambiguity: "Amongst the vertebrata the degree of intellect and an approach in structure to man clearly come into play" (C. Darwin 1961, 221). This sets us up for the preparation and reception of the *Descent of Man*, the work that occupied the last decades of Darwin's life. The theme of progress, particularly at the absolute level, was endorsed and driven home further, as Darwin tried to show how humans represent the high point of the evolutionary picture—and how certain white Anglo-Saxon males represent the high point of the human picture.

In basic theory, there was not a great deal of difference between progress in the *Descent* and Darwin's general treatment of progress; although there was a sense that, toward the end of human evolution, the action started to switch almost completely from the physical to the psychosocial. Enormously impressed by an article written by Wallace in 1864, Darwin wrote: "The great leading idea is quite new to me, viz.

that during late ages the mind will have been modified more than the body; yet I had got as far as to see with you that the struggle between the races of man depended entirely on intellectual and moral qualities" (letter to Wallace, May 28, 1864, in Wallace 1916, 2:127). This emphasis on the mind got reflected right into the *Descent of Man*, as did all sorts of Victorian racial and sexual sentiments. "Man is more courageous, pugnacious, and energetic than woman, and has a more inventive genius" (C. Darwin 1871, 2:316). Woman, on the other hand, has "greater tenderness and less selfishness" (C. Darwin 1871, 2:3326). There has simply been more intense sexual selection between men than there has been between women. It is men who have to battle with the elements— and with each other. Naturally, here as elsewhere in life, the effort pays off. "The female . . . ultimately assumes certain distinctive characters, and in the formation of her skull, is said to be intermediate between the child and the man" (C. Darwin 1871, 2:326).

Early Influences

Let us move to our second question. Is there reason to think that Darwin was a progressionist and that this inclined him to biological progressionism? The first part is answered readily. Charles Darwin was virtually predestined to be a progressionist. He came from the family that worshiped at the shrine of progressionism. Both of his grandfathers were Lunatiks, and we know what an obsession progress was with Erasmus. Robert apparently followed very much in the tradition—interested in science, free-thinking, committed to change. And Uncle Jos (Josiah Wedgwood the second), later Darwin's father-in-law, a man who was much respected by Darwin for his integrity, was an ardent Unitarian and so far committed to the philosophy of progress that he became one of the reforming members of parliament after the bill of 1832.

Then, perhaps more influential than any, there was Darwin's older brother Erasmus. It was he who inducted Charles into the joys of chemistry—hardly a theory-neutral enterprise, given the importance of that particular discipline in the Industrial Revolution. It was he who went before Charles in shucking the trimmings of conventional Christianity (and may well have had a strong influence on his younger brother, judging from worried comments by Emma Darwin). And it was he who, after the *Beagle* voyage, introduced Charles to intellectual London, including such luminaries as Harriet Martineau and Thomas Carlyle.

Darwin's intellectual and social background was one of progressionism. Moreover, this was a major factor feeding into his becoming an evolutionist. Evolutionism for Charles Darwin meant progress, linked to progressionism. In the 1830s, that was the very heart of the doctrine. And we see Darwin, whose introduction to evolutionism came precisely through the ideas of Lamarck and Erasmus Darwin and others, accepting that background without question. Indeed, at times Charles Darwin went so far as to echo his grandfather. Thus Erasmus, in a passage scored by Charles: "Perhaps all the productions of nature are in their progress to greater perfection! an idea countenanced by modern discoveries and . . . consonant to the dignity of the Creator of all things" (Erasmus Darwin 1803, 1:54). . . .

Progress was the philosophy of the day; Darwin was submerged in it; and the indications are that it seeped over—more precisely, flooded—into his science. Moreover, at that time people were thinking hard about causes of progress, and this too is Darwin's concern, especially as reflected by the distinctively Darwinian, relativistic notion of progress. Here we have competition between organisms, striving to stay one step ahead, with consequent divergence and development into all kinds of specializations. Although it took years to develop fully, the idea is there from the beginning in Darwin's thought, and it surely has Malthusian connections since complexification emerges. Significantly, countering Malthus's pessimism would have been the optimists who argued that competition leads to specialization and specialization leads to progress. I refer, of course, to the members of the highly influential Adam Smith school of political economy. And, just as Darwin was influenced in his thinking about upward progress—that is, of a more traditional absolutist kind—by views about an absolute form of Progress, so we find he was influenced in his thinking about relativistic progress by thoughts on a relativistic notion of Progress.

Later Influences

Let us turn the calendar to the period between the discovery of natural selection and the appearance of the *Origin*. Here, tied in with the "principle of divergence," the notion of splitting or branching was crucial.

Organisms compete against each other and success in the competition frequently comes to that which is somewhat more specialized—that is, success comes less often to the all-purpose being, and more often to the being that has specially suited adaptations. Darwin connected this view with the then-current view of progress, which saw specialization as a mark of advance. And the end result is competitive highness or relative progress, or, more precisely, a rethinking and refinement of this notion. Note incidentally that there is an ambiguity at some point here between specialization within a single organism and specialization of individual organisms within a group. The principle of divergence requires the latter, but Darwin seemed to think that the former obtains also. . . .

We come to the time after the *Origin*, when Darwin was leading up to the work on humans at the beginning of the 1870s. Relativistic progress took a back seat to absolute progress, as Darwin tried to show that we humans—especially Anglo-Saxon humans—are the apotheosis of the evolutionary process. Once again, extra-biological views on progress were crucial (John Greene 1977). For instance, on his copy of Wallace's important essay, Darwin carefully highlighted some typical Victorian sentiments on progress: "The better and higher specimens of our race would therefore increase and spread, the lower and more brutal would give way and successively die out" (Wallace 1864, clxv). Consequently: "The red Indian in North America, and in Brasil; the Tasmanian, Australian and New Zealander in the southern hemisphere, die out, not from anyone special cause, but from the inevitable effects of an unequal mental and physical struggle" (Wallace 1864, clxv). This got double line marking in the margin by Darwin and a scribbled comment confirming the point, quite apart from a letter to Wallace saying that this part of the paper was "grand and most eloquently done" (Wallace 1916, 2:127).

The Evidence

We come to the question about the evidence on which Darwin based his theorizing. First, we ask on what foundation rested the belief that biological progress of some kind actually occurred: in Darwin's case that there was a kind of branching process with some branches proving more significant than others, and that ultimately one branch—that leading to humankind—triumphed above all others. Here, obviously, paleontology is the most significant potential source of evidence. Although Darwin tended to use fossils for geological ends (proving the Lyellian steady state), he always kept well abreast of modern research. Yet, it seems not unfair to say that the Darwin of the 1830s was making houses without stones, or rather progressive speculations only with stones (Bowler 1976). We know that there was some evidence of a sequential succession of fossils—Darwin himself was impressed by the fossils of South America but there was not so much connection that Lyell could not argue against any kind of progress or evolution. Certainly, it was not primarily the fossils that made Darwin an evolutionist, or a progressionist.

Empirically more pertinent was comparative morphology, particularly as represented by Darwin's massive study on barnacles. One must move with caution here since, given that this study occupied the years at the end of the 1840s and the beginning of the 1850s, publicly Darwin was working in a non-evolutionary mode.... Yet it is clear that Darwin was using the barnacle work to articulate and confirm his progressionist beliefs, though not so much at the general level.... "On the whole, I look at a Cirripede [barnacle] as a being of a low type, which has undergone much morphological differentiation, and which has, in some few lines of structure, arrived at considerable perfection,—meaning, by the terms perfection and lowness, some vague resemblance to animals universally considered of a higher rank" (C. Darwin 1854a, 20).

Through the 1850s there was a growing realization (from and through people like Owen) that perhaps the fossil record is not so very antithetical to evolutionism—which, remember, everyone equated with progressionism in one form or another. It is true that the gaps in the fossil record remained. Yet, some holes were plugged. The greatest find came after the *Origin*, when the bird/reptile *Archaeopteryx* was discovered and identified (J. Owen 1863, T. H. Huxley 1867–68). Of course, more significant than any of this was (by the 1850s) the unambiguously branching nature of the fossil record, even if it was branching with—if not a purpose—at least with a rhythm. One starts with a fairly general form and then goes off in various directions, adding adaptive specializations. Furthermore, the early forms tend to be more embryo-like, a point that Darwin seized on: "If it should hereafter be proved that ancient animals resemble to a certain extent the embryos of more recent animals of the same class, the fact will be intelligible" (C. Darwin 1859, 267).

What of our own species? There was some evidence, recognized as such by the end of the 1850s, that the human species is very old—far older than allowed by the traditional six thousand years of Genesis (Oakley

1964, J. W. Gruber 1965). Human remains had been discovered along with the remains of extinct organisms, and people appreciated that this points to an early human beginning. The fossil evidence of human evolution was less helpful, and really did not make the case at all. Neanderthal man was discovered (in Germany) in the 1850s, but the opinion even of Darwin's supporters was that it is part of our own species rather than a genuine "missing link" (Huxley 1863).

For Darwin this absence of human fossil evidence was hardly devastating to his general case for human evolution; it merely pointed (as he and everyone thought) to our comparatively recent arrival on this globe. Darwin's main argument for human evolution, and for our close relationship with the apes, was grounded in comparative anatomy and in embryology and so forth. He did indeed speculate (in discussion with Lyell) about the actual course of human evolutionary history; but the details of this were never that significant for his overall position. Yet, as we swing next to look at Darwin's thinking on the causes of progress, we should not let him escape too easily here. Qua progressionist, he held to the claim that humans come top. If we think purely in terms of the evidence available to Darwin about paths and histories, it is clear that his human-centered progressionism transcended the available data. Even if it is agreed that we humans are among the final products of the evolutionary process, this in itself does not prove progress. Degenerate barnacles are also among the last in the process. At most, therefore, Darwin could make a rough claim for consistency between the evidence and his theory. There was certainly no overwhelming proof.

What about the question of causes? Darwin thought that evolution results in relative progress, marked in terms of the division of labor (and flagged by such things as complexity), and that this somehow shifts over into absolute progress. Humans—particularly white, European, male, capitalist humans—come top. The issue here is whether natural selection, with a little help from secondary causes, particularly sexual selection, can do all that is asked of it. Or, rather, whether the evidence behind selection can do all that is asked of it. And the candid answer is that, judged by today's standards and even more by the standards of Darwin's day, to arrive at the conclusions that were reached there has to be a huge amount of (shall we say) "stretching" of the evidence to meet the theory. Beginning with the idea of a struggle, going on to selection, speaking to adaptation, and triumphing through a division of labor—all this meant going beyond the given, in dramatic ways. I am not criticizing Darwin from today's perspective and even less from that of his own day. The mark of the genius is to see a pattern among the scattered pieces. I am just stating a fact.

"I should premise that I use the term Struggle for Existence in a large and metaphorical sense, including dependence of one being on another, and including (which is more important) not only the life of the individual, but success in leaving progeny" (C. Darwin 1859, 62). ... If cautionary notes are needed in treating of the evidential basis for the struggle, they are required even more when we turn to the variation Darwin postulated to make the struggle biologically meaningful and (especially) to provide an ongoing supply of fuel or "raw stuff" for evolution. By the time of the writing of the *Origin*, thanks especially to his work on barnacles, Darwin was starting to feel quite confident that large-scale variation of the needed type does indeed exist; but, again, this was hardly a simple function of reading off the "facts" from the book of nature. Nor was natural selection itself such a simple function. Indeed, Darwin had virtually no direct evidence of the process. And, as numerous critics pointed out, the analogy from artificial selection was shaky (to say the least). No matter how much you select, you are not about to turn a cow into a pig.

Relativistic progress may be an important aspect of reality. We shall learn that many today believe that it is. But, whatever its current status, Darwin gives us little reason to believe it a common empirical phenomenon. My hunch is that his faith in its existence was primarily a function of his knowledge of what occurs in the breeders' world, where competition can lead to ever exaggerated features. His experience with breeding was spliced with an understanding of the course of manufacturing and technology, where again one finds the "evolution" of ever developed contrivances—combined of course with an ever greater reliance on a division of labor. ...

Darwin and National Strands of Progress

As in all things, with respect to progress Darwin wove his rope from many strands. First, from France: although Darwin kept Lamarck at arm's length, one should not underestimate the influence of that particular author. Darwin took in Lamarck's ideas indirectly from Grant and from the *Principles of Geology* (despite Lyell's intentions), and more directly from personal reading of Lamarck and others, like Geoffroy.

Apart from anything else, Darwin always accepted the key Lamarckian mechanism of the inheritance of acquired characters. And this is not to mention additional French influences, like Comte, Milne-Edwards, and others. Likewise, although Darwin was no great Germanophile—he always had great trouble with the language—socially and scientifically there were influences. People in brother Erasmus's set, like Thomas Carlyle, were great enthusiasts for German culture. Also there was Whewell and his almost Hegelian view of the history of science, not to mention Owen and, either directly or through him, the important work of such scientists as Von Baer. The unity of type, so important in Darwin's thought, clearly owes much to German idealism—as do many other elements that found their way into the *Origin*. In fact, at times Darwin wrote as though he were a full-blown recapitulationist: "As the embryo often shows us more or less plainly the structure of the less modified and ancient progenitor of the group, we can see why ancient and extinct forms so often resemble in their adult state the embryos of existing species of the same class" (C. Darwin 1959, 704).

When all is said, other contributions to Darwin's ideas pale beside the British element. Whatever the foreign sources, the notion of relative progress was deeply embedded in his own culture. Darwin's picture was one of hard, slow, physical grind—of a real groping gradually through much sweat and toil, often unsuccessfully, toward a better state. And this state, ultimately as absolute as one could wish, was mid-Victorian middle-class society. In the fullest sense of the word, Charles Darwin was the heir of the eighteenth-century British Enlightenment—David Hume, Adam Smith, and above all others, Erasmus Darwin and his circle. As Marx famously remarked in a letter to Engels, Darwin's genius was to transfer the British industrialist's philosophy—which had at its heart a faith in progress—right into the biological world (letter of June 18, 1862, in Marx and Engels 1965, 128).

Chapter 6, Reading 5

Evolutionary Directionality: No Direction to Evolution[1]

THE DISCUSSION IS GOING to take a bit more of a theological turn. I am not a theologian and don't pretend to be. However—and this is good advice to anyone, however dignified and important he or she may be—I am happy to turn to the textbooks for information and this is what I have done now. As you will see as the discussion proceeds, the tremendously important ideas being discussed in this chapter quickly take us out and beyond the theological to the scientific and the philosophical. So, in a sense, all of us are going to be underqualified in one way or another. This is nothing to be ashamed of and in fact is what makes this whole topic so incredibly interesting and exciting. In this chapter, I combat the whole idea that the universe displays a directionality intended and orchestrated by God. In particular, I argue from evolutionary science that the biological realm was not somehow preparation for the intended appearance of Homo sapiens as special beings.

---

1. Ruse 2017h. Reprinted with permission of Oxford University Press.

Bottom-Line Demands

Rather than starting straight off with science, let me turn things around for a moment and start with Christianity. Acknowledge one basic fact. In the Christian scenario human beings are not a contingent add-on—I take it that this is true of any theistic religion. One presumes that God did not have to create at all and one presumes that, having decided to create, God did not have to create humans or humanlike beings. But God did, and in the Christian (theistic) picture we have a starring role. God made humans "in his own image" to love and to have us love and worship him in return. When we fell into sin, he went on loving us so much that he was prepared to die in agony on the cross to pave the way for our eternal salvation, whatever that might be. It isn't that God doesn't care about other organisms—he knows when every sparrow falls—but we have a special place. We have an obligation to care for other organisms, but one has a strong sense that they are there for our benefit. Jesus told us to care about the sick and the needy, even those in prison, but I don't remember any sermons on the virtues of being a vegan.

What does it mean to say that we are made "in the image of God"? A lot of theological ink has been spilled on this one. Generally—and I will go with this here—"image" focuses in on our rationality. . . . At any rate, we humans are able to think and reason, to be self-aware, to have memories that we can articulate, and so forth. This is also an essential part of God's nature, and inasmuch as we are made in God's image it speaks of these sorts of things. Where does morality fit into all of this? The obvious answer is that it is all part and parcel of the image. . . . Recognizing that there is a huge amount of theology that needs unpacking here, about the will of God and the binding nature of moral rules, we can surely say that in some sense God is a moral being and so also are we humans. This is only in a sense, of course, because there is at least one influential strand of theological thinking that claims that God is beyond or above morality—he could never be subject to external norms. This thinking comes through in places like the book of Job, where God tells man that he, God, makes the rules. "Shall he that contendeth with the Almighty instruct him?" (Job 40:2 KJV). But whether making or following, God is good. The tricky part is that God is a supremely good being—God would never do wrong—whereas it is an essential part of Christian theology that because of the fall we are anything but supremely good beings. We know what we should do, but we don't always do it.

One way around this (because of Irenaeus of Lyons) is to distinguish being made in God's likeness from being made in his image, with the former referring to our original unfallen state when we did and thought only good. I think it enough for us here to say that the important thing is that humans have a moral sense in somewise, unlike, say, lions and tigers, and that this reflects God in important respects.

Does Evolution Go Anywhere?

We must produce humans. Rather less cryptically, because obviously humans have been produced, we must show that the arrival of humans was no chance. . . . It is not enough to pull some fancy Thomistic shoe shuffle about primary and secondary causation, saying that theologically we need humans—this is primary causation and we know that God being all powerful can do this—and scientifically this is not his problem—this is secondary causation and it is up to nature to find a solution. We must ask: "What solution?" We cannot have God left empty handed by the evolutionary process. But is this possible? I have mentioned the history of life and let me flesh this out a bit, because it does seem promising.

First, life is approximately or, rather, more than 3.5 billion years old. It was primitive, and more complex cells had to wait until about 2 billion years ago. Coming down, less than a billion years ago things started to pick up. The so-called Cambrian explosion didn't occur until more than half a billion years ago, and it was then that we started to get the forms that were the ancestors of organisms living today. Increasingly, however, the pre-Cambrian is being uncovered and increasingly it is clear that the organisms then were of a kind that one would expect would explode into being in the Cambrian. On the other hand, there are prevertebrates. After that, it all seems rather plain sailing. In the animal world—there is a similar story for the plant world—we have fish, amphibians, reptiles, dinosaurs (actually fancy reptiles), birds, mammals, monkeys, apes, humans.

Today, thanks both to paleontology and to molecular biology, the human lineage is well known (Ruse 2012). Living in Africa, we got up on our hind legs about five or less million years ago, left the jungles to live on the plains, and increasingly turned to activities that required bigger and better brains. At some points in the past half million years or so, gangs of humans left Africa and started to spread around the

EVOLUTIONARY DIRECTIONALITY: NO DIRECTION TO EVOLUTION

world. For a while, it was thought—a bit like the soldier with the tinderbox who kept emptying his pockets every time he came across better booty—each new outward migration spelled doom for those who had gone before. But now, thanks to fabulous DNA studies, we know that there was some interbreeding. Don't run down those Neanderthals. Grandma and Grandpa sometimes had wandering eyes in that direction, and it wasn't just the eyes that wandered.

All of this sounds like a smooth and predetermined progression up to Homo sapiens. Early evolutionists thought so. Charles Darwin's grandfather, Erasmus Darwin, was given to expressing his thoughts in (rather bad) verse, and he had no doubts that humans are the necessary end point, monad to man, as they used to say.

> Imperious man, who rules the bestial crowd,
> Of language, reason, and reflection proud,
> With brow erect who scorns this earthy sod,
> And styles himself the image of his God;
> Arose from rudiments of form and sense,
> An *embryon* point, or microscopic *ens*!
> (E. Wilson 1803, 1:11)

Although he was no Christian, Erasmus Darwin was a committed deist, thinking that God started everything off and then let it unfurl through unbroken law. As noted, this was probably the position of his grandson Charles through most of his life, and although in his last decade or so he moved toward agnosticism, he remained committed to the progress of the evolutionary process from the first primitive forms of life to the apotheosis of history, Homo sapiens—actually, more precisely, upper-middle-class Homo sapiens from a small island off the coast of Europe. Remember the final words of the *Origin*:

> Thus, from the war of nature, from famine and death, the most exalted object which we are capable of conceiving, namely, the production of the higher animals, directly follows. There is grandeur in this view of life, with its several powers, having been originally breathed into a few forms or into one; and that, whilst this planet has gone cycling on according to the fixed law of gravity, from so simple a beginning endless forms most beautiful and most wonderful have been, and are being, evolved. (C. Darwin 1859, 490)

But here's the rub! The Darwinian process seems to knock the stuffing out of the inevitable climb to humankind. There is no necessary high point to the process, and indeed Darwin warned himself about such thinking. In the margin of his copy of an early evolutionary tract by the Scottish publisher Robert Chambers (*Vestiges of the Natural History of Creation*) Darwin scribbled, "Never use the word higher & lower—use more complicated." Why is this? There are two reasons. First, although natural selection is no tautology—those that survive are those that survive—it is relativistic. What succeeds in one case may well not succeed in another. There is simply no good reason to think that large brains and intelligence are always better than any alternatives. In the immortal words of the late Jack Sepkoski (one of the leading paleontologists of his day), "I see intelligence as just one of a variety of adaptations among tetrapods for survival. Running fast in a herd while being as dumb as shit, I think, is a very good adaptation for survival" (Ruse 1996b, 486). Second, as we saw in the previous chapter, the raw building blocks of evolution, the variations on which selection works, are random, not in the sense of being uncaused, but in the sense of not appearing according to need. They are certainly not directed toward the production of human beings. Evolution through selection is opportunistic, not directed. And if further confirmation is needed, there is the case of the "hobbit," *Homo floresiensis*, discovered in 2003 on one of the islands of Indonesia. About three and a half feet tall with a brain the size of chimpanzees' (about four hundred cc), it nevertheless showed advanced, toolmaking behaviors. It went extinct only about twelve thousand years ago, and so it might plausibly have survived to the present. What price inevitable progress were it around, or had it alone survived?

Causes

Expectedly, in the light of this challenge, there has been a century and a half of effort trying to show that progress to humankind can in fact be expected given evolution through selection. Darwin was always stone-cold certain that selection applies to humans and not just to us physically. Indeed, in 1838, in his private notebooks, humans are the focus of the first unambiguous articulation of natural selection. "An habitual action must some way affect the brain in a manner which can be transmitted.—this is analogous to a blacksmith having children with strong arms.—The other principle of those children. Which chance? produced with strong arms, outliving the weaker one, may be applicable to the formation of instincts, independently of habits.—?" (C. Darwin 1987, N42). But will we win? Darwin started one of the major optimistic strategies, suggesting that the best will win out and we are the best.

> If we look at the differentiation and specialization of the several organs of each being when adult (and this will include the advancement of the brain for intellectual purposes) as the best standard of highness of organization, natural selection clearly leads towards highness; for all physiologists admit that the specialization of organs, inasmuch as they perform in this state their functions better, is an advantage to each being; and hence the accumulation of variations tending towards specialization is within the scope of natural selection. (C. Darwin 1861, 134)

This kind of thinking has found supporters, right down to the present day. Julian Huxley, the grandson of Thomas Henry Huxley (and the older brother of the novelist Aldous Huxley), put the idea in military garb, suggesting that there are biological "arms races," with lines of evolving organisms competing against each other (1912). Just as with battleships—Huxley was writing just before World War I, when Britain and Germany were competing by building ever larger naval behemoths—where every increase in gunpower is matched by an increase in armor, so we find that every adaptation of one line (great speed, thicker shell) is matched by an adaptation of the other line (even greater speed, stronger boring power). Richard Dawkins is a great enthusiast for evolutionary progress. "Directionalist common sense surely wins on the very long time scale: once there was only blue-green slime and now there are sharp-eyed metazoa" (Dawkins and Krebs 1979, 508). In support of his intuition, he has wholeheartedly embraced arms race–type thinking, arguing that in military races we have seen a move to ever greater reliance on electronic equipment and that this has implications for the world of organisms (Dawkins 1986). Those beings with the biggest onboard computers are at the top. The fact that we are twenty-three times brighter than your average hippopotamus may not tell you everything, but it does tell you "something."

I don't know whether you regard this line of thinking as flawed or merely wishful, but either way I don't see humans guaranteed. Big brains really are expensive to maintain. You need ongoing supplies of protein, meaning the bodies of other animals, and one can think of 101 reasons why these might not be readily available. Again and again the winner in battles goes with something simple, cheap, and reliable. In World War II, German tanks at their best were terrific; but the Nazi regime never really twigged on to mass manufacture, using just a few models with interchangeable parts. It made all the difference on the Allied side. Perhaps humans are likely, but Christians need more than this. Faced with this failure, other Darwinians have tried an approach that makes much of ecological niches. This was a favorite of the late Stephen Jay Gould (1985) and has been taken up with gusto by Simon Conway Morris (2003), justly famed for groundbreaking work on the fossils of the Burgess Shale. Conway Morris notes that there are many instances of evolutionary convergence, meaning that organisms that are unrelated sometimes take similar paths. There are, for instance, unrelated species of saber-toothed tiger, indeed, some being placental mammals and others being marsupial. This suggests that there is, independent of organisms themselves, a niche for predators with the kinds of shearing teeth that the saber-tooth possesses. It is just a question of an evolving line finding that niche and moving in. Perhaps, then, there is a niche for intelligence and it was wanting to be found, as predictably happened.

> If brains can get big independently and provide a neural machine capable of handling a highly complex environment, then perhaps there are other parallels, other convergences that drive some groups towards complexity. Could the story of sensory perception be one clue that, given time, evolution will inevitably lead not only to the emergence of such properties as intelligence, but also to other complexities, such as, say, agriculture and culture, that we tend to regard as the prerogative of the human? We may be unique, but paradoxically those properties that define our uniqueness can still be inherent in the evolutionary process. In other words, if we humans had not evolved then something more-or-less identical would have emerged sooner or later. (Conway Morris 2003, 196)

This is a nice idea, but again, probably more hopeful than established. Many ecologists and evolutionists question the basic assumption that niches just wait there objectively to be found. They argue, rather, that niches are created as much as anything. There was, for instance, no niche for high-flying insects until trees evolved. Now the niche is filled and busy, but it hardly existed at all at one point. The same with culture. It is agreed that such a niche now exists, but it was not necessary that it existed at all before we came along. Contradicting himself on these matters, it is Gould who stresses the contingency of things. Making a joking reference to the asteroid that hit the Earth sixty-five million years ago and wiped out the dinosaurs, making possible the age of mammals, Gould wrote, "Since dinosaurs were not moving toward markedly larger brains, and since such a prospect may lie outside the capabilities of reptilian design . . . we must assume that consciousness would not have evolved on our planet if a cosmic catastrophe had not claimed the dinosaurs as victims. In an entirely literal sense, we owe our existence, as large and reasoning mammals, to our lucky stars" (Gould 1989, 318).

In any case, theologically, the problems are horrendous. If God is willing to get involved occasionally, why not on every occasion? Every day children are dying in horrendous pain because of random mutations. Why does God not prevent these? Whether or not one thinks that God should have prevented them anyway, one can at least understand why he doesn't prevent them in individual cases if he has a general policy of not intervening or at the least only intervening when salvation matters are at stake. Better to stay out of the directed variation business entirely. . . . Let me offer another suggestion. The Darwinian agrees that, however improbable, natural selection could produce humans because it has in fact produced humans. I take it that what this means is that no matter how unlikely, so long as you are prepared to roll the dice billions and billions of times, you are going to come up eventually with human beings, or, to help the odds a little, with beings that are rational and with a sense of morality. I don't think the Christian story cares too much if we all have green skin and are bald. Suppose the multiverse really is a possibility, billions and billions of other universes with no correlation to our own time and space. At some point you are going to get humans and the Christian story can get underway. It is true that God is now lumbered with billions and billions—less one—worthless universes, but remember that traditionally he is outside time and space so it is not as if he is sitting around in heaven waiting anxiously for humans to appear. In his eyes, a thousand years are as a day, and a day is as a thousand years (2 Pet 3:8).

Is God Too Complex?

Richard Dawkins complains that the more complex you make the universe, the more complex you make God because he has to keep up with his creation (2006). And the more complex you make him, the more improbable you make him. I am not sure that this is entirely so. The whole point of science and mathematics is that you start with the very simple—like Euclid's axioms—and quickly get a lot more complex and counter-intuitive—like Pythagoras's theorem. This said, however, it does start to seem that God has fewer and fewer dimensions of freedom. Perhaps he did not have to create through law, although if he did not, one wonders what kind of world one would have. Would humans have all of the marks of evolution, even if they were not so? This starts to sound like the world of the pre-Darwinian naturalist Philip Gosse, who suggested that God might have put the marks of the past on the world, although thereby he was deceiving us (1857). Not surprisingly, people were not happy with that suggestion. But if he did create by law, then it does seem that God was forced to be, shall we say, a little bit exuberant to achieve his ends. A Being who had to create billions and billions of universes to get his goal strikes me as a Being not entirely in control of things.

I am not concluding that Darwinian theory makes impossible Christian claims about life's direction and the appearance of humans, but I am saying that there are bigger difficulties than many realize.

7

Design, Telos, and Purpose in the Natural World

Chapter 7, Reading 1

The Argument from Design: A Brief History[1]

THE ARGUMENT FROM DESIGN for the existence of God—sometimes known as the teleological argument—claims that there are aspects of the world that cannot be explained except by reference to a creator. It is not a Christian argument as such, but it has been appropriated by Christians. Indeed, it forms one of the major pillars of the natural-theological approach to belief—that is, the approach that stresses reason, as opposed to the revealed theological approach that stresses faith and (in the case of Catholics) authority. This chapter is a very brief history of the argument from design, paying particular attention to the impact of Charles Darwin's theory of evolution through natural selection, as presented in his *Origin of Species*, published in 1859.

From the Greeks to Christianity

Note that here there is a two-stage argument. First, there is the claim that there is something special about the world that needs explaining—the fact of growth and development, in Plato's example. To use modern language, there is the claim that things exist for certain desired ends, that

---

1. Ruse 2003a. Reprinted with permission of Cambridge University Press.

there is something "teleological" about the world. Then, second, there is the claim that this special nature of the world needs a special kind of cause, namely, one dependent on intelligence or thinking.... For Plato, it was the second stage of the argument—the argument to design—that really mattered. He was not that interested in the world as such, and clearly thought that design could be inferred from the inorganic and the organic indifferently. His student Aristotle, who for part of his life was a working biologist, emphasized things rather differently. Although, in a classic discussion of causation, he argued that all things require understanding in terms of ends or plans—in terms of "final causes," to use his language—in fact it was in the organic world exclusively that he found what I am calling organic complexity. Aristotle asked: "What are the forces by which the hand or the body was fashioned into its shape?" A woodcarver (speaking of a model) might say that it was made as it is by tools such as an axe or an auger. But note that simply referring to the tools and their effects is not enough. One must bring in desired ends. The woodcarver "must state the reasons why he struck his blow in such a way as to effect this, and for the sake of what he did so; namely, that the piece of wood should develop eventually into this or that shape." Likewise, against the physiologists he argued that "the true method is to state what the characters are that distinguish the animal—to explain what it is and what are its qualities—and to deal after the same fashion with its several parts; in fact, to proceed in exactly the same way as we should do, were we dealing with the form of a couch" (Aristotle 1984b, 641a7–17, 997)....

Aristotle certainly believed in a god or gods, but these "unmoved movers" spend their time contemplating their own perfection, indifferent to human fate. For this reason, whereas Plato's teleology is sometimes spoken of as "external," meaning that the emphasis is on the designer, Aristotle's teleology is sometimes spoken of as "internal," meaning that the emphasis is on the way that the world—the organic world, particularly—seems to have an end-directed nature. Stones fall. Rivers run. Volcanoes erupt. But hands are for grasping. Eyes are for seeing. Teeth are for biting and chewing. Aristotle emphasizes the first part of the argument from design. Plato emphasizes the second part. And these different emphases show in the uses made of the argument from design in the two millennia following the great Greek philosophers. Someone like the physician Galen was interested in the argument to organized complexity. The hand, for instance, has fingers because "if the hand remained undivided, it would lay hold only on the things in contact with it that were of the same size that it happened to be itself, whereas, being subdivided into many members, it could easily grasp masses much larger than itself, and fasten accurately upon the smallest objects" (Galen 1968, 1:72). Someone like the great Christian thinker Augustine was interested in the argument to design.

The world itself, by the perfect order of its changes and motions, and by the great beauty of all things visible, proclaims by a kind of silent testimony of its own both that it has been created, and also that it could not have been made other than by a God ineffable and invisible in greatness, and ineffable and invisible in beauty (Augustine 1998, 452–53). As every student of philosophy and religion knows well, it was Saint Thomas Aquinas who put the official seal of approval on the argument from design, integrating it firmly within the Christian *Weltanschauung*, highlighting it as one of the five valid proofs for the existence of God.

The fifth way is taken from the governance of the world. We see that things that lack intelligence, such as natural bodies, act for an end, and this is evident from their acting always, or nearly always, in the same way, so as to obtain the best result. Hence it is plain that not fortuitously, but designedly do they (the things of this world) achieve their end. Then from this premise (equivalent of the argument to organization)—more claimed than defended—we move to the creator behind things (argument to design). "Now whatever lacks knowledge cannot move towards an end, unless it be directed by some being endowed with knowledge and ... intelligence; as the arrow is shot to its mark by the archer. Therefore some intelligent being exists by which all natural things are directed to their end; and this being we call God" (Aquinas 1952a, 1:26–27).

After the Reformation

Famous though this "Thomistic" argument has become, one should nevertheless note that for Aquinas (as for Augustine before him) natural theology could never take the primary place of revealed theology. Faith first, and then reason. It is not until the Reformation that one starts to see natural theology being promoted to the status of revealed theology. In a way, this is somewhat paradoxical. The great Reformers—Luther and Calvin, particularly—had in some respects less time for natural theology than the Catholics from which they were breaking. One finds God by faith alone (*sola fide*), and one is guided to him by Scripture alone (*sola scriptura*). They were putting pressure on the second part of the argument from design. At the same time, scientists were putting pressure on the first part of the argument. Francis Bacon (1561–1626), the English philosopher of scientific theory and methodology, led the attack on Greek thinking, wittily likening final causes to vestal virgins: dedicated to God but barren (Blane 1819, 69)! He did not want to deny that God stands behind his design, but Bacon did want to keep this kind of thinking out of his science. The argument to complexity is not very useful in science; certainly the argument to complexity in the nonliving context is not useful in science. And whatever one might want to say about the argument to complexity for the living world, inferences from this to or for design (a Mind, that is) have no place in science. Harshly, Bacon judged: "For the handling of final causes mixed with the rest in physical inquiries, hath intercepted the severe and diligent inquiry of all real and physical causes, and given men the occasion to stay upon these satisfactory and specious causes, to the great arrest and prejudice of further discovery" (Bacon [1605] 2000, 119).

But there was another side, in England particularly. Caught in the sixteenth century between the Scylla of Catholicism on the Continent and the Charybdis of Calvinism at home, the central Protestants—the members of the Church of England, or the Anglicans—turned with some relief to natural theology as a middle way between the authority of the pope and the Catholic tradition and the authority of the Bible read in a Puritan fashion. . . . After Hume, how was Paley able to get away with it? More pertinently, after Hume, how did Paley manage to influence so many of his readers? Do logic and philosophy have so little effect? The philosopher Elliott Sober points to the answer (2000). Prima facie, Paley is offering an analogical argument. The world is like a machine. Machines have designers/makers. Hence, the world has a designer/maker. Hume had roughed this up by suggesting that the world is not much like a machine, and that even if it is, one cannot then argue to the kind of machine-maker/designer usually identified with the Christian God. But this is not really Paley's argument. He is offering what is known as an "inference to the best explanation." . . .

Charles Darwin

Darwin was not the first evolutionist. His grandfather Erasmus Darwin put forward ideas sympathetic to the transmutation of species at the end of the eighteenth century, and the Frenchman Jean-Baptiste de Lamarck did the same at the beginning of the nineteenth. But it was Charles Darwin who made the fact of evolution secure and who proposed the mechanism—natural selection—that is today generally considered by scientists to be the key factor behind the development of organisms (Ruse 1999b): a development by a slow natural process from a few simple forms, and perhaps indeed ultimately from inorganic substances. In the *Origin*, after first stressing the analogy between the world of the breeder and the world of nature, and after showing how much variation exists between organisms in the wild, Darwin was ready for the key inferences. First, an argument to the struggle for existence and, following on this, an argument to the mechanism of natural selection. . . .

With the mechanism in place, Darwin now turned to a general survey of the biological world, offering what the philosopher William Whewell had dubbed a "consilience of inductions" (1840). Each area was explained by evolution through natural selection, and in turn each area contributed to the support of the mechanism of evolution through natural selection. Geographical distribution (biogeography) was a triumph, as Darwin explained just why it is that one finds the various patterns of animal and plant life around the globe. Why, for instance, does one have the strange sorts of distributions and patterns that are exhibited by the Galápagos Archipelago and other island groups? It is simply that the founders of these isolated island denizens came by chance from the mainlands and, once established, started to evolve and diversify under the new selective pressures to which they were now subject. Embryology, likewise, was a particular point of pride for Darwin.

Why is it that the embryos of some different species are very similar—man and dog, for instance—whereas the adults are very different? Darwin argued that this follows from the fact that in the womb the selective forces on the two embryos would be very similar—they would not therefore be torn apart—whereas the selective forces on the two adults would be very different—they would be torn apart. Here, as always in his discussions of . . . evolution, Darwin turned to the analogy with the world of the breeders in order to clarify and support the point at hand. "Fanciers select their horses, dogs, and pigeons, for breeding, when they are nearly grown up: they are indifferent whether the desired qualities and structures have been acquired earlier or later in life, if the full-grown animal possesses them" (C. Darwin 1859, 446).

All of this led to that famous passage at the end of the *Origin*: "There is a grandeur in this view of life, with its several powers, having been originally breathed into a few forms or into one; and that, whilst this planet has gone cycling on according to the fixed law of gravity, from so simple a beginning endless forms, most beautiful and most wonderful have been, and are being, evolved" (C. Darwin 1859, 490). But what of the argument from design? What of organized complexity? What of the inference to design? Darwin's evolutionism impinged significantly on both of these stages of the main argument. With respect to organic complexity, at one level no one could have accepted it or have regarded it as a significant aspect of living nature more fully than Darwin.

To use Aristotle's language, no one could have bought into the idea of final cause more than the author of the *Origin of Species*. This was Darwin's starting point. He accepted completely that the eye is for seeing and the hand is for grasping. These are the adaptations that make life possible. And more than this, it is these adaptations that natural selection is supplied to explain. Organisms with good adaptations survive and reproduce. Organisms without such adaptations wither and die without issue. Darwin had read Paley and agreed completely about the distinctive nature of plants and animals. At another level, Darwin obviously pushed adaptive complexity sideways somewhat. It was very much part of his evolutionism that not everything works perfectly all of the time. And some features of the living world have little or no direct adaptive value. Homology, for instance—the isomorphisms between organisms of very different natures and lifestyles—is clearly a mark of common descent, but it has no direct utilitarian value. What end does it serve that there are similarities between the arm of humans, the forelimb of horses, the paw of moles, the flipper of seals, the wings of birds and bats? There is adaptive complexity, and it is very important. It is not universal....

To sum up: Darwin stressed the argument to adaptive complexity, even as he turned it to his own evolutionary ends. He made—or, if you prefer, claimed to make—the argument to design redundant by offering his own naturalistic solution, evolution through natural selection. In other words, Darwin endorsed internal teleology to the full. He pushed external teleology out of science. He did not, and did not claim to, destroy all religious belief, but he did think that his theory exacerbated the traditional problem (for the Christian) of design. How could a good God allow such a painful process of development?

From Darwin to Dawkins

Scientifically speaking, after the *Origin*, evolution was a great success. Natural selection was not. In certain respects, it was Herbert Spencer rather than Charles Darwin who came to epitomize the evolutionary perspective. Adaptation was downplayed, and selection was sidelined. In its stead, evolution was linked to the popular ideology of the day—progress, from blob to vertebrate, from ape to human, from primitive to civilized, from savage to Englishman. From what Spencer termed the uniform "homogenous" to what he termed the differentiated "heterogenous." ... It was not until the 1930s and the coming of Mendelian genetics (as generalized across populations) that Darwinism in the strict sense really took off as a functioning professional science (Ruse 1996b). ... In the 1960s and 1970s, the Darwinian program became, if anything, even more intense and successful as researchers turned increasingly to problems to do with behavior, especially social behavior as exhibited by such organisms as the hymenoptera (the ants, the bees, and the wasps).

The argument to adaptive complexity has thrived, although one should not think that ultra-Darwinians have always had it their own way. A sizeable minority of evolutionary biologists—the most notable being the late Stephen Jay Gould—has always warned against seeing function and purpose in every last feature of organic nature. Gould, particularly, argued that if one looks at the long-term results of evolution as shown in the fossil record ... one finds good reason to think that selection is but one player in the causal band (Gould 2002). He is well known for having formulated, with fellow paleontologist Niles Eldredge, the theory of punctuated equilibria, where change goes in jumps and where selection and adaptation have but lesser roles (Eldredge and Gould 1972, Ruse 1999c).

In recent years, those who warn against seeing too much (natural) design in nature have been joined by people working on development, trying to integrate this area of biological inquiry fully into the evolutionary picture (so-called "evo-devo"). The mantra of these enthusiasts is that of constraint. It is argued that nature is constrained in various ways—physical and biological—and that, hence, full and efficient adaptation is rarely, if ever, possible. Some of the most remarkable of recent findings are of homologies that exist at the micro (even the genetic) level between organisms of very different kinds (humans and fruit flies, for instance).... There is much more to the story. Any full history, from the years after Darwin down to the present, would need to extend the story from Protestants to Catholics, although in fact in this case probably (for all that the Catholic Church gives a special place to the philosophy of Saint Thomas) the tale is not so very different. Certainly, the most important reconciler of science and religion in the twentieth century—the French Jesuit paleontologist Pierre Teilhard de Chardin—made progress the backbone of his evolution-based natural theology (1955).

Again, any full history would need to acknowledge the fact that in the early twentieth century—thanks to Karl Barth, the greatest theologian of the century—natural theology as a very enterprise in itself took a severe beating. Barth complained, with some reason, that the God of the philosophers bears but scant resemblance to the God of the Gospels (1933). And although natural theology has recovered somewhat, there seems to be general recognition among theologians that old-fashioned approaches—supposedly proving God's existence beyond doubt—are no longer viable enterprises. What one must now aim for is a theology of nature, where the world informs and enriches belief, even though it cannot substitute for faith. If the god of the Bible is the creator of the universe, then it is not possible to understand fully or even appropriately the processes of nature without any reference to that God. If, on the contrary, nature can be appropriately understood without reference to the God of the Bible, then that God cannot be the creator of the universe, and consequently he cannot be truly God and be trusted as a source of moral teaching either (Pannenberg 1993).

Yet for all the reservations and qualifications—perhaps because of the reservations and qualifications—in major respects natural theology (or whatever you call it now), as represented by the argument from design, has not moved that significantly from where it was in the years after Darwin. On the one hand, we have those (in the tradition of Hodge) who want to maintain the good old-fashioned emphasis on organized complexity, on adaptation, and who think that this leads to a definitive proof of a creator—potentially, at least, the Christian creator—and who think that evolution by blind law (especially Darwinian evolution by natural selection) denies all of this. These are the people who have embraced, with some enthusiasm, the position known today as intelligent design

(Behe 1996, Dembski 1998a). The flip side to these believers are those (in the tradition of Darwin himself) like Richard Dawkins (1995) and the philosopher Dan Dennett (1995), who are no less ardent about the argument to organized complexity, yet who feel that Darwin made any further inference unnecessary and who feel that the problem of pain makes any further inference of a theological variety impossible. . . .

Then, on the other hand, we also have today the more mainstream Christians (in the tradition of Temple and Beecher), who accept evolution, who continue to be wary of natural selection, who downplay adaptation and play up progress, and who argue that this mix gives at least an adequate theology of nature, even if a traditional natural theology is no longer possible. Entirely typical is the Colorado State University professor Holmes Rolston III, a philosopher of the environment, an ordained Presbyterian, a trained physicist, and a recent Gifford lecturer. He takes as his text a line from one of my earlier books: "The secret of the organic world is evolution caused by natural selection working on small, undirected variation" (Ruse 2003b, 309). And he makes clear that (whatever today's scientists might say), although one must opt for science rather than miracles, a Darwinian approach is not adequate. The direct problem is that of design; but in a more important sense, the issue is one of direction.

The troublesome words are these: chance, accident, blind, struggle, violent, ruthless. Darwin exclaimed that the process was "clumsy, wasteful, blundering, low and horribly cruel" (Darwin 2013). None of these words has any intelligibility in it. They leave the world, and all the life rising out of it, a surd, absurd. The process is ungodly; it only simulates design. Aristotle found a balance of material, efficient, formal, and final causes; and this was congenial to monotheism. Newton's mechanistic nature pushed that theism toward deism. But now, after Darwin, nature is more of a jungle than a paradise, and this forbids any theism at all. True, evolutionary theory, being a science, explains only *how* things happened. But the character of this *how* seems to imply that there is no *why*. Darwin seems anti-theological, not merely nontheological (Rolston 1999, 91).

Continuing, Rolston focuses on the need for a mechanism of progress. We have to have some way in which nature guarantees order and upward direction. Apparently we need (and have) a kind of (highly non-Darwinian) rachet effect, whereby molecules spontaneously organize themselves into more complex configurations, which then get incorporated into the living being and thus move the living being up a notch, from which it seems there is no return....

This yields a kind of upward progress to life. For all that every "form of life does not trend upslope," finally, as expected, we move up to humankind, God's special creation (Rolston 1999, 116–17). At this point, Rolston (in fact echoing many earlier Christian progressionists) ties in the upward struggle of evolution with the upward struggle of humankind. What theologians once termed an established order of creation is rather a natural order that dynamically creates, an order for creating. The order and newer accounts both concur that living creatures now exist where once they did not. But the manner of their coming into being has to be reassessed. The notion of a Newtonian architect who from the outside designs his machines, borrowed by Paley for his watchmaker God, has to be replaced (at least in biology, if not also in physics) by a continuous creation, a developmental struggle in self-education, where the creatures through "experience" becomes increasingly "expert" at life. Do not be perturbed by this.

This increased autonomy, though it might first be thought uncaring, is not wholly unlike that self-finding that parents allow their children. It is a richer organic model of creation just because it is not architectural-mechanical. It accounts for the "hit-and-miss" aspects of evolution. Like a psychotherapist, God sets the context for self-actualizing. God allows persons to be imperfect in their struggle toward fuller lives . . . , and there seems to be a biological analogue of this. It is a part of, not a flaw in, the creative process (Rolston 1999, 131)....

I draw to the end of my survey of twenty-five hundred years of the argument from design. Deliberately, I have tried to be nonjudgmental, merely telling the story of the ideas as they appeared in history. But, as I conclude, I cannot resist drawing an obvious inference from my history. Intelligent design theorists and atheistical Darwinians cannot both be right, but they are both surely right in thinking that they are more in tune with modern evolutionary biology than are the mainstream reconcilers of science and religion. For all the qualifications, adaptation—organized complexity—is a central aspect of the living world, a central aspect of the work and attentions of the professional evolutionist today. Any adequate natural theology—or theology of nature—must start with this fact.

One must indeed cherish this fact and make it a strength of one's position rather than something to be acknowledged quickly and rather guiltily, and then ignored. I say this irrespective of related questions about whether or not progress is truly the theme of life's history, although I should note that its existence is a highly contentious . . . question in evolutionary circles. Highly contentious in theological circles also, for many believers deny the idea that unaided humans can in some sense improve their lot. Progress is seen to be in conflict with the Christian notion of providence, where God's unmerited Grace alone gives salvation. But these are topics for another discussion. Here, I simply leave you with the reflection that the argument from design has had a long and (I would say) honorable history, and that this history seems still to be unfinished.

Chapter 7, Reading 2

Two Thousand Years of Design[1]

Plato's Argument from Design

PLATO EXPRESSED HIS IDEAS in the form of dialogues, many purporting to be conversations between Socrates and his young male followers, aristocratic Athenians who would gather at Socrates's feet to listen, debate, and learn. The early dialogues are probably authentically Socratic—fairly accurate reports of actual discussions between Socrates and his followers and opponents. But at some point, although Plato still used the dialogue form, he started to insert his own ideas. Socrates himself probably formulated a version of the argument, but it was Plato's distinctive thinking on the problem of design in these later dialogues that was to influence posterity.

First some crucial background on Plato's beliefs about the nature of existence or being. Plato's ontological theory centers on the eternal Forms: patterns or templates representing universal paradigms, of which the things of this world are mere temporal copies or reflections. Belonging to some meta-world of ultimate rationality, shared only with the laws of mathematics, the Forms reach the peak of reality in the Form of the Good, which gives life and illumination. In our world, this Form is represented by the Sun. Since our domain of experience is only partly real—real only inasmuch as it "participates" in the world of the Forms—ours is

---

1. Ruse 2003d. Reprinted with permission of Harvard University Press.

the domain of becoming and decay, of change and time, of wrong as well as right. Yet despite these flaws, this domain of ours is not one of mere chance and chaos—we have order, as seen in the motions of the heavenly bodies. In the Platonic heavens (in some sense the true abode of human souls, which are themselves eternal), the stars forever trace their paths and patterns as they move in perfect, endless circles.

In Plato's example of growth, eating and drinking cause future growth and development, but because these are things we prize, it is proper to go further and say that eating exists for the purpose of bringing about growth and development. If we did not want to grow and develop and were content to stay six forever and ever, then we would not speak and think in terms of the purpose of eating. It would be enough to give the first kind of explanation about how eating and drinking cause growth. There would be no more call for further explanation than there is when we say that stubbing your toe causes the nail to turn black and fall off. You did not stub your toe so that the nail would fall off. You stubbed. It fell. That is all.

According to Plato, this is the problem we face when we try to explain causes. It may seem that we are facing a funny kind of causation, where cause and effect are reversed in time. The cause is the thing in the future, bringing about the effect in the present or past. The growth brings on the eating and drinking, and so forth. But this is not the essence of the kind of causation that we are considering now. Values are what really count, and where values come into play—things that people desire—we need a different kind of explanation, an explanation that refers to the ends and purposes of things. Wants and desires imply consciousness, intention; values imply a mind. The stone in the brook does not value the water flowing over it, but I do value growing big and strong. Here we reach the controversial part: Whose mind is it that puts everything in motion and orders things for their own future good? It is hardly our own minds—at least, it is hardly our minds once we look beyond our intentions and desires. We did not decree that eating would be of importance in achieving growth and maturity. It is rather, according to Plato, the Mind of God. Not necessarily the Jewish or Christian God, in the sense of one who created a world from nothing. But certainly a designer God, whose nature Plato nailed down in his later dialogue about the workings of the universe, the *Timaeus*. The detailed speculations of this work may or may not have been endorsed by Plato himself, but its general philosophy was surely congenial. . . .

Another term, the argument to complexity, founders, I think, on the various meanings of complexity. A thunderstorm might be thought complex, and yet it calls for no special understanding. Richard Dawkins speaks in terms of organized complexity or adaptive complexity (1983), yet even here we are hardly getting away from anthropomorphic terms. Who did the organizing of organized complexity? Perhaps we would do better to say apparent organized complexity, and probably this is about as good as we are going to do. Apparent organized complexity or adaptive complexity can be abbreviated simply to order or complexity, so long as the context is understood. The important point here is that the first move in Plato's argument is to recognize the distinctive nature—I would call it the complexity—of certain things whose existence requires some kind of special explanation.

After the argument to complexity (the argument that moves us to a recognition that complexity exists), Plato's second move is from the complex, distinctive nature of things to an explanation of that distinctive nature. This is sometimes referred to as the argument from design—the move from acknowledgment of design toward acknowledgment of a designer of some sort. I am even more uncomfortable with this second term, because the move itself strikes me as almost trivial. If we all agree that literal design has occurred, there must, by definition, be a designer (although the nature of that designer is, of course, another matter). But has design actually occurred, just because complexity exists? That is the question that concerns us in this book—whether, having made the move to complexity, one is committed to making the subsequent move to design.

I prefer to retrieve the term "argument" to "design" for this second move, meaning an argument that gets us from recognizing the distinctive complexity of the world to acknowledging that this complexity entails design (and, by definition, a designer of some sort). Can one block this second move, either by blocking the first or by allowing the first but cutting off the second? That is, can we argue convincingly that there is no complexity? If not, can we argue convincingly that complexity exists, but that it does not imply design? The second part of the argument, the move from the complexity of the world to the mind behind it—the designing mind—was Plato's true obsession. He spent little time on the first part of the argument, the part that looks at the distinctive nature of the world and what makes it distinctive. Purpose in Plato's teleology was an external purpose—imposed upon things by outside factors, in this case the designer God.

Aristotle's Final Causes

Aristotle took so much from Plato, and yet he differed from him on many points—nowhere more so than over ideas about design and purpose in the universe. Aristotle's views came as part of his overall analysis of causation. He claimed that there are four different senses in which phenomena can be said to bring about, or to cause, or to be the causes of, other phenomena. The one of interest to us is the fourth, final causes, where things occur for the sake of desired goals. In Aristotle's *Physics*, just as in Plato, we find human intentionality: we ourselves do something, or we make instruments to do something, with our own ends in view. But Aristotle was a practicing biologist for part of his life, and, although he used a human model to explain what he was about, in his biological discussions he introduced final causes without direct reference to intentionality at all.

Criticizing the atomists' belief that everything just happens by chance and that there is no need for end-related thinking, Aristotle drew an analogy between a piece of furniture and the features, or characteristics, of an organism. Just as we think of furniture as made by certain actions of a craftsman for some end, so also we should think of the organic part as made by certain actions for some end. Distinguishing a model hand . . . from a real hand, and criticizing physiologists who think that all they need to do is to refer to the immediate causes of features, Aristotle chided: "What are the forces by which the hand or the body was fashioned into its shape?" A woodcarver (speaking of a model) might say that it was made as it is by an axe or an auger. But simply referring to the tools and their effects is not enough. One must bring in ends. The woodcarver "must state the reasons why he struck his blow in such a way as to effect this, and for the sake of what he did so; namely, that the piece of wood should develop eventually into this or that shape." Likewise against the physiologists, "The true method is to state what the characters are that distinguish the animal—to explain what it is and what are its qualities—and to deal after the same fashion with its several parts; in fact, to proceed in exactly the same way as we should do, were we dealing with the form of a couch" (Aristotle 1984b, 641a7–17, 997).

We see that Aristotle had in mind the model of a craftsman, as did Plato, but his was not an argument intended to prove a designing mind. For him, the end direction, the purpose, was more naturalistic—it is more part of the way that nature works. His emphasis was on the argument to complexity rather than the argument to design. He stressed the end-directed nature of the things under discussion. But what about the designer, the obsession of Plato? What about the argument to design? Does it make sense to talk of interests or wants without a consciousness behind them? Can we have values without a valuer? Clearly, in a sense, Aristotle thought we can. The good or the final cause is the well-being of the individual organism. An organic feature exists for its possessor's wellbeing. It has its value in this sense, rather than in any value desired and imposed by an external being. An organic feature is part of the nature of things. It is constitutive of objects. It belongs to their ontology. In this sense, Aristotle's explanation of final causes was internal as opposed to the external teleology of Plato.

It is for this reason also that, despite his general discussion in the *Physics*, Aristotle's philosophy of purpose and design focused much more directly on the world of organisms. Indeed, some have argued that his teleology was focused exclusively on organisms. Whereas Plato saw a purpose in the whole universe, Aristotle worked at the individual, physical level. Inanimate objects seem not to have a purpose or an end—in the *Meteorology*, for instance, there is no mention whatsoever of final causes. Organisms, however, do have ends or final causes in his system.

Not that Aristotle was entirely naturalistic in the sense of thinking that final causes have no metaphysical underpinnings. You cannot simply drop consciousness out of the picture and then believe that value-based thinking remains untouched. Ultimately, Aristotle thought that the overall picture is one of parts functioning for the sake of wholes (organisms), which in turn are able to flourish and reproduce and thus participate in the eternal, which for Aristotle is identified with the deity or the divine. None of this is terribly helpful or consoling, however, because Aristotle's ultimate deity—the unmoved mover—has no knowledge of us and spends its time contemplating its own perfection. What is helpful and consoling is that Aristotle has put in place another part of the puzzle about purpose and design. Plato had explicated the value component and had seen that this ties in somehow with mind and forethought. Aristotle—a hands-on biologist as well as a philosopher—saw that a designing mind as such can have no place in science. Yet, we cannot do without purposelike understanding, at least in the realm of organisms. In some way, this purpose or design in the living world must be understood right down at the level of the individual, with all its adaptations. Aristotle left much labor for others to do in reconciling the external and the internal, but the groundwork was laid. More provocatively, one might say that the paradox was proposed. Thinking in terms of ends means thinking in terms of values, and values imply a consciousness. Yet science has no place for such a consciousness. What is one to do?

The Christians

As we leave Athens for Jerusalem, we must move the clock forward, past the birth of Jesus and into the Christian era. But as we do, we should note that Greek thought did not die or go absent. Indeed, as the power center of the ancient world moved west toward Rome, we find that final cause thinking—both the argument to complexity and the argument to design, which taken together I prefer to call the argument from design—was picked up, cherished, and elaborated. The great Latin orator Cicero (106–43 BC) argued in his *De natura deorum* for the intricate nature of the living world, for the artifact or craftsman-produced nature of things like the eye, and concluded that such end- or purposelike complexity could not have come about blindly or by chance. Likewise, two centuries later, the highly influential anatomist and physiologist Galen (AD 129–ca. 200) took an explicitly Aristotelian attitude toward the living world. "Aristotle is right when he maintains that all animals have been fitly equipped with the best possible bodies, and he attempts to point out the skill employed in the construction of each one" (Galen 1968, 1:108). Following his own prescription, Galen looked at all anatomical parts from a final cause perspective. The hand, for instance, has fingers because "if the hand remained undivided, it would lay hold only on the things in contact with it that were of the same size that it happened to be itself, whereas, being subdivided into many members, it could easily grasp masses much larger than itself, and fasten accurately upon the smallest objects" (Galen 1968, 1:72).

What about the Christian thinkers? Conveniently for our purposes, we can focus on their contributions for the first millennium and a half after Christ, breaking with them only when we reach the sixteenth century, with its Renaissance flowering of art and literature (much of it reaching back to the Greeks), the Reformation of the Catholic Church and the rise of Protestantism, and the Scientific Revolution, when Copernicus put the Sun at the center of the universe and a host of other scientists refined and extended his discoveries. The greatest theologians of the earlier era—the first fifteen hundred years of Christianity—were practitioners of a religion that grew out of Judaism. Yet, they were much influenced by the Greeks and struggled to synthesize the reasoned conclusions of the philosophers with the revealed truths of their Judeo-Christian faith. First there was Saint Augustine (354–430), much taken with the philosophy of Plato, and then eight centuries later there was Saint Thomas Aquinas (1225–74), who devoted his life to reconciling Christianity with the newly discovered works of Aristotle. Both Augustine and Aquinas saw in the insights of the ancients a way toward an appreciation of the existence and nature of the Christian God.

For both of these men the argument to design was the really important idea. The argument to complexity left them relatively unmoved; they were not practicing scientists, as Aristotle had been, or close to science and mathematics, as was Plato. They were theologians first and philosophers second. Thus, teasing apart the intricacies of the empirical world held little fascination. Whatever high rank these men gave "natural" theology (that is, reason-based argumentation for a divine designer), it necessarily played a secondary role to "revealed" understanding or belief. For Plato, the argument to design stood on its own; it led one to a belief in a god or gods. For Augustine and Aquinas, Christian revelation, particularly as presented in the Holy Bible, was the essential starting point for the believer.

For Aquinas, the real force of the argument from design—one of five famous proofs that he gave for the existence of God—was not really that it proved God's existence in the face of unbelief. Aquinas did not live in a society of skeptics and agnostics and atheists. Rather, as one who had argued strenuously against the argument that the very idea of God proves necessarily that God exists (the so-called ontological argument), Aquinas now needed the design argument to show that one could nevertheless reach a knowledge of God's existence through reason.

But not reason unaided. Reason that starts with the world of experience—experience of the entire world, not just the world of organisms. "The fifth way is taken from the governance of the world. We see that things that lack intelligence, such as natural bodies, act for an end, and this is evident from their acting always, or nearly always, in the same way, so as to obtain the best result. Hence it is plain that not fortuitously, but designedly do they achieve their end." Then from this premise (the argument to order), more claimed than defended, he moves to the creator behind things (the argument to design). "Now whatever lacks knowledge cannot move towards an end, unless it be directed by some being endowed with knowledge and intelligence; as the arrow is shot to its mark by the archer. Therefore some intelligent being exists by which all natural things are directed to their end; and this being we call God" (Aquinas 1952a, 1:26-27).

The Scientific Revolution

The great Christian thinkers of late antiquity and the medieval period absorbed the Greek concerns with ends—the argument to complexity and especially the argument to design. During the intellectual and social upheavals that occurred in the middle of the last millennium, these issues resurfaced with explosive force. The Protestant Reformation emphasized direct access to God by ordinary people—and not just learned individuals. One finds God by faith alone (*sola fide*), and one is guided to him by Scripture alone (*sola scriptura*). What need have we of the argument to design when it is enough to read the word of God and to open our hearts to his message? In a sense, reason can only get in the way of this direct channel.

Not just in the Protestant churches but in science also, enthusiasm for final causes started to decline. The world of Copernicus (who published his major work, *De Revolutionibus Orbium Caelestium*, in the year of his death, 1543) is a world of efficient cause, not final cause. We want to know what makes the planets go around the Sun, not what purpose the endless cycling serves. The metaphor of the world as an organism went into steep decline, and a new metaphor—the world as a machine, unthinking, uneventful, simply going blindly through the motions—started to take over. In the mid-seventeenth century the British chemist and general man of science Robert Boyle was a major spokesman for this vision. But he was not alone, for his attitude fit nicely with the views of the methodologists describing and prescribing the new ways of doing science.

Francis Bacon (1561–1626) had led the attack on Greek teleology, wittily likening final causes to vestal virgins: dedicated to God but barren (Blane 1819, 69). He did not want to deny that God stands behind his design, but he did want to keep this kind of thinking out of science. The argument to complexity is not very useful in science, especially in the nonliving context, Bacon asserted. And whatever one might want to say about the argument to complexity for the living world, inferences from this to a mindful designer have no place in science. "For the handling of final causes mixed with the rest in physical inquiries, hath intercepted the severe and diligent inquiry of all real and physical causes, and given men the occasion to stay upon these satisfactory and specious causes, to the great arrest and prejudice of further discovery" (Bacon [1605] 2000, 119).

The French physicist, mathematician, and philosopher René Descartes (1596–1650) felt much the same way. Introducing an argument that was as much theological as philosophical or scientific, he warned that "when dealing with natural things, we will, then, never derive any explanations from the purposes which God or nature may have had in view when creating them and we shall entirely banish from our philosophy the search for final causes. For we should not be so arrogant as to suppose that we can share in God's plans." Stick rather to efficient causes, "starting from the divine attributes which by God's will we have some knowledge of, we shall see, with the aid of our God-given natural light, what conclusions should be drawn concerning those effects which are apparent to our senses" (Descartes 1985, 202, principle 28). Wary of the power of the church as always (he was writing at the time of Galileo), Descartes warned that we should never presume to go against the truths of revelation. But ultimately, like Bacon, Descartes could see no place for the design argument in our scientific understanding of the world of experience. Leave it to the theologians and the philosophers to make the overall argument for design.

David Hume

What then were theologians and philosophers to make of final causes? By the end of the eighteenth century the answer would seem to be "not very much." The Scotsman David Hume (1711–76), once wittily described by a fellow countryman (David Brewster) as "God's greatest gift to the infidel," was an empiricist and a skeptic. He reduced all knowledge to sensation and then doubted that certain knowledge was ever attainable. The best consolation we have is that our psychology will not let us believe our philosophy, and so we are able to go about the business of our daily life. Sentiment and emotion drive us, and reason can never be anything other than the slave of the passions. Hume saw religion as little more than something to stave off the trials of life and fears of death and the unknown. He was himself no out-and-out atheist—absolute nonbelief, he thought, requires a commitment no less justified than Christianity—but he was deeply skeptical of any claims to belief. What then of the arguments for the existence of God: what in particular of the design argument? Hume's *Dialogues Concerning Natural Religion* is the most sustained attack ever penned against theology and religious belief of any kind. First, through one of the participants in the encounter, Cleanthes, Hume sets up the argument in its classic form. He starts by arguing for the special nature of the world (the argument to complexity), and from this goes on to infer that something creative stands behind it (the argument to design).

> Look round the world: contemplate the whole and every part of it: You will find it to be nothing but one great machine, subdivided into an infinite number of lesser machines, which again admit of subdivisions, to a degree beyond what human senses and faculties can trace and explain. All these various machines, and even their most minute parts are adjusted to each other with an accuracy, which ravishes into admiration all men, who have ever contemplated them. The curious adapting of means to ends, throughout all nature, resembles exactly, though it much exceeds, the productions of human contrivance; of human designs, thought, wisdom, and intelligence. Since therefore the effects resemble each other, we are led to infer, by all the rules of analogy, that the causes also resemble; and that the Author of Nature is somewhat similar to the mind of man; though possessed of much larger faculties, proportioned to the grandeur of the work, which he has executed. By this argument a posteriori, and by this argument alone, do we prove at once the existence of a Deity, and his similarity to human mind and intelligence. (Hume 1947, 115–16)

Then, through another of the participants, Philo, Hume knocks it all down like a house of cards. If we liken the world to a machine, he asks, then are we opening the way for multiple planners of machines and many prior less adequate worlds? "If we survey a ship, what an exalted idea must we form of the ingenuity of the carpenter, who framed so complicated, useful, and beautiful a machine? And what surprise must we feel, when we find him a stupid mechanic, who imitated others, and copied an art, which, through a long succession of ages, after multiplied trials, mistakes, corrections, deliberations, and controversies, had been gradually improving?" More generally:

> Many worlds might have been botched and bungled, throughout an eternity, ere this system was struck out: much labour lost: many fruitless trials made: and a slow, but continued improvement carried on during infinite ages in the art of world-making. In such subjects, who can determine, where the truth; nay, who can conjecture where the probability, lies, amidst a great number of hypotheses which may be proposed, and a still greater number which may be imagined. (Hume 1947, 140)

This is a counter to the second phase of the design argument, from complexity to a creator, that we might want to take seriously. Hume also went after the argument to complexity itself, which sets out to prove that something special exists that is in need of explanation. In Hume's opinion, we should be careful about making any such inference. We might question whether the world really does have marks of organized, adaptive complexity. For instance, is it like a machine or is it more like an animal or a vegetable, in which case the whole argument collapses into some kind of circularity or regression? It is certainly true that we seem to have a balance of nature, with one part's change affecting and being compensated by a change in another part, just as we have in organisms. But this seems to imply a kind of non-Christian pantheism. "The world, therefore, I infer, is an animal, and the Deity is the SOUL of the world, actuating it, and actuated by it" (Hume 1947, 143–44).

Unneeded in science, riddled with paradox in philosophy, and obstructive to genuine belief in religion—proving, if anything, the existence of just the kind of god one does not want to have around—teleological thinking seemed destined for the slag heap of discredited ideas, along with phlogiston, which was also making a deserved exit at the end of the eighteenth century. No good case had been made for something special about the world, no need to infer a creative intelligence behind everything was generally felt, and no way was seen to infer it, even if a need existed. After a run of more than two thousand years, understanding in terms of final causes, purpose, or design seemed to have come to a disastrous end.

Chapter 7, Reading 3

Design[1]

THERE IS NO STANDARD Christian position on the role of reason in religion. Catholics think that "natural theology" has a significant and full role to play: "Illumined by faith, reason is set free from the fragility and limitations deriving from the disobedience of sin and finds the strength required to rise to the knowledge of the Triune God" (John Paul II 1998, 43). While there are Protestants who accept and even welcome natural theology, the "neoorthodox" (like Barth) think not only that it fails but also that it is pernicious in its effects and promises. A true faith needs no proofs and indeed is destroyed by such proofs. Our radical freedom to accept God's gift of grace would be compromised were it possible to give logical proof of Christian claims. Obviously we must discuss the interaction of Darwinism and natural theology, but equally obviously the Christian's own stand will have to be considered in any overall assessment of these issues.

The Teleological Argument

Arguments for the existence of God lie at the heart of natural theology. Some such arguments touch but slightly or not at all on the Darwinian system. The "teleological argument" or the "argument from design,"

---

1. Ruse 2000b. Reprinted with permission of Cambridge University Press.

however, is right on the front line. Many people, Richard Dawkins most vocally recently, claim that here above all Darwinism and Christianity come into conflict, precluding belief in both systems. By going back in history, let us see why this opinion might be held.

Notwithstanding Hume's criticisms—pointing to conclusions that he himself was not prepared to accept in full—the argument from design flourished right through to the nineteenth century. Interestingly, its most important base was Protestant Britain rather than Catholic Europe, mainly because—given the nonprofessional status of British science as opposed to that found on the Continent, in France especially—British scientists had to work particularly hard to justify their activities to the outside (nonscientific) world (Appel 1987). Burnishing the argument from design was a perfect antidote to the worry that studying nature might put undue pressure on tenets of revealed religion. Its most famous formulation occurs in *Natural Theology*, by Archdeacon William Paley in 1802:

> I know of no better method of introducing so large a subject, than that of comparing a single thing with a single thing: an eye, for example, with a telescope. As far as the examination of the instrument goes, there is precisely the same proof that the eye was made for vision, as there is that the telescope was made for assisting it. They are made upon the same principles; both being adjusted to the laws by which the transmission and refraction of rays of light are regulated. (Paley 1802, 1)

A watch demands a watchmaker. Hence an eye demands an eye maker—or rather, an eye designer. Call this "God": the God of the Christian, moreover, since the eye and other organic characteristics attest to a designer of great skill and power. The popularity of this argument makes understandable one of the most important points about Darwinism: the author of the *Origin* accepted completely and utterly the initial premise of the teleological argument, namely that organisms are design-like (Ruse 1999b). Indeed, this is the problem to which natural selection speaks: the explanation of adaptations like the eye and the hand. It is here that Darwinism distinguishes itself from almost all other evolutionary theories. Darwin argued that, thanks to natural selection, we will have the formation or evolution of features like the hand and the eye, those very organs of which the natural theologians made so much. Darwin regarded the features as adaptations, as did the theologians. They were not just idle bodily parts or appendages, but things with a purpose or end or function. This is the reason that the *Origin* incorporates all of the teleological language of the theologians. If you like, put it this way: the metaphor of design is just as much a feature of Darwin's *Origin* as it is of Paley's *Natural Theology*.

Does Darwin Exclude Real Design?

Now what does all of this imply? Some people think that Darwin spelled the end to the argument from design. Before Darwin, one had no choice but to accept a designer. After Darwin, the designer was finished and the way was open for atheism.

> Paley's argument is made with passionate sincerity and is informed by the best biological scholarship of his day, but it is wrong, gloriously and utterly wrong. The analogy between the telescope and the eye, between watch and living organism, is false. All appearances to the contrary, the only watchmaker in nature is the blind forces of physics, albeit deployed in a very special way. A true watchmaker has foresight: he designs his cogs and springs, and plans their interconnections, with a future purpose in his mind's eye. Natural selection, the blind, unconscious, automatic process which Darwin discovered, and which we now know is the explanation for the existence and apparently purposeful form of all life, has no purpose in mind. It has no mind and no mind's eye. It does not plan for the future. It has no vision, no foresight, no sight at all. If it can be said to play the role of watchmaker in nature, it is the blind watchmaker. (Dawkins 1986, 5)

Because he did not know about evolution through selection, Hume hesitated before the final leap into nonbelief. Now such a leap is nigh obligated: "Although atheism might have been logically tenable before Darwin, Darwin made it possible to be an intellectually fulfilled atheist" (Dawkins 1986, 6). But surely the Christian has a counter to this? One might argue that although selection makes redundant—closes off the option of—an intervening and designing God, it still leaves open the option of God's designing at a distance. Perhaps God put his design into action through the medium of unbroken law. Indeed, as Baden Powell argued in the years just before the *Origin*, perhaps a God who works this way is superior to a God who has to intervene personally and miraculously: "Precisely in proportion as a fabric manufactured by machinery affords a higher proof of intellect than one produced by hand; so a world evolved by a long train of orderly disposed physical causes is a higher proof of Supreme intelligence than one in whose structure we can trace no indications of such progressive action" (Powell 1855, 272).

Dawkins will have none of this. He regards Darwinism not simply as proving that the argument from design does not work, but as proving that atheism is true. Natural selection explains adaptive complexity. God simply cannot do this, because apart from anything else, one would then have the burden of explaining God.

> Any God capable of intelligently designing something as complex as the DNA/protein replicating machine must have been at least as complex and organized as that machine itself. Far more so if we suppose him additionally capable of such advanced functions as listening to prayers and forgiving sins. To explain the origin of the DNA/protein machine by invoking a supernatural Designer is to explain precisely nothing, for it leaves unexplained the urging of the Designer. (Dawkins 1986, 141)

But are we not being a little unfair to Dawkins at this point, missing the real force of his argument? His basic objection is that whether you think that God designed through miraculous intervention or through the medium of natural selection, you are still leaving unexplained the very existence and nature of this wonderful God who is supposedly capable of doing all of this. Which point of course is true and in the opinion of many is a good reason for nonbelief. Ultimately, assuming the existence of God really solves and explains nothing. Yet this surely is a problem for Christian belief generally and not something brought on by Darwinism specifically. There are of course various responses one can make to the problem, which may or may not be judged adequate. For instance, traditionally, God is thought to exist necessarily, so the question of his beginnings is ruled irrelevant. To which critics object that the idea of necessary existence is a conceptual confusion. At which point we can pull back gracefully and let the disputants argue among themselves. Their premises have nothing to do with evolutionary theory. Dawkins has not shown that being a Darwinian denies, or even exacerbates the difficulties of, Christian commitment. In the spirit of Baden Powell, one might think that God's magnificence is confirmed as one realizes that he does so much with so simple a mechanism as natural selection.

Because of our Darwinism—confident that it can, that it must, explain all—might we not turn away from precisely those things which theologically are the most significant? This is the fear which underlies the thinking of biochemist Michael J. Behe, author of *Darwin's Black Box: The Biochemical Challenge to Evolution*, a man who believes that he has made a breakthrough where "the result is so unambiguous and so significant that it must be ranked as one of the greatest achievements in the history of science. The discovery rivals those of Newton and Einstein, Lavoisier and Schrodinger, Pasteur and Darwin" (Behe 1996, 232–33). Perhaps so, but moving to the arguments, let us see why he gives us reason to fear Darwinism. Behe's key notion is something he labels "irreducible complexity." Some organic phenomena are just so complex that they cannot have been produced by blind unguided law. That is just a fact of nature.

> By irreducibly complex I mean a single system composed of several well-matched, interacting parts that contribute to the basic function, wherein the removal of any one of the parts causes the system to effectively cease functioning. An irreducibly complex system cannot be produced directly (that is, by continuously improving the initial function, which continues to work by the same mechanism) by slight, successive modifications of a precursor system, because any precursor to an irreducibly complex system that is missing a part is by definition nonfunctional. (Behe 1996, 39)

Behe adds, surely truly, that an irreducibly complex biological system has to be a major challenge to a Darwinian mode of explanation. Darwinism insists on gradualism, and this is precisely what is not on offer. "Since natural selection can only choose systems that are already working, then if a biological system cannot be produced gradually it would have to arise as an integrated unit, in one fell swoop, for natural selection to having anything to act on" (Behe 1996, 39). Which essentially means that natural selection is redundant.

As a matter of fact, Behe does not want to rule out a natural origin for all irreducible complexities, but we learn that as the complexity rises, the likelihood of getting things by any indirect natural route "drops precipitously" (Behe 1996, 40). As a physical example of an irreducibly complex system, Behe instances a mousetrap: something with five parts (base, spring, hammer, and so forth), any one of which is individually necessary for the mousetrap's functioning. It could not have come into being naturally in one step, and it could not have come about gradually. Any part-piece would not function properly alone, and any part missing would mean failure of the whole. It had to be designed and made by a conscious being—a fact that is true also of organisms. "The purposeful arrangement of parts" is the name of the game (Behe 1996, 193).

Irreducible Complexity Challenged

Now what are we to say about this claim? Obviously, if Behe's overall argument is well taken, then Darwinism is in trouble and will surely strike back at Christianity. But are we to accept Behe? As it happens, Behe's choice of a mousetrap as an exemplar of intelligent design is somewhat unfortunate. All sorts of parts can be eliminated or twisted and adapted to other ends. There is no need to use a base, for example. You can just attach the units directly to the floor, a move that at once reduces the trap's components from five to four. But even if the mousetrap were a terrific example, it would hardly make Behe's point. No evolutionist ever claimed that all of the parts of a functioning organic feature had to be in place at once, nor did any evolutionist ever claim that a part used now for one end must always have had that function. Ends get changed, and something that was introduced for one purpose might well take on another purpose. It might be only later that the new purpose gets incorporated in such a way that it becomes essential.

Here, although Behe himself is in as much trouble in the realm of philosophical theology as he was in the realm of biological science, let us see how others try to haul him from the hole into which he has pitched himself. The mathematician-philosopher William Dembski (1998a, 1998b) recognizes that one must find some way to separate such things as mal-mutations from such things as highly complex functioning entities, else the whole new anti-Darwinian revival of the design argument (what its proponents call "intelligent design") comes crashing down. To this end, Dembski proposes something he calls an "explanatory filter." The essence of this idea is that you always explain things at the most economical or plausible level of understanding, and you only go on down to another level if the first level fails. So, faced with some (biological) phenomenon, you explain if you can through regular unbroken law. If that works, then the cheering can begin. Your job is finished. If it does not work, then you go to the next level: chance. If that works or is plausible, again your work is over. But if it does not work, then you must go on to another level: design. "To attribute an event to design is to say that it cannot plausibly be referred to either law or chance. In characterizing design as the set-theoretic complement of the disjunction law-or-chance, one therefore guarantees that these three modes of explanation will be mutually exclusive and exhaustive" (Dembski 1998a, 98).

The sad truth is that Behe is in the same boat as those physicists we dismissed earlier. He has offered us a freshened-up version of the old "God of the gaps" argument for the Deity's existence. A Supreme Being must be invoked to explain those phenomena for which I cannot offer a natural explanation. But such an argument proves only one's own ignorance and inadequacy. It tells us nothing of beings beyond science. In the words of the Christian theologian and martyr Dietrich Bonhoeffer: "We are to find God in what we know, not in what we don't know" (Bonhoeffer 1979, 311).

Darwinism Explaining Christianity

I conclude this chapter by considering an argument which goes the other way. Could it not be that the Darwinian approach to function and design really does prove too powerful for the Christian? Could it not be that Darwinism shows that religion itself is just a part of the adaptive design of human nature, and that once we recognize this it will be seen that religion, including Christianity, falls to the ground?

This is certainly the position of Edward O. Wilson. Wilson does not want to belittle religion in the fashion of Dawkins. He sees it as an important and significant aspect of human culture. But he wants to turn precisely this importance and significance back on itself. For him, religion exists purely by the grace of natural selection. Those organisms which have religion survive and reproduce better than those which do not. Religion gives ethical commandments, which are important for group living. Also, religion confers a kind of group cohesion, something which is a very important element of Wilson's picture of humankind. "A kind of cultural Darwinism ... operates during the competitions among sects in the evolution of more advanced religions. Those that gain adherents grow; those that cannot, disappear. Consequently religions are like other human institutions in that they evolve in directions that enhance the welfare of the practitioners" (E. Wilson 1978, 174–75). Although Wilson writes here about cultural evolution, in fact he thinks that religion is ingrained directly into our biology. Thanks to our genes, it is part of our innate nature. "The highest forms of religious practice, when examine more closely, can be seen to confer biological advantage. Above all they congeal identity" (E. Wilson 1978, 188).

Religious enthusiasm is part of the human condition. We can explain religion. We can never eliminate it. At best, we can promote biology as an alternative secular religion: "The final decisive edge enjoyed by scientific naturalism will come from its capacity to explain traditional religion, its chief competition, as a wholly material phenomenon. Theology is not likely to survive as an independent intellectual discipline" (E. Wilson 1978, 192).

Explaining Religion Away?

Wilson's writings are rooted as much in his own childhood experiences of Baptist fundamentalism in the American South, as in any knowledge or study of empirical reality (Ruse 1999). But, taking his position at face value, let us ask about its implications for Christianity. In Wilson's own mind, what is happening is that Darwinism is explaining religion (including Christianity) as a kind of illusion: an illusion that is necessary for efficient survival and reproduction. Once this explanation has been put in place and the illusion exposed, one can see that Christianity has no reflection in reality. In other words, epistemologically one ought to be an atheist. Since Wilson still sees an emotive and social power in religion, he would replace spiritual religion with some kind of secular religion. That secular religion, as it turns out, happens to be Darwinian evolutionism. A Darwinian cannot be a Christian, but a Darwinian should be a Darwinian! We are dealing with a "myth"; but, when all is said and done, "the evolutionary epic is probably the best myth we will ever have" (E. Wilson 1978, 201).

In short, Wilson's Darwinism in itself does not prove the inadequacy of Christian belief; rather, his Darwinism shows why one might have a Christian belief, if evolution be true. Try again. Could one not argue that Darwinism shows that there is something wrong with religion, since Darwinism is indifferent as to the form of religious belief? . . . Religion, however, might be effective in achieving group cohesion, even though it takes on very different forms: monotheism, polytheism, animism, and so forth. All of which suggests that, given this range of biologically adequate options, Darwinism is more corrosive of religious belief than one suspected at first.

The conclusion is clear. Christians surely ought to consider seriously the empirical claims that Wilson and fellow thinkers are making about their religion. The theological implications being extracted are another matter. No sound argument has been mounted showing that Darwinism implies atheism. The atheism is being smuggled in, and then given an evolutionary gloss. This is no good reason for giving a negative answer to our title question.

Chapter 7, Reading 4

Darwin and Design: Darwin Destroys Design[1]

IN A WAY, THIS is the key chapter in our debate. We are going to be raising fundamental questions about the nature of organisms. Note, however, that neither Michael Peterson nor Michael Ruse is a biologist. Although they both take the science seriously, and we lay ourselves right open to criticism if we get the science wrong, they don't even pretend to be biologists. One nice thing about our engagement is that, as Ruse explained earlier, they are not arguing about the science. Without being too overwhelmingly condescending, Ruse suspects that both Peterson and himself feel rather smugly satisfied that unlike some of our good friends—those more in the center of things, like the Unitarians—they don't feel the need to argue about which parts of science are right and which are wrong. That, for them, is not their job and not really their prime interest—and certainly not their prime need. They look at things from a theological and philosophical perspective, and don't feel the need to slip in solutions through the key chapter in their debate. They raise fundamental questions about the nature of organisms. Note again, however, that neither Michael Peterson nor Michael Ruse is a biologist.

Following Malthus, at most, resources can increase only at an arithmetic rate (one, two three, four . . .). Hence, not all can survive or, what

1. Ruse 2017a. Reprinted with permission of Oxford University Press.

is even more important in the biological world, not all can reproduce. There is going to be a struggle, although as Darwin stressed, this struggle need not be actually physically violent:

> I should premise that I use the term Struggle for Existence in a large and metaphorical sense, including dependence of one being on another, and including (which is more important) not only the life of the individual, but success in leaving progeny. Two canine animals in a time of dearth, may be truly said to struggle with each other which shall get food and live. But a plant on the edge of a desert is said to struggle for life against the drought, though more properly it should be said to be dependent on the moisture. A plant which annually produces a thousand seeds, of which on an average only one comes to maturity, may be more truly said to struggle with the plants of the same and other kinds which already clothe the ground (C. Darwin 1859, 62–63).

From this, Darwin goes on to natural selection. The success of organisms in the struggle will, on average, be a function of their different features. Over time, this will lead to change:

> Let it be borne in mind how infinitely complex and close-fitting are the mutual relations of all organic beings to each other and to their physical conditions of life. Can it, then, be thought improbable, seeing that variations useful to man have undoubtedly occurred, that other variations useful in some way to each being in the great and complex battle of life, should sometimes occur in the course of thousands of generations? If such do occur, can we doubt (remembering that many more individuals are born than can possibly survive) that individuals having any advantage, however slight, over others, would have the best chance of surviving and of procreating their kind? On the other hand, we may feel sure that any variation in the least degree injurious would be rigidly destroyed. This preservation of favorable variations and the rejection of injurious variations, I call Natural Selection (C. Darwin 1859, 80–81)

In one major respect, Darwin's theory was incomplete. For evolutionary change to occur, one needs a steady supply of new variations. His knowledge of animals and plants in domestication, combined with an eight-year systematic study of barnacles, had convinced Darwin that such variation does occur. He was also totally convinced that such variation does not occur to speak to the needs of organisms but in this sense is "random." But Darwin had little idea about why and where such variation occurs and, as important, little idea about how (without being swamped out by already existing forms) variation can gain a foothold and persist in a population. Supplying this part of the story had to wait until the twentieth century when work by Darwin's contemporary, the Moravian monk Gregor Mendel, was uncovered and biologists were rapidly able to put together a full and adequate theory of heredity. It was in the context of this theory, generalized to populations, that the Darwinian mechanism of selection came into its own and in the 1930s the amalgamated "synthetic theory of evolution" or "neo-Darwinism" was formulated. To use a somewhat hackneyed term, evolutionary studies now had its "paradigm," and this has held sway for the past seven or eight decades. Of course there have been advances and changes, many linked to the coming of molecular studies, but it is the neo-Darwinian approach that guides today's researches.

The Problem of Final Causes

The machine metaphor has triumphed. In the felicitous language of Richard Dawkins, "We are survival machines, but 'we' does not mean just people. It embraces all animals, plants, bacteria, and viruses." Continuing,

> We are all survival machines for the same kind of replicator—molecules called DNA—but there are many different ways of making a living in the world, and the replicators have built a vast range of machines to exploit them. A monkey is a machine which preserves genes up trees, a fish is a machine which preserves genes in the water; there is even a small worm which preserves genes in German beer mats. DNA works in mysterious ways. (Dawkins 1976, 22)

The story is a bit more interesting than this. (This is one of the reasons why Ruse loves working on evolutionary theory. It is always a bit more interesting.) Aristotle didn't think in terms of little men in the future reaching back to the present and altering things. But he did think it made sense to talk in terms of the future for present understanding. That is what the organic metaphor is all about. Of the newborn boy, you can ask about the function of the penis. One part of the answer will obviously refer to reproduction—something very much in the future and possibly even something that will never happen. He might become a Catholic priest!

But it makes sense to say that the boy has a penis because he will need it for reproduction. He has it to reproduce or for the purpose of reproduction. In the physical world after the Scientific Revolution, this kind of thinking—"final cause" thinking—was dropped. You just focus on the machine as something going around and around according to unbroken law. You don't ask about the function, the purpose of the moon. Or if you do, you are making a joke: "It is to light the way home for drunken philosophers." But everyone agreed that in the biological world this dropping simply isn't possible. The eye obviously does exist to see and the hand to grasp—now and in the future (Ruse 2003).

The point about Darwin's natural selection is that he spoke to this issue of final cause, realizing that if the machine metaphor were to triumph in the biological world—ultimately it is all just blind law cycling endlessly—he had to deal with final causes. One way might have been to ignore them or deny that they are really that significant. Robert Chambers ignored them, and Darwin's great supporter Thomas Henry Huxley always downplayed their significance. This was not Darwin's way. He believed in final causes. Natural selection does not just bring about change. It brings about change in the direction of adaptive complexity. Organisms that have well-functioning hands and eyes do better in the struggle than organisms that do not have such characteristics. Of course, this is not always the case. A falling rock might wipe out the better organism—what evolutionists call the "fitter" organism—but, on average, quality will tell. So we have a natural explanation of final causes.

Note, and this is important, that all of the final cause work is done by natural selection. The variations that selection works on are random, although not in the sense of being uncaused. Darwin always thought that there were causes, and today (especially with the coming of molecular biology) we know a lot about them. Rather, as Darwin stated without equivocation, variations are random in the sense of not occurring according to need. A new predator arrives, and the prey needs a new adaptation. Unless the prey already has something in its tool kit, it is probably going to be out of luck. New variations, what today we call *mutations*, rarely do anything to help their possessors with respect to their immediate needs. You might think that this in itself is something of a refutation of natural selection. If the right variations are so rare, how can an organism survive, let alone improve? To use an analogy, think of a class assignment where you are asked to write an essay on dictators, discussing one example in detail. The way most people think of the availability of usable variation is as if your only source material were provided by the Book of the Month club, that is to say a book every thirty days chosen by others without regard to your immediate needs. No doubt within, let us say, ten years the club would choose some book on Hitler or Stalin or Castro or some like person, but by the time it arrived on your doorstep, the course would be long over, and you would have failed! Note, and this is important, that all of the final cause work is done by natural selection. The variations that selection works on are random, although not in the sense of being uncaused.

Likewise in the favorite. Darwin himself always embraced it as a secondary mechanism. However, we now know that Lamarckism—incidentally not original with Lamarck and only a side cause for him—simply isn't true. Every now and then you see claims that it has been revived, but they always turn out to be some fairly trivial, short-term effect, if that. Another mechanism (if you can so call it) that is popular in some circles today is that the unaided laws of physics and chemistry can produce complex functioning—life, we are told, is "self-organizing" (Kauffman 1993). A popular example of such a phenomenon—cutely characterized as "order for free"—is phyllotaxis, meaning the organization of petals on plants and other fruits and seeds, including, famously, the pine cone. Non-Darwinians argue that it is simply a function of the mathematics of producing flowers and seeds (Goodwin 2001). The parts are produced in a sequential fashion and by the laws of nature they must follow a fixed pattern, a pattern, incidentally, that is governed by the so-called "Fibonacci series," made famous by the thriller *The Da Vinci Code* (D. Brown 2004).

Expectedly, Darwinians don't buy into alternatives to natural selection. With respect to self-organization, it is argued that at most it could cover only a fraction of the known adaptations of organisms and that investigation invariably shows that selection has been at work. No one denies that there will be constraints put on the operation of selection by the laws of physics and chemistry—you cannot make a cat the size of an elephant because whereas the legs increase linearly, the weight increases as a function of the cube—but this does not mean that selection is not at work. In the case of plants, it will be orienting petals in one way rather than another and so forth. It is not the laws of physics and chemistry *or* natural selection but the laws of physics and chemistry *and* natural selection.

This all said, there is still a worry here about the power of natural selection, and although he never broke faith either with selection or with the nondirected nature of new variations, Ruse is not sure that Darwin himself ever got on top of this problem. Fortunately, today's evolutionists have a ready answer, suggesting that concerns of this kind are misplaced. Most new variations don't just appear on the scene, as it were. Thanks to the Mendelian system of heredity, where two units (genes) work to produce characteristics (thanks to the paired system of chromosomes), even if new variations are deleterious, if they are generally masked by their mates (in geneticists' language, if they are recessive), they can be carried on and on in populations. This means that if they meet up with an identical mate (allele), they express their physical (phenotypic) characteristic. In other words, all of the time in a reasonably large population, features stored in the genotype are reappearing in the population and can be used if they are of worth or needed (Lewontin 1974). In fact, the situation is even better than this. Apart from the fact that most mutations tend to be recessive and hence not wiped out immediately (for the simple reason that most mutations stop some biochemical processes, and the mate is there to pick up the slack), selection often works to keep variations in populations. One common reason is that an organism with just one variant may do better than either of the two possible other organisms, those with no or two variants, and so selection works to keep the variants in the population. (This is known as "balanced heterozygote fitness.")

To return to our analogy, if you were dependent solely on the offerings of the Book of the Month club, you would indeed be in trouble. Suppose, however, you have a library at your disposal. If there were nothing on Hitler or Stalin or Castro on the shelves, and there probably would be, then surely there would be something on someone else or perhaps on someone that could be used for the purpose. Perhaps there would be something on Churchill and, although no one would say he was a dictator, one could write on his wartime powers and how they were dictatorial in one sense but not in another. You might even get an A for ingenuity! And the same is true in evolution. You might think that, with a new predator, a simple change of color to provide more effective camouflage would be best. But perhaps there is no variation speaking to color, but there is one that makes a move to a different ecological niche possible (let us say digestion of a hitherto inedible foodstuff), and with this move the prey does even better than it did before. It gets an A for ingenuity! Just as in class there is no one predetermined best answer, so in nature there is no one predetermined best answer. Life is graded on a curve. It is not how well you do absolutely, but how well you do compared to your fellows. With a large genetic tool kit at your disposal, you are at least in the game.

Does Darwinian Selection Refute Christianity?

Now turn the story the other way. Let's continue with the focus on final cause for a moment. You might want to say that as well as having the machine metaphor, Darwin brings back or retains the organic metaphor. Ruse is happy if you want to say that, although since presumably you are no longer looking at the inorganic world as an organism, he is not sure that we should speak of "metaphor" here since now everything seems literal. For himself, he would rather work entirely in terms of the machine metaphor. After all, machines do have ends—the automobile is for transport—it is just that in the inorganic world we drop that aspect of the metaphor as misleading or unhelpful. If Ruse says his love is a rose, he is probably not going to spend a great deal of time talking about her root system. He would say, therefore, that when it comes to organisms, we find it helpful to use the machine metaphor more broadly when trying to understand their parts—he can ask not only how the heart works but also what it is for—but then obviously when it comes to organisms as a whole, we have the truncated metaphor. What is the ultimate purpose of organisms? In the scientific world, they have no more purpose than the rivers of the world. Organisms, of course, do things and change the environment. But so also do rivers. It is just that neither is in place to do what it does—except in special cases that rather prove the general points, as when an ant uses an aphid to provide food for its young. Overall, however, it is just blind law in action.

So where now does this leave Christianity? Something important is being said here, because second only to the cosmological argument for the existence of God—if indeed second—is the argument from design. The organic world is as if designed, and the reason why it is as if designed is because it is designed—by God! Eyes are like telescopes. Telescopes have telescope designers and makers. Therefore, the eye has a designer and maker—the Great Optician in the Sky. After the Scientific Revolution, this type of argument became somewhat of a staple in English theological circles, as Peterson notes—partly because it fit in with the scientific temperament and partly for political reasons, as the Anglican Church trod a via media between the authority of Catholicism and the *sola scriptura* of the Calvinists. Probably the most famous expositor was Archdeacon Paley at the beginning of the nineteenth century (1802), and a major reason why Darwin took final cause so seriously is because he was soused in Paley as an undergraduate at Cambridge. But the argument in some form goes back to Plato in the *Phaedo*, if not before.

Although there is debate about the different formulations of the argument, whichever way, the Darwinian story has implications for it. Some think the argument is a straight analogy. The eye is like a telescope and so forth. If this is the case, then although you don't really need Darwinism to point this out for you—Paley himself takes note of it—a weakness in the argument comes if you can show that the organic world really doesn't seem all that well designed. Those of us growing old, says this writer with feeling, know only too well the troubles of aging eyesight. God might have spared a thought for weary, elderly readers. I think it fair to say that Darwinism picks up on this point and runs with it—and this, picking up from the point left dangling at the end of the previous chapter, is a big reason why Ruse is not inclined to give Michael Peterson even the deist God. Since everything is done by blind law, and since it is of the essence of Darwinism that you cannot go back to the beginning and start anew but must work with what you have, you are going to get an awful lot of what the English call Heath Robinson machines and what Americans call Rube Goldberg machines—ludicrous contraptions that do the simplest things in the most complex of ways. The late George Williams used to make much of the male reproductive system, where nothing seems to go from A to B in the quickest, most efficient manner, but is rather meandering and looped in almost bizarre ways (Williams 1966). How could something like this be the production of a good God who is also all powerful and all knowing? Even a trainee human physiologist could have done better.

Some think the argument from design is what is known as an "argument to the best explanation"—an approach Peterson favors. The general rational order of the world—a world in which biological life occurs—cannot be by chance. Remember Sherlock Holmes speaking to Dr. Watson: "You will not apply my precept," he said, shaking his head. "How often have I said to you that when you have eliminated the impossible, whatever remains, however improbable, must be the truth?" You are starting with the fact that organisms are highly intricate machines, exhibiting organized complexity. Despite the optimism of the order-for-free crew, you know that normally such complexity just doesn't happen. The world is ruled by Murphy's law—if something can go wrong, it will. Nicely functioning machines don't come through chance. They never get started and, if they do, they break down. So there must be some reason for the hand and the eye, and all else eliminated, it has to be God.

Although Hume was at most a deist and not a theist, general opinion is that he was not entirely indifferent to Paley's design argument. At the end of his *Dialogues Concerning Natural Religion* ([1779] 1947), having done the world's greatest hatchet job on any system at any time, somewhat sheepishly Hume admits that there might be something—or Something. Now, however, Darwinism steps forward, and you no longer need God. The design argument collapses. Assuming that the design-like nature of organisms is the one thing holding you from falling into nonbelief, that barrier has now been removed and you are on your way. In the words of Richard Dawkins, "Although atheism might have been logically tenable before Charles Darwin, Darwin made it possible to be an intellectually fulfilled atheist" (Dawkins 1986, 6).

So, What's the Answer?

Take note of where we are now. We have not disproven the existence of the Christian God. Even given the inadequacy of male plumbing, you can still speak of God as a creator. It is just that, assuming you want to stay with him as loving and don't think he spends his days laughing at poor old men with urinary problems, you have to agree that speaking of God as omnipotent—able to do anything—has to be circumscribed. Of course, you knew that already. God cannot make two plus two equal five, although Ruse thinks Descartes thought that might be a possibility. Now you have to recognize that there are a lot more things that God cannot do. In the same spirit, it may now be that you cannot force belief in God on people, but you can still speak of God as existing and even of God as a designer. We saw that at the time of the *Origin*, Darwin wanted to do this. Admittedly, by then Darwin was probably a deist and not a theist, but there are those who were undoubtedly theists who felt the same way. The great Catholic convert John Henry Newman wrote to a correspondent, "I believe in design because I believe in God; not in a God because I see design." This wasn't just a matter of being backed into a corner. He continued, "Design teaches me power, skill and goodness—not sanctity, not mercy, not a future judgement, which three are of the essence of religion" (Newman 1973, 97).

The main point is that Darwinism does have major implications for natural theology but that they are in a sense muted. If you are a creationist, thinking that God created everything miraculously, then Darwinism is an immediate threat. Ruse desires to leave it to others to fret about whether you ought to interpret the initial breath of life as in some sense referring to a miraculous intervention but simultaneously accept the subsequent creation of animals and plants as a natural phenomenon. Personally, although Ruse sees no contradiction in such a stance, it strikes him as a little odd. Ruse mentions the interesting point that today's creationists seem prepared to accept this in somewise, because they think that God created "types" and that after these left the ark they evolved rapidly (thanks to natural selection) into the forms we have today. (If you don't believe Ruse, go and see the excellent discussion of natural selection in the creationist museum in Northern Kentucky.)

At this point, the theist may be hugging him or herself—although Ruse insists to leave it to others to decide whether with joy or with relief. Going back to the disjunction expressed by Aubrey Moore, you might say that Darwinism was indeed a friend because now Christianity is freed from the tyranny of natural theology. Peterson will differ from Ruse and object to the use of the word "tyranny" or to a blanket statement about all natural theology. Here, he and Ruse will have to differ. Or rather, he and I plus many of the major Christian theologians will have to differ. Ruse suspects that someone like Aquinas might be able to live with the collapse of natural theology as the basis for God belief as much as Kierkegaard and Barth. Aquinas was a naturalist—his inspiration, Aristotle, Ruse suspects would have loved Darwin—and we have seen that Aquinas ever had a nuanced view of the proofs of reason. He and the saintly Newman would have found much common ground. Design is not being denied, it is just being reframed. However, do note the other side to Moore's disjunction. Dawkins is right. A nasty roadblock has been cleared from the route to atheism.

Chapter 7, Reading 5

Design as a Metaphor[1]

AT THE HEART OF modern evolutionary biology is the metaphor of design, and for this reason function-talk is appropriate. Organisms give the appearance of being designed, and thanks to Charles Darwin's discovery of natural selection we know why this is true. Natural selection produces artifact-like features, not by chance but because if they were not artifact-like they would not work and serve their possessors' needs.

Still, is it a concern that we have a metaphor here, a human-based metaphor? Are we not being unduly anthropomorphic? Remember Ernst Mayr's fourth worry: "The use of terms like purposive or goal-directed seemed to imply the transfer of human qualities, such as intent, purpose, planning, deliberation, or consciousness, to organic structures and to subhuman forms of life" (Mayr 1988, 40). Well, yes it does! But as Darwin pointed out, we use metaphors all the time in science, and many are based directly on human emotions and actions. Force, pressure, attraction, repulsion, work, charm, resistance are just a few of these many metaphors. Without metaphors of this kind, science would grind to a halt (to use a well-worn metaphor!). Darwin himself wrote: "The term 'natural selection' is in some respects a bad one, as it seems to imply conscious choice; but this will be disregarded after a little familiarity. No one objects to chemists speaking of 'elective affinity'; and certainly an acid has no more choice in combining with a base,

---

1. Ruse 2003c. Reprinted with permission of Harvard University Press.

than the conditions of life have in determining whether or not a new form be selected or preserved" (C. Darwin 1868, 1:6).

Frankly, I am not sure that Darwin ever entirely convinced himself on this issue. He kept scratching away at it. Certainly, he did not convince all of his contemporaries or subsequent evolutionists. The design metaphor—its putative existence and significance—is still a matter of much debate, so let us look at some of the issues.

The Metaphor of Design

Start with the most basic question. Does evolutionary biology really have at its heart the metaphor of design? Do evolutionists truly regard organisms as artifactual? Well, yes: we have the backing of history, we have the words of evolutionists themselves, and we have the examples of their work (remember the trilobite eye). Yet, some people, for instance the philosopher Colin Allen and the biologist Marc Bekoff, worry that there are dis-analogies between artifacts and organisms. Hence, they deny that we have a true metaphor. "Functional claims in biology are fully grounded in natural selection, and are not derivative of psychological notions such as design, intention, and purpose" (Allen and Bekoff 1995, 612). In the human case, they write: "Many people use natural objects (driftwood, seashells, heads of game animals, etc.) for decorating rooms and buildings. These objects are clearly not designed for that purpose (although they are presumably placed in strategic locations by design, in the sense of intent design). A rock on a desk may function as a paperweight, but unless the rock has had a flat base chiseled into it, or other similar modification, it is not appropriate to say that this object was designed for the purpose of holding down papers." Hence, "function does not entail design for that function" (Allen and Bekoff 1995, 614).

This is not a strong point, although it does hint at important facts about the use of concepts and metaphors. With respect to the use of rocks and like objects for certain human functions, while the rocks themselves are not necessarily designed, they are put by design in certain places in particular ways, after making choices (we would not choose a ten-ton rock for a paperweight). At the very least we have borderline cases, which, outside mathematics, is what we encounter in real life all of the time anyway. Whether we think that a rock used as a paperweight is an instance of design is just such a borderline case. As are the kinds of borderline cases that we find in biology.

Suppose a characteristic has been produced by selection for one particular end—it is designed for this end and has such a function. Suppose now that the characteristic starts to get exploited for other ends. Bones may have started their evolutionary careers as calcium banks, and only later were used for the rigid supports of vertebrates. Does one want to say that the bones were designed as banks and have (and can only have) the function of banks? Or does one want to allow that the bones may have now a new function without necessarily having been designed for such an end? I have a geranium planter in my back garden that started life as a beer barrel. It changed functions, and, whether you tie design into its every moment or whether you say that it has a function for which it was not designed, the change does not negate a connection between function and design. By using the barrel as a planter, I am starting to change not just the function but the design—especially when I paint the barrel in bright colors. By using the bones as supports, nature changed not just the function but the design of bones—especially when selection started to work on the support virtues of bones. . . .

George Williams's example of the male reproductive system in humans makes this point precisely. And before him, of course, Darwin himself stressed the improvised nature of organic design: "Throughout nature almost every part of each living being has probably served, in a slightly modified condition, for diverse purposes, and has acted in the living machinery of many ancient and distinct specific forms" (C. Darwin 1862, 348).

Goal-Directed Systems Again

Grant that the reason we think it appropriate to use functional language in evolution is the metaphor—the Paley/Darwin metaphor—of design. Organisms, produced by natural selection, have adaptations, and these give the appearance of being designed. This is not a chance thing or a miracle. If organisms did not seem to be designed, they would not work and hence would not survive and reproduce. But organisms do work, they do seem to be designed, and hence the design metaphor, with all the values and forward-looking, causal perspective it entails, seems appropriate.

Now, let us think back to goal-directed or directively organized systems. I would say that humans are the apotheosis of goal-directed systems and that this points to two interrelated notions of teleology in the human world. On the one hand, we have human intentions and goals—the things that we want for ourselves and others, and also the thoughts and actions that we have and make and take to achieve these ends and goals. On the other hand, we have the objects—the artifacts—that we design and make to realize those intentions. In a way, therefore, an artifact's purpose is conferred on it by humans in order to achieve their own ends. The knife itself hardly has the end of cutting. We have the end of cutting, and so we design and make the knife.

Frans de Waal's studies of caged chimpanzees and Jane Goodall's studies of chimps in the wild contain example after example of all sorts of elaborate, end-directed behavior. When males are obstructed from their goals by competition from other males, they enlist the help of females in alliances and the like. And since the females have their own goals, getting their help can require a great deal of manipulation and persuasion and sheer hard work. I see no reason why one should deny the label of "intentionality" to such cases as these. Moreover, I think one can speak of the manufacture or use of "artifacts" by these brutes—sticks for poking or reaching and the like. They are simple and crude compared with human artifacts, but they are artifacts nonetheless.

Wrong Focus

Once you have the metaphor of design in play, then of course you can ask questions about borderline instances and extensions and so forth. The real question, though, is whether, in the first place, the metaphor itself is an appropriate one. The question is not whether metaphors should ever be used at all but whether the specific metaphor of design should be used to explain evolution.

Darwinians argue strenuously that it must be used. Richard Dawkins speaks to precisely this issue, asking what job we expect an evolutionary theory to perform. Allowing that some people get excited about issues (such as the nature of species) that leave him cold, Dawkins agrees with John Maynard Smith that the "main task of any theory of evolution is to explain adaptive complexity, i.e. to explain the same set of facts which Paley used as evidence of a Creator." Jokingly he refers to himself as a "neo-Paleyist," concurring with the natural theologian "that adaptive complexity demands a very special kind of explanation: either a Designer as Paley taught, or something such as natural selection that does the job of a designer. Indeed, adaptive complexity is probably the best diagnostic of the presence of life itself" (Dawkins 1983, 404).

Formalists would beg to differ. They would raise the objection that adaptation and the associated design metaphor have their roots in British natural theology of the pre-Darwinian era, and they would ask again why, in the secular world of the twenty-first century, we should be bound by this retro thinking. At the very least, they would like to see the metaphor used a lot less frequently. This worry has been voiced explicitly by the philosopher Peter Godfrey-Smith, who wondered if our interest in design in nature might not be little more than a reflection of something about our psychology. He distinguished among three concepts: empirical adaptationism, the belief that everything is a result of selection and has adaptive value; explanatory adaptationism, the view that adaptationism is the really big question facing evolutionists; and methodological adaptationism, the belief that one should look for adaptations when one is faced with organisms, especially new and unfamiliar ones. The first and third concepts are relatively unproblematic. No one holds the first in a strict form, and few deny the third. It is the second, explanatory adaptationism, that raises interesting and difficult issues. . . .

Both sides of this debate about the design metaphor—speaking now specifically about explanatory adaptationism—seem to agree that an element of taste is at work here. An evolutionary biologist does not have to spend his or her time focused exclusively—or even chiefly—on adaptation. Other problems, like working out life's histories and the nature of variation and the reasons for speciation and more, compel. But this is true of any science. Some physicists like to work on galaxies, others on quarks. Yet, just as in physics, some issues in biology are fundamental, in that their solutions seem to have ramifications for much more. Working out the theory of forces has implications for so many fields in physics, and in a similar vein one might say that adaptation is just such an issue in biology.

The critic might respond that one has here a circular situation: Darwinians make searching for adaptation central to their program, and then when they find the adaptations they so fervently seek, they use them as support for Darwinism. But a better term than "circularity" might be "self-reinforcement." Darwinism is a successful theory—our scientific examples show that—and at the moment (and for the foreseeable future, whatever the qualifications) it is the only game in town, on its own merits. Fruit flies, dunnocks, dinosaurs, fig wasps—this is a theory on a roll. It has earned the right to set the agenda.

Adaptation is not the only evolutionary problem one can work on. A lot of the best evolutionary biologists today are working on development, where the functional issues are not foremost. But adaptation lurks in the background of most if not all evolutionary issues. Take the question of variation. Richard Lewontin has pointed out a huge amount of naturally occurring variation at the molecular level, as we have seen. In itself, one might say that these findings have little or nothing to do with adaptation. But in another sense, they have everything to do with adaptation. The search for variation was sparked by Dobzhansky's question about where the necessary fuel for natural selection could come from. Lewontin's finding of massive variation immediately sparked questions about whether it is held in populations through the workings of natural selection or whether it is the result of genetic drift or other nonadaptation-producing forces. These are adaptation issues. In physics today, we cannot get away from force. In biology today, we cannot get away from adaptation.

All of this is quite apart from questions about the extent to which the adaptation of organisms is ubiquitous and whether other putative causal factors should be getting more attention than they do. We have certainly seen no arguments that adaptation is nonexistent or that it is not widespread. More important, we have seen no arguments that adaptation is not a, if not the, fundamental issue. Dawkins is surely right that life is complex and that it does need explaining. If you are a biologist, you have made a commitment to explaining the complexity of life. And this holds true whether you are a Darwinian or not. So perhaps, until things start to grind to a halt, until the font of questions and answers dries up, perhaps we can put on hold Godfrey-Smith's worry that, from a broader historical perspective, the design problem has been "oversold." We have good scientific reasons for taking adaptation as the fundamental issue. Of course, Lewontin and his school do not much care for many of the findings of the adaptationists. But to say that we should not play the game at all, or that we should count all as equal, requires some persuasive arguments. Better than arguments would be examples. Let those who worry about explanatory adaptationism show their dunnocks and dinosaurs and fig wasps. When they demonstrate that they can do science that explains and predicts without invoking adaptation even implicitly, then we can start to take their position seriously.

A Return to Reduction

We come back to the worries that started this philosophical part of the discussion. If we appeal to or use the concept of purpose or end-directedness, are we not caught in some kind of unacceptable vitalism or finalism? Does all hope of a satisfying and legitimate reduction go out the window? Is the status of physics and chemistry forever barred to evolutionary biology? Certainly, the worst of these worries has long been laid to rest. At the ontological level—the level that raises questions of existence—nothing we have seen has given any hint of appeal to special forces or impetuses or the like. Looking at organisms as though they were designed does not mean that they have vital forces. Anything but. They were made by natural selection, in a good old-fashioned mechanical way.

But when we come to the theoretical level of reduction—the level that raises questions about the deducibility of biology from physics and chemistry—matters are rather different. Here is surely an impossible block. Our analysis of biological functionality does not imply that we have reverse causality. No one is saying that the future-focused trilobite eye causes the present lens. But the analysis does imply that we are thinking in reverse fashion about causes—understanding what is happening in terms of what we expect will happen. The lens is understood in terms of what we think it will do—because, causally, we have a repeating circular situation, where lenses have done in the past what we think this one will do in the future. Moreover, we have a value component, as we judge entities—the parts of organisms—in terms of the overall benefit or good of the organism. It is hard to see how any amount of deduction (which is what a theoretical reduction is all about) is going to eliminate or explain away these factors. We may perhaps talk about eliminating the metaphor of design, but we are not going to translate it into non-design-type language or understanding. Biology brings a different perspective to things, and that is a fact. Should we be looking, if not to the translation of teleology in biology into non-teleology, to its elimination altogether? I am not now thinking so much of those who dislike adaptationism and who would like to see end-directed thinking diminished by doing away with adaptationism but rather of those who are adaptationists and who nevertheless think talk of purpose or function should be done away with.

I refer now to Darwinians who are ashamed of what they take to be the weaknesses of their science. Responding to the Gould and Lewontin charges about overdoing the adaptationist strategy, Michael Ghiselin would have us "reject such teleology altogether." He would have us change the very questions we ask. "Instead of asking, What is good? we ask, What has happened? The new question does everything we could expect the old one to do, and a lot more besides" (Ghiselin 1983, 362–63). Less drastically, we have those like Ernst Mayr who would like to hide teleology by calling it a different name, "teleonomy." Some—philosophers mainly—would argue that since the obsession with ends comes in through the metaphor of design, a more mature science will emerge which drops the metaphor altogether and with it talk of purpose. "When you actually start to do the science, the metaphors drop out and the statistics take over" (Fodor 1996, 19–20). And even the philosopher Morton Beckner, a student of Nagel, likewise seemed to think that talk of purpose and design can go. "Even though teleological language cannot be translated away, there is a sense in which teleological language is eliminable. The sense is this: given any single case of an activity which is describable in teleological language, that case is also describable in non-teleological language. By this I mean that every observable aspect of the activity which in fact serves as the basis for our application of teleological concepts can also be described by means of a conceptual apparatus that is not teleological in character" (Beckner 1969, 162).

Of course, Aristotle, Kant, Cuvier, and Whewell would disagree with all of this. They would argue that if one is to think about organisms, one must think in terms of purpose and design. I am sympathetic to this older viewpoint, although I am wary of philosophical claims insisting that ways of thought must necessarily be as they have been. Perhaps, if making artifacts comes with being rational, then it follows that, as a matter of fact, serious thought about organisms will always be focused on end-directedness. But necessary or not, biology as we know it today would be dreadfully impoverished without a perspective that asks "what for"—and this includes molecular biology as much as traditional evolutionary biology, for the genetic code is as much part of the design metaphor as is the trilobite lens.

Indeed, for all that the thought of elimination might be possible, this point about impoverishment was appreciated by Beckner: "Suppose we are watching a tank in which there is a single anchovy, and that we introduce a barracuda into the tank. At first nothing happens; then, when it would be reasonable for us to suppose that the anchovy spots the barracuda, the anchovy behaves as follows: he turns sharply, swims quickly to the surface, leaps out of the water, reenters, and repeats the sequence." This description (call it B) is entirely nonteleological, which certainly says more but in another sense says very much less than the teleological statement (call it A): "On sighting the barracuda, the anchovy engaged in an escape reaction." There is no translation. As Beckner points out: "A says less than B since B offers details not mentioned in A. It says more, since calling the anchovy's behavior an 'escape reaction' implies that it serves the function of escape, whereas this is not implied by B" (Beckner 1969, 162–63).

And that really is the nub of things. If we want a biology that is not interested in the reason why the stegosaurus has such a funny display on its back, is not intrigued by the peculiar shape of the trilobite lens, does not care why some butterflies mimic other butterflies, is unconcerned about the spirals of the sunflower, then presumably something can be done. Some low-grade biological activities like classification, perhaps, can indeed be done better than usual if we refrain from asking why: as we have seen, one of the big complaints of taxonomists is that adaptation makes the job difficult and that assumptions about evolution only cloud the picture. I presume we could also do a certain amount of embryology and physiology and other sorts of biological activity without any questions about adaptation—the sorts of things for which we saw Lauder and Amundson promoting a capacity-type analysis of function. But it is all surely going to be very limited—a classic case of cutting off your scientific nose to spite your philosophical face. Dobzhansky's most famous claim was that nothing in biology makes sense except in the light of evolution—that is, Darwinian evolution. And for many evolutionary biologists active today, the search for causes is here to stay.

But, to return to the worry that has haunted evolutionists from Wallace to Mayr: Is not the very use of a metaphor a sign of weakness? Should not evolutionists be striving to avoid it? I am not sure why they should. Here I stand with Charles Darwin. Avoiding metaphor is not something that physicists and chemists feel called upon to do. Of course, the logical positivists would probably have had us avoid metaphor, but we need not be bound by their strange demands any more than we need be bound by Paleyian natural theology. Today, the consensus is that, far from being a sign of weakness, metaphors are an essential part of thought—including scientific thought. They are a prime force in causing people to think in new, imaginative, and fertile ways. They are a key ingredient in the heuristic side of thought, one of the most prized virtues of great science. Metaphors do not provide answers so much as they raise questions. Precisely because one thinks (metaphorically) of organisms as if they were designed, one has many, many questions thrown up to be answered. Why are there plates on the back of stegosaurus? Why is the trilobite lens such a strange shape? Why does one butterfly mimic another? Why are pine cones and sunflowers patterned as they are? And much, much more. Without the metaphor, the science would grind to a halt, if indeed it even got started. The plates are there; the lens is there; the mimics are there; the swirls are there. Brute statements, which probably we would not be led to make in the first place because there would be no reason to even notice. Who would care about the trilobite lens? Why bother to describe it rather than something else? Whatever the ultimate logic of the case, at the practical, real-life level, Kant got it right. Biologists "are, in fact, quite as unable to free themselves from this teleological principle as from that of general physical science. For just as the abandonment of the latter would leave them without any experience at all, so the abandonment of the former would leave them with no clue to assist their observation of a type of natural things that have once come to be thought under the conception of physical ends" (Kant 2007, 25). . . .

Purpose in evolution is obviously alive and well and mixing in the best circles! But what about the value component in the design metaphor? One does not have to be a logical positivist to insist that science ought to be value free, as opposed to (say) religion and politics. Hence, inasmuch as the design metaphor brings in values—and this has been a significant conclusion of our analysis—we might fear that something has to be wrong with it. Unless, of course, we agree with those people today who would deny the very possibility of anything being value free, including—especially including—science. People like Lewontin, who would argue that part of the oppressive power of science is that it pretends to be value free when truly it represents all of the values of Western capitalist, patriarchal, racist society.

Fortunately, we do not need to open this particular can of worms here; we could point out that even the logical positivists agreed that science can be shaped by pragmatic values, aesthetic values, and the like. I have argued that methodological adaptationism is not merely a matter of taste, but taste certainly comes into the equation. Darwinism does show an interest in organized complexity, pushing it to the fore. Just as formalism shows its interests. Remember Thomas Henry Huxley: "What I cared for was the architectural and engineering part of the business, the working out of the wonderful unity of plan in the thousands and thousands of diverse living constructions" (L. Huxley, 1900, 1:7–8). So there is certainly a value component at this level, in Darwinism as in all sciences. . . .

However we make the cut, the value component introduced by the metaphor of design is relative—relative to an organism's well-being. What is the function of the nostril at the front of the face? To ensure good airflow, so that the brain can be heat regulated in an efficient manner. What is the function of the trilobite's strangely shaped lens? To focus sharply in the aqueous environment in which it finds itself. Those trilobites without such lenses, or with such lenses only working partially, were at a survival and reproductive disadvantage. Science does not ask the question whether the trilobite's survival and reproduction is an ultimate or absolute good. Parasites such as viruses have parts that function well, only too well from the standpoint of some hosts, but no scientist says that a parasite is an ultimate or absolute good.

Absolute values about what is "good" are the problematic ones, the ones that we want to keep out of science. Relative values are commonplace. The difference is reflected in the distinction between "valuation" and "evaluation," the latter a phenomenon within science that even as conservative a philosopher as Ernest Nagel acknowledges and respects. Take the notion of "anemia." This refers to an animal having insufficient red blood corpuscles, which means that among other things it cannot keep the same constant internal temperature as normal animals in the species. But what is "normal" in this context? We have to make some kind of judgment call. An investigator or a physician will have to decide if the condition makes much difference to the animal's well-being. Can it function as well as other animals? If the animal is human, is he or she as comfortable as other people? We all have our ups and downs, but is this person outside the acceptable bounds? And so forth. The point is that some kind of valuing is going on here. "When the investigator reaches a conclusion, he can therefore be said to be making a 'value judgment,' in the sense that he has in mind some standardized type of physiological condition designated as 'anemia' and that he assesses what he knows about his specimen with the measure provided by this assumed standard" (Nagel 1961, 492).

The key thing to note is that this is not an offensive kind of valuation—one is judging against a standard. One is making an "evaluation." And this is precisely what is happening in evolutionary biology. One is not making an absolute valuation. One is making a judgment against the standard of success in survival and reproduction. So can one truly say that the trilobite lens is of value to the trilobite—even of value in a relative sense—when the trilobite almost certainly did not have full consciousness and an awareness of its needs? The serrated knife for cutting my bread is of value to me because—and precisely because—I have interests. That is where we came in. The trilobite has no interests. It just is. But that is surely the point at issue. We are seeing the trilobite as if it had interests. As if certain ends were of value to it. And that leads to understanding—our understanding. Whether or not we want to buy into all that he claimed about necessity, Kant was right in seeing that we do the science, and we try to make sense of the trilobite and its lenses. And we are doing this through metaphor—looking at the trilobite as if it had intentions and interests. As if it had values. The interests and values are not there in reality, but they are part of the way in which we map reality in order to make sense of it.

This is an uncomfortable conclusion for those who think the aim of science is to give an unvarnished report on reality.... Darwinian evolutionary biology is different. Because of the nature of organisms—their distinctive complexity—the design metaphor is appropriate and highly fruitful. The argument to organized complexity is precisely that—an argument to organization, to design (taken as a metaphor). This gives rise to a forward-looking kind of understanding—understanding in terms of final causes. A powerful kind of understanding that raises interesting and fruitful questions and stimulates important answers. It should be celebrated, not regretted.

8

Naturalism, Sociobiology, and Their Entailments

Chapter 8, Reading 1

Naturalism[1]

WE MOVE ON NOW to the central Christian drama. God, seeing us in a state of sin, became incarnate in the human form of Jesus Christ, lived and preached and then was crucified for our benefit, rising again on the third day. Darwinian evolutionary theory is simply irrelevant to much of this story. How we should interpret God's death, for instance: as a sacrifice, as a substitute, as a ransom, or what? But Darwinism does impinge on the story in very important respects. Most obviously, there is the problem of miracles. The Christian story tells of Jesus born of a virgin, turning water into wine and feeding the five thousand and raising the dead (and of some of his disciples being able to do some of these things also), and most significantly coming back from the dead himself just a short while after he had been taken down from the cross and buried. Darwinism is a theory committed to the ubiquity of law. In the language of the philosophers, it is a "naturalistic" theory. How can it be reconciled with a world picture so obviously committed to the breaking of law?

Miracles

As always in philosophical discussions, a lot depends on the meaning of terms. By "law" in this context we mean scientific law, and this means

1. Ruse 2000d. Reprinted with permission of Cambridge University Press.

a universal statement referring to a regularity of the empirical world, which in some sense is both true and necessary (Nagel 1961; Hempel 1966) ... Newton's law of gravitational attraction, about all bodies feeling an attractive force inversely proportional to the square of the distance between them, is a scientific law. If we have an apparent exception like a boomerang not falling at once to the earth, we look for intervening factors, because without such factors we know that the object must fall. And Darwinian theory is certainly intended to be a body or network of such laws. Natural selection is something which is going to hold when you have populations of organisms, and Mendelian/molecular genetics does not fluctuate as the seasons wax and wane (Ruse 1973, 1975a). Miracles are generally taken to be in some sense violations of or exceptions to law, brought on by divine desire and agency (Swinburne 1970). But notice the key qualifiers "generally" and "in some sense." If Lazarus, whom Jesus raised from the dead, was stinking rotten with maggots crawling through him and then got up and walked, this is a straightforward use of miracle, no question. The laws of nature have been broken. Likewise if Jesus really did walk on water, we have such a miracle. But not all miracles, including not all Christian miracles, have been of this nature. Remember the Catholic doctrine of transubstantiation. The belief is that the bread and the wine of the communion service turn, literally, into the body and blood of Christ, the "host." In Aristotelian terms it is the essence, the substance, that changes; the accidents, the contingent attributes remain untouched. The host remains bread-like and wine-like, all the way through to the naked eye and even under the most intense microscopic investigation. Something is happening outside of the laws of nature, perhaps, but no laws as we know them are being broken. ...

The everyday miracles of the Gospels—turning the water into wine and feeding the five thousand and even raising Lazarus—can be explained as the enthusiasm of the moment. People's hearts were so filled with love by Jesus's talk and presence that spontaneously and out of character they shared their food. To think otherwise—to think that Jesus actually turned loaves and fishes into a banquet—is if anything a bit degrading, making the Redeemer a kind of high-class caterer. Lazarus and the ruler's daughter were more than likely brought back from trances. They may have been dead to all intents and purposes, and Jesus's actions were highly significant, but one should not suppose that Lazarus and the girl were necessarily clinically dead.

In fact, even the supreme miracle of the resurrection requires no lawbreaking return from the dead. One can think Jesus in a trance, or more likely that he really was physically dead but that on and from the third day a group of people, hitherto downcast, were filled with great joy and hope. That a psychologist or sociologist might be able to explain all of this by natural laws is totally irrelevant—something of a relief, actually. What counts is that it happened and that it was unexpected and that it mattered. Conjuring tricks are beside the point. It is from this regeneration of spirit that true Christianity stems, not from some law-defying physiological reversals in the early hours of a Sunday morning (O'Collins 1993, 553–57).

If the virgin birth of Jesus were natural, presumably it would start with the spontaneous dividing of an ovum of Mary, which would mean that the child would be an XX chromosome bearer and hence female. One feels, therefore, that the birth of Jesus had to be more than natural or nonlawbreaking. But in any case, for such a person there is a strong theological objection to explaining away the miracles as natural. The whole point is that Jesus was well and truly dead and then rose miraculously back to life. His conquering of death was precisely that which makes sense of the whole story of the incarnation and atonement. (This is the position of the important German theologian Wolfhart Pannenberg [1968]).

Indeed, as for the liberal thinker, there are strands of theological thought that rather reinforce one's position, welcoming the scientific background against which miracles (sensu law violations) supposedly occur. In the first place, one might say that the whole point of miracles is that they are miraculous. If they are occurring all of the time, then the miracles of Jesus are hardly that exciting or significant. It is precisely because they do not occur as a matter of course—that the world is so law bound—that they become particularly significant. In the second place, complementing the first point, one can make a traditional distinction between the order of nature and the order of grace. That is, between what is known as cosmic history and what is known as salvation history. "The train of events linking Abraham to Christ is not to be considered an analogue for God's relationship to creation generally. The incarnation and what led up to it were unique in their manifestation of God's creative power and a loving concern for the created universe." Because and precisely because we as free beings had sinned, a special intervening act was required of God. "Dealing with the human predicament 'naturally,' so to speak, would not have been sufficient on Gods part" (McMullin 1993, 324). It goes without saying that the creation of animals and plants was an entirely different matter and that there was no call here for miraculous intervention.

Atheism

Not everyone will be happy with this synthesis or attempt at harmony. There are both Darwinians and Christians who argue that if one starts using law, becoming a naturalist, this is the slippery slope that ends at the bottom with materialism: meaning at this point that nothing supernatural at all exists, which means atheism, which means that Christianity is ruled out as false. Hence, Darwinism, as a supreme manifestation of the naturalistic philosophy, ends in the falsity of Christianity.

On the one side, we find the historian of evolutionary biology William B. Provine. He does not hesitate to label as "intellectually dishonest" those who think that law and God are compatible. His is a kind of theology by stipulative fiat: "A widespread theological view now exists saying that God started off the world, props it up and works through laws of nature, very subtly, so subtly that its action is undetectable. But that kind of God is effectively no different to my mind than atheism" (Provine 1988, 170). Having made this judgment, Provine has little trouble finishing the case, concluding that this new God "does nothing outside of the laws of nature, gives us no immortality, no foundation for morals, or any of the things that we want from a God and from religion." The honest Darwinian is an atheist, in practice if not in name. On the other side, we find Phillip Johnson. Speaking of naturalists, he notes that they "concede that some problems are not yet solved, but they are confident that science will solve them by proposing natural mechanisms because science has so often been successful in the past. Bringing God or intelligent design into the picture is giving up on science by turning to religion (miracle) and invoking a 'God of the gaps/The Creator belongs to the realm of religion, not scientific investigation" (Johnson 1995, 208). From here it is but an easy step to atheism.

Naturalism is a metaphysical doctrine, which means simply that it states a particular view of what is ultimately real and unreal. According to naturalism, what is ultimately real is nature, which consists of the fundamental particles that make up what we call matter and energy, together with the natural laws that govern how those particles behave. "Nature itself is ultimately all there is, at least as far as we are concerned. To put it another way, nature is a permanently closed system of material causes and effects that can never be influenced by anything outside of itself—by God, for example. To speak of something as 'supernatural' is therefore to imply that it is imaginary, and belief in powerful imaginary entities is known as superstition" (Johnson 1995, 37–38).

What Is Darwinism? It Is Atheism!

Surely there is a response here. As a scientist, as a Darwinian, one is committed to the rule of empirical law. But this is aside from whether one thinks that there is a reality beyond this law. As a Christian, one thinks that there is more. Yet, even if one wants to argue for rule-breaking miracles imposed by grace on top of the order of nature, one is trying deliberately to keep these beliefs separate. Let us therefore speak of "methodological naturalism" and of "metaphysical naturalism." The metaphysical naturalist is the person who is an atheist, who does deny that there is anything beyond blind law working on inert matter. The methodological naturalist, who may well be an ardent Darwinian, is one who states that for the purpose of doing science nothing but law will be entertained, but who recognizes that there might be more, in fact or meaning.

Augustinian Science

Plantinga has a number of arguments, and it is important to understand their overall intent. He is not—at least, he claims he is not—against science as such. His objection is to science which makes central the rule of law. His objection is to naturalistic science, meaning at the least what we are now calling methodologically naturalistic science. Darwinism is highlighted as a paradigmatic example of such a science. Plantinga himself would substitute something that he calls "Augustinian science," which allows for miracles as well as laws—specifically, which allows for Christian miracles—and which apparently is rightfully so labeled because this is the science that Saint Augustine would endorse were he alive today. Because Augustinian science is not even methodologically naturalistic, there is no temptation, much less compulsion, to slip over into God denial, what we are calling metaphysical naturalism.

What is the objection to Augustinian science? One major complaint made by Plantinga is that Augustinian science is simply ruled out by definition: in a quite arbitrary fashion, something like Darwinism unfairly delimits the boundaries of real science. By fiat, science is characterized as something produced according to a methodologically naturalistic philosophy. Christianity is therefore put beyond the bounds of science. Even if it is not declared logically impossible, it is made into second-class knowledge. It is in some sense belittled. As a Darwinian, I myself am named as a major culprit in this respect, and my Arkansas testimony is highlighted as particularly offensive. "Science" is defined as "miracle excluding," and then it is triumphantly announced that Christianity is not science!

> One thinks this would work only if the original query were really a verbal question—a question like Is the English word "science" properly applicable to a hypothesis that makes reference to God? But that was not the question: the question is instead Could a hypothesis that makes reference to God be part of science? That question cannot be answered just by citing a definition. (Plantinga 1997, 146)

At one level it is easy to answer Plantinga. It would indeed be very odd were I and others simply trying to characterize "science" as something that, by definition, is based on a (methodologically) naturalistic philosophy and hence excludes God, and then simply leaving things like that. Our victory (if denial or exclusion of Christianity is our aim) would be altogether too easily won. We would indeed simply be ruling religion out of science by fiat. But this is not quite what is happening. There is no attempt to offer an analytic definition of what one means by "science," as one might offer an analytic definition of "straight line" as meaning "shortest distance between two points." This is a definition that is analytic or stipulative. What is going on—what I was trying to do in Arkansas—is the offering of a lexical definition; that is to say, we are giving a characterization of the use of the term "science." What Plantinga in the passage quoted above calls giving an answer to a "verbal" question. And the suggestion is simply that what we mean by the word "science" in general usage is something that does not make reference to God and so forth, but which is marked by methodological naturalism. Moreover, whether one likes this fact or not, it is true. Since the Scientific Revolution, the professional practice of science has been marked by an ever greater reluctance to admit social or cultural beliefs, including those of religion. Plantinga may promote Augustinian science as "science," but it would be he who was making a stipulative definition.

There is another level to what is going on, and here one has to confess to a certain sympathy for Plantinga's complaint. He is surely correct that if all we had was a dispute about words, then the debate would be trivial. Whether one is defining "science" stipulatively or lexically, in itself nothing much rests on it. But there is more. The fact is that, having set the boundaries to science, many do go on immediately to claim that what lies beyond the boundaries is wrong or misguided or nonsensical. In the language we have been using, whatever people may say that they are doing—and many are proudly open in their actions—there is often a slide from methodological to metaphysical naturalism. The logical positivists used to claim that everything outside logic and science is meaningless, and this would certainly include Christianity. Plantinga is absolutely right that there is a tendency to characterize science on the basis of subjects like Darwinism and then to denigrate everything that does not fit the pattern. But note that this is surely only a tendency, and if one is indeed a committed Christian then there is nothing in Darwinism, or in the notion of science that it supports, that says that your commitment is wrong or stupid. Yours is not a scientific commitment, but you knew that already. If scientists and philosophers persist in saying that your position is meaningless simply because it is not science, then it is they who are guilty of arbitrary stipulative definitions.

Plantinga would spurn this proffered help. He switches from simply pointing out that, whatever the logic of the situation, Darwinians do slide into atheism. Now he takes a more philosophical tack, showing that methodological naturalism is more than just a heuristic. It forces you into some unacceptable ontological conclusions. If you are a Darwinian you are committed to the rule of law, and this enters into your definition of science. But, complains Plantinga, this means that your science cannot accept or treat of the unique or unrepeatable. Laws are universal. And this means that at some level you are denigrating or dismissing the unique and unrepeatable as beyond the reach of respectable or legitimate explanation. Which means that again Christianity suffers, because virtually by definition it deals precisely with the unique and unrepeatable. Jesus was not just another prophet. He was the Christ, and uniquely he suffered, was crucified, and rose from the dead. So you can say what you like. At some deep level, Darwinism is anti-Christian. Again there is a ready response. On the one hand, one can draw attention to the distinction between cosmic history and salvation history.

Even if the story of Jesus be a story of unique events, this is salvation history—a religious story—set against the background of law-governed repeatable events, cosmic history—a scientific story. On the other hand, repeatability and uniqueness are somewhat relative terms. In a sense, everything is unique and, in a sense, everything is repeatable. Take the demise of the dinosaurs at the end of the Cretaceous period. This was in itself a unique phenomenon and unrepeatable; but, uniqueness notwithstanding, the demise was made up of many factors which can individually be brought under lawful understanding. It seems most probable that the event was triggered by an asteroid or a comet or some such thing hitting the Earth (Alvarez et al. 1980). This was no unique phenomenon, nor was the hitting of the Earth by the asteroid or comet such that the normal laws of nature—that is to say, Galileo's laws of motion—could not be applied. It is believed that there was a huge dust cloud raised, and the Earth became dark. Again, even if this was a unique phenomenon—and the dust cloud in the last century after the explosion of Krakatau makes one doubt this—one can still apply laws. One has all sorts of experience of dust causing darkness; then of darkness cutting off photosynthesis of plants and of the consequent dying of plants; and then of the consequent starvation of animals, which are part of the ecological food chain depending on plants. In other words, although the dinosaurs existed only once and will never reappear—so their demise was certainly something unique—the various components involved in the extinction of the dinosaurs are such that they can be brought under regularity. In principle, we have nothing different from any frequently repeatable phenomenon, like the death of annual plants at the end of every growing season.

Conversely and interestingly, one might also say that the Christian story is made up of many nonunique elements. Great healers do have a dynamic energy to bring people back from the brink, restoring them to health. Men and women of immense charisma do have the ability to fill groups with love and compassion toward each other, in ways barred to the rest of us—just as the Hitler-featuring Nazi rallies showed that such people can also turn crowds to vile and wicked ends. And it is not unknown for downcast groups to be filled with hope and joy and renewal, even in the face of the worst circumstances. Whether or not one thinks that law-breaking miracles were involved in this drama, the real point of this Christian story is that these phenomena all came together in a unique and overwhelming pattern. It is the story of order and meaning. Hence, in a way Plantinga helps us to see that science (including Darwinism) and religion (including Christianity), even if different, can come together in shared patterns of understanding. Because of its components and the way they are ordered, the Cretaceous extinction is not just any old event. And, because of its components and the way they are ordered, Christianity is not just any old event.

Plantinga is not finished. Harking back to some of the themes we discussed in an earlier chapter, he claims that if you are a Christian, then you ought to accept that the Jesus story involves miracles outside the bounds of law. But this acceptance is weakened if you deny a similar status to the miracles—including the creation story miracles—of the Old Testament. Hence, you should accept these earlier miracles also, and Darwinism obviously goes against this. This is why we need Augustinian science rather than Darwinian (naturalistic) science. We need an approach that recognizes God's actions as basic and not just incidental.

> Natural laws are not in any way independent of God, and are perhaps best thought of as regularities in the ways in which he treats the stuff he has made, or perhaps as counterfactuals of divine freedom. (Hence there is nothing in the least untoward in the thought that on some occasions God might do something in a way different from his usual way—e.g., raise someone from the dead or change water into wine.) (Plantinga 1997, 149)

One final argument by Plantinga. He claims that the Darwinian scientist has pragmatic as well as theoretical reasons to be against Christianity. Many scientists, including Darwinians, would defend the naturalistic philosophy that lies behind their science, because even though there are gaps in our scientific understanding—the origin-of-life question for the Darwinian, for instance—enough solutions have been found in the past that one would be foolish to deny that they will continue to be found in the future. This will be pressed on the Christian, who will be encouraged (to say the least) to take a liberal attitude on miracles, interpreting them as explicable by the laws of nature and so forth. Plantinga challenges this. While he agrees that giving up on methodological naturalism is in some sense what he calls a "science stopper"—something that brings methodologically naturalistic science to an end—as Christians, we have no reason to think that such science-stopping events do not happen.

Naturalism Self-Refuting

We are not yet finished with Plantinga. In an argument that has achieved some considerable notoriety, he has been claiming that Darwinism collapses in on itself, either in contradiction or in a circle that is intellectually stifling. Hence no Darwinian can be a true Christian, for as such an evolutionist one is forced into a radical skepticism that cannot be reconciled with genuine religious commitment. Everyone, and most especially the Christian, has to take a stand outside Darwinism (Plantinga 1991a, 1993).

The argument is as follows. If naturalism (in some sense) be true, then we should be evolutionists, Darwinian evolutionists of a kind. If we are evolutionists, then this must extend to our reasoning and cognitive powers. But, Darwinian evolution cares nothing for truth, only for survival and reproductive success. Hence, there is really no reason why our reasoning and cognitive powers should tell us the truth about the world. They just tell us what we need to believe to survive and reproduce, which information (although effective) could as easily be quite false. Plantinga tells the story of an overly rich dinner in an Oxford College, where Richard Dawkins spoke up for atheism before the philosopher—A. J. Ayer—a classic case of preaching to the converted, I should have thought—and then goes on to draw a philosophical moral. Perhaps none of our thoughts can tell us about reality.

None of this is to deny or downplay this very important fact that all our knowledge starts with sensation and understanding and reasoning done with evolution-produced adaptations. And one consequence of this is that one cannot get directly in touch with what philosophers often call "metaphysical reality": the kind of ultimate being (or Ultimate Being) that exists whether anyone is around to sense it or not. (The tree that falls in the forest when no one is around, in the popular example.) But apart from the fact that there are serious doubts about whether this kind of reality truly makes sense, ignorance of its nature hardly tips one into the kind of radical skepticism that Plantinga threatens. The Darwinian simply denies that truth can mean correspondence between one's ideas and reality, arguing rather that truth means (as I have been arguing in the preceding paragraphs) a coherence between all the parts that we hold important and significant. Unless challenged, one accepts the touchstones and tries to make a comprehensive, consistent, and meaningful overall picture (Ruse 1986b).

In any case, the kind of coherence about which I am talking is hardly one to make Christian faith impossible. As Descartes showed in his *Meditations*, logically one cannot exclude the possibility of an evil demon who is corroding our most certain of beliefs, even those of mathematics and logic! If the Christian says that he or she has a direct line to God—faith or some such thing—the rest of us still have the right to ask whether this direct line is any more secure than any other. And the answer is surely that it is not. I am not now saying that one cannot be a Christian, whether or not one is a Darwinian. I am saying that we all start with ourselves and our powers and abilities. Plantinga is being naive or arrogant if he truly thinks that the Christian has an impregnable foundation of belief not shared by those of us who start from empirical evolutionary premises.

The basic problem is that we are not really arguing. Ultimately, nothing is going to make Alvin Plantinga sympathetic to Darwinian theory or to its underlying naturalistic philosophy. If he knew of the success of the science and the power of the philosophy, he would realize how his Christian faith cannot remain static and untouched but must move forward creatively to meet the challenges. As it is, he does his religion no great service.

Chapter 8, Reading 2

Naturalism and the Scientific Method[1]

WHAT DO WE MEAN by "naturalism"? I presume that it is something set off against "supernaturalism," and that this latter refers to a God or gods and their intervention in this world of ours. The physical Jesus rising from the dead on the third day, I take to be a paradigmatic example of a supernatural event. So therefore I take it that naturalism means an approach to, or understanding of, our world that makes no reference to a God or gods. By "our world" I assume something fairly unproblematic, that would include not just the world of physical experience, but also consciousness and things like mathematics and love and hate and literature and art and so forth. Whatever the Euler identity may be, it is not supernatural, and neither is the Mona Lisa or my love of my wife or my sense that it is wrong to be unkind to small children.

Metaphor

Let's start at the beginning with what in fact it means in practice, especially in practice as a scientist, to be a methodological naturalist. No gods, and certainly no gods intervening, so what is the alternative? I

---

1. Ruse 2013c. Reprinted with permission of Oxford University Press.

presume it is explaining (and doing the rest of science, like predicting) on the basis of unbroken, unguided (blind) law. One assumes that the world runs according to certain regularities and that is it. So if you want to explain why the planets go in ellipses you appeal to Kepler's laws, and then further back to Newton on the subject. If you want to explain why the pressure of the gas went up when you squashed it down, you appeal to Boyle's law, and then on to gas theory. If you want to explain why the child of two blue-eyed parents is also blue eyed, you appeal to Mendel's first law, and more recently you get into the DNA molecule and its structure. If you want to explain why some chap is gay rather than straight, you point to his hostile father and domineering mother, and beyond that to issues like bisexuality and how it manifests itself in child development; although, obviously as we move toward Freud's theories of sexuality we are in very choppy water and I am certainly not assuming that his ideas are well taken. (In this particular case, at the most charitable, he got things backward.)

Now notice, we are already going beyond the fact of the regularities, the laws. At least, we are if we are worth our salt as scientists. Sure, Kepler tells us that planets go in ellipses, but we want to know why. And this is where Newton comes in, telling us about the gravitational forces between bodies and the effects that they have. Likewise, with the gases. Boyle's law gives us a formula, but we want to know why. This is where gas theory, with its assumptions about gases being collections of buzzing elastic balls, comes into play. And so the story goes. But what are these bigger pictures, these theories or (as Thomas Kuhn called them) these paradigms? In one sense, they are collections of laws, but in another important and informative sense they are ways of looking at things. They are models or metaphors—thinking of the world "as if." Think of gases "as if" they were collections of buzzing balls, bouncing off the sides of their containers. Think (to take the case of Darwinian evolutionary theory) as if the mechanism of change were some chap constantly selecting for one form rather than another. Think (to take cognitive science) as if the brain were a computer. It isn't, but pretend that it is.

Cutting to the quick, there has been one dominant metaphor—what is known in the lingo as a "root" metaphor—that has dominated and guided science since the Scientific Revolution in the sixteenth and seventeenth centuries (Hall 1983, Dijksterhuis 1961). This is the metaphor of the world as a machine, that is, a mechanism. But note a machine without purpose, one that grinds endlessly on, simply following the laws of nature. It is true that the early mechanists were believers, to a person—Christians or deists. But they came to see that God or purposes or (what Aristotelians called) "final causes" had no role in science. Francis Bacon, the philosopher of the revolution, joked that final causes are like vestal virgins—decorative but sterile (Blane 1819, 69). In the words of one of the greatest of the historians of the Revolution, E. J. Dijksterhuis, God became "a retired engineer" (1961, 40).

So, we see first that physics and chemistry were brought under or into the metaphor. As Robert Boyle pointed out, the heavens circulating endlessly are like one of those clocks (his example was the clock made in the late sixteenth century and installed in the cathedral at Strasbourg) that not only tell time, but have phases of the moon showing appropriately and other celestial happenings all exhibiting themselves, literally by clockwork (Boyle [1686] 1996)! Then, thanks to Darwin (1859), came biology. The purposes of organic parts—hands, teeth, eyes, and so forth—were the products of the blind workings of natural selection brought on by the struggle for existence. Organisms, in the felicitous phrase of Richard Dawkins, were seen as "survival machines" (Dawkins 1976, 2). Read Darwin on the contrivances of orchids for reproduction (1862), if you want to see the metaphor in full flight. And finally, as noted just above, we humans fell before the metaphor. As it has been put, brains are nothing but computers made of meat (Ruse 2010b).

What Follows From the Incompleteness of Science?

But grant that we do have some genuine issues and that these are not explained by the machine metaphor. Not even attempted by the machine metaphor. And so these escape the grasp of methodological naturalism. What follows? One response, and it would be mine, would be: "I simply don't know." I am a skeptic on these sorts of matters. I profess ignorance—although somewhat arrogantly I think others should do likewise! Is this a legitimate response for the methodological naturalist? I would argue that it is. Darwinian evolutionary theory is one of the triumphs of naturalistic science. It stresses that there are no absolutes. Organisms have adaptations to help them to survive and reproduce. But to retell the old joke: When escaping from a bear you don't need to be a great runner, just faster than the chap next to you! Brains are obviously great adaptations, but they come with a cost, namely particularly they need lots of energy (usually found in the form of protein, meaning other animals) and sometimes this is not readily available. If you are living on grasslands, then in the immortal words of the paleontologist Jack Sepkoski: "I see intelligence as just one of a variety of adaptations among tetrapods for survival. Running fast in a herd while being as dumb as shit, I think, is a very good adaptation for survival" (Ruse 1996b, 486). The point I am making is that there is nothing sacrosanct about brains and their computing power (to use the metaphor). And in particular, there is no reason why our brains should be able to peer into all of the mysteries of the universe. As Richard Dawkins has rightly said: "Modern physics teaches us that there is more to truth than meets the eye; or than meets the all too limited human mind, evolved as it was to cope with medium-sized objects moving at medium speeds through medium distances in Africa" (Dawkins 2003, 19). So I am happy to remain with "I just don't know."

What I don't think is justified is a jump to metaphysical naturalism, at least not in the sense that what has been shown means that one must jump to metaphysical naturalism. I think my skepticism already shows this. It is not that I think that a God or gods must exist, but rather that I am leaving a blank space here. What of the person who would fill it up, let us say with the Christian God? Obviously I don't think you have to and for me there are good reasons why I would be unwilling to fill it in this fashion. If asked, I would start with the problem of evil. I don't want any part of a God who let Anne Frank die at Bergen-Belsen. Moreover I would point out that anyone who does invoke the Christian God is still going to have major challenges.

For example, if God is responsible for the very fact of existence—why there is something rather than nothing—then he (she or it) is going to have to exist necessarily. Otherwise one gets caught on sophomoric questions about "What caused God?" I am not saying that the notion of necessary existence is impossible—Christian theologians like Aquinas have written extensively on the topic—but it is a tough one to which to give a fully convincing answer (see Davies 2013, Grayling 2013). However, if you persist through all of this, then I do not see how the methodological naturalist can object if the Christian does try to explain existence and ethics and consciousness and meaning from his or her perspective. You cannot give a scientific answer, but if you want to talk about creations and meanings and purposes and so forth, then I don't see that the methodological naturalist has ruled these out of court—that is, not in the role of methodological naturalist. This is not an argument against metaphysical naturalism, just not one in favor of it.

Miracles

This has been one line of argument. The questions that methodological naturalism does not attempt to answer. Let us turn now to another line of argument. Could it not be that because of the positive workings of or through methodological naturalism one is directed at least to the plausibility of metaphysical naturalism? Let us look at three issues here, starting with the tensions between understanding through law and invocation of miracles. Now in a sense you might say that this question has been decided before we start. The whole point of methodological naturalism is that we don't allow miracles, certainly not in the sense of breaking or violating laws. So how can there be any tensions? However, surely at this point it is legitimate to raise matters of pragmatic significance and success. If methodological naturalism keeps getting stumped then surely one might think that something more is going on, and that something more is miraculous. This is the position of today's so-called intelligent design theorists. They argue that certain biological phenomena are "irreducibly complex"—they could not be explained by a slow, natural process like natural selection. Hence, one must invoke the agency of an intelligent designer (Behe 1996, Dembski and Ruse 2004). Admittedly in theory this designer might be a natural superintelligence on Andromeda, but truly everyone is supposed to think that this designer is supernatural and in fact looks remarkably like the Christian God.

Free Will

Turn to another worrisome problem, namely that of free will. I presume that in some sense the methodological naturalist is committed a priori to determinism. Things don't just happen. A white rabbit appears out of the hat, and you know that there has to be a law-bound answer. That after all is the whole point about machines. Like the Strasbourg clock, they just keep grinding on indefinitely and with enough information you can tell exactly what the time (or its equivalent) will be at any point in the future (or retrodicting backward for that matter). Does this then mean that there is no such thing as free will? And if this is so, then what price God (the God of theism that is)?

There are those who answer affirmatively and conclude that any talk of free will is at best an illusion and at worst self-deception (Coyne 2012). None of this proves metaphysical atheism obviously, and there are those who might say that indeed such a stance strengthens the case for the Calvinist claims about predestination. God knows exactly what is going to happen at any point in time and the reason is precisely because we are dealing with a machine situation. It goes without saying that not everyone, including believers, is entirely happy with this theology. It is felt that in order to get a sense of responsibility—a notion at the heart of faith—some real kind of freedom must be preserved. In any case, many (myself included) would be inclined to invoke some form of G. E. Moore's argument for commonsense realism (1939). He said he would throw out any argument that disproved the existence of his own hand before him! We feel that the sense that we have of free will is so overwhelming that we would throw out any argument that disproved its existence. It is about as basic as anything could be. Those with slightly more philosophical sophistication, that is those like me (!), who think that free will is real, tend to divide into those known as *libertarians* (not in the social sense) believing that even though determinism may hold true, this still leaves room for the exercise of a genuine sense of free will, and those known as *compatibilists* believing that there are two dichotomies at work here—determinism vs. indeterminism, and freedom vs. restraint—and that not only is determinism compatible with free will but it is required for free will (Fischer et al. 2007). If indeterminism rules then our actions are random and not subject to evaluation, which is the very essence of freedom.

Did he choose right or did he choose wrong? The point is that most who have thought on these issues believe that being a methodological naturalist does not preclude some kind of freedom. Parenthetically, a point about quantum theory, something that is often invoked in contemporary discussions about free will. The consensus opinion is that ultimately it is not as significant as some seem to think that it is. If we have genuine random events (at least, from our perspective) then we have indeterminacy. At once the usual inferences apply. Randomness is not to be equated with free will. If I suddenly start to tear off my clothes in public for no reason whatsoever, then I am a lunatic, not a person showing free will. On the other hand, if it is pointed out that although individual events may be random, on a statistical basis we have very tightly controlled phenomena—the half-lives of decaying substances are as exact and predictable as (let us say) the motions of the planets—then all of the arguments made in the previous paragraphs apply without need of further comment.

Humans: Did They Appear Necessarily?

We can conclude then that, with respect to the freedom issue, while methodological naturalism does not speak for or against the existence of a God or gods of some kind, it does not rule out the existence of the Christian God or indeed of any god of the theist. Let us turn now to one final topic of this ilk, remembering that we are concerned not with whether methodological naturalism opens up a way for the metaphysical nonnaturalist, but whether there is that in the exercise of methodological naturalism that precludes the God or gods that people do actually believe in and so in some sense points to metaphysical naturalism. Thus far we have talked about the attributes of humans, like free will, but let us turn now to the very existence of humans. What does a methodological naturalistic approach have to say about this topic and what are the implications within the context of our discussion?

Let us stay with biology, specifically Darwinian evolutionary theory, recognizing that a full account of humankind would need an extensive discussion of culture also. I think it fair to say today that no evolutionary biologist, specifically those concerned with the evolution of humankind—*paleoanthropologists*—thinks that our species offers insurmountable problems in explanation (Ruse 2012b). Not everything is yet explained, but we have a large (and growing) amount of fossil evidence and this is backed by increasing understanding (thanks especially to new techniques) of the causal factors involved in our appearance. We split from the apes (in Africa) about five million years ago, came out from the jungle onto the plains, moved to our hind legs, and our brains started their growth in size, accompanied obviously by ever greater social and intellectual skills. At some point in the not-too-distant past (a hundred or so thousand years ago) we moved out of Africa, spreading right across the globe, and as we did so we developed the distinctive racial characteristics that are so obvious in our species. Agriculture and the like is a pretty recent invention of the past ten thousand years or so.

Obviously none of this is very compatible with a religion committed to a literalistic understanding of (what Thomas Carlyle called) "Jewish old clothes." There was no miraculous creation of organisms, no unique Adam and Eve, no universal flood through which only one extended family survived. This of course puts greater or lesser constraints on God, depending on where one started from. If one has long taken Genesis in a metaphorical or allegorical fashion, then there are no great tensions. Although it should be understood that some fairly significant revisions in theology may be required. The Augustinian view of original sin has all humans tainted because of the actual actions of Adam and Eve. Like carriers of a bad gene, we all suffer because of what went before. This kind of thinking no longer holds water, a matter of serious concern for those in the Augustinian tradition (like Calvinists). There are available other theologies that fit more readily with the evolutionary story, for instance the views attributed to Irenaeus of Lyon, who argued (before Augustine) that original sin should be understood in terms of human incompleteness (Schneider 2010). Thus the coming of Jesus should not be seen an improvised "Plan B," something that God did when things went wrong, but as an always-intended part of the completion of creation. The idea that we are incomplete, a mixture of altruism and selfishness, fits well with the modern biological understanding of human nature. (Obviously biologists would not use the word "incomplete," but they do stress our conflicting natures—altruistic to work well in the group and selfish to look after number one.) What about the actual appearance of human beings? One thing we can say with certainty is that given Darwinian theory—the relativism of the selective process and the non-guidedness of the variations (mutations)—no one would say that humans absolutely must have evolved. This was certainly the view of pre-Darwinian evolutionists like Erasmus Darwin (Charles's grandfather) and Jean-Baptiste de Lamarck—monad to man—but is simply no longer tenable (Ruse 1996, 2012). Not that this has stopped people from offering suggestions as to how our evolution was very probable.

One, originating with Darwin but endorsed most recently by (of all people) Richard Dawkins, seizes on arms races, translated into biological terms. Lines of organisms compete and their adaptations improve. The prey gets faster and so the predator gets faster. Perhaps, as in military arms races, things turn electronic and the beings with the best onboard computers emerge and win. Referring to something known as an animal's EQ, standing for "encephalization quotient" (Jerison 1973)—a kind of cross-species measure of IQ that factors out the amount of brainpower needed simply to get an organism to function (whales require much bigger brains than shrews because they need more computing power to get their bigger bodies to function), and that then scales according to the surplus left over, Dawkins writes: "The fact that humans have an EQ of 7 and hippos an EQ of 0.3 may not literally mean that humans are 23 times as clever as hippos! But the EQ as measured is probably telling us something about how much 'computing power' an animal probably has in its head, over and above the irreducible amount of computing power needed for the routine running of its large or small body" (Dawkins 1986, 189). No prizes for guessing the nature of that something!

Paleontologist Simon Conway Morris (echoing an argument of Stephen Jay Gould) seizes on the notion of convergence. Evolution does not happen randomly, but rather as a series of moves toward already existing ecological niches, bringing on remarkable similarities in physical nature. The saber-toothed tiger niche for instance was occupied by both marsupial and placental mammals. Conway Morris (a sincere Christian) draws a happy conclusion about the cultural niche: if brains can get big independently and provide a neural machine capable of handling a highly complex environment, then perhaps there are other parallels, other convergences that drive some groups toward complexity. Could the story of sensory perception be one clue that, given time, evolution will inevitably lead not only to the emergence of such properties as intelligence, but also to other complexities, such as, say, agriculture and culture, that we tend to regard as the prerogative of the human? "We may be unique, but paradoxically those properties that define our uniqueness can still be inherent in the evolutionary process. In other words, if we humans had not evolved then something more-or-less identical would have emerged sooner or later" (Conway Morris 2003, 196). There are other suggestions, including the somewhat non-Darwinian notion that complexity is bound to increase over time simply from random factors and that this could spell the prospect of humankind arriving to order (McShea and Brandon 2010). (Darwin does toy with some of these ideas in an early notebook, but the real author of such thinking was Darwin's fellow Englishman Herbert Spencer.)

But the fact is that none of them actually guarantee the arrival of humans. This is surely a problem for theists for if such people claim anything it is that existence of humankind is not a contingent phenomenon. By this is meant not that we have necessary existence (as is presumed of

God) but that our being here is necessary. Gould suggested that even if humans did not necessarily have to appear here on Earth—joking about the asteroid that wiped out the dinosaurs and made possible the rise of the mammals, he said that literally we owe our existence to our lucky stars (Gould 1989, 318)!—given the vastness of the universe we might reasonably expect that somewhere, sometime, humanoids (beings with the necessary human features like intelligence) would have evolved. Perhaps so, although to be completely safe one might want to invoke the notion of multiverses, infinitely many of them, so that (given that our existence shows that human evolution could have happened) without doubt the evolution of human beings would have happened.

Many will disagree with the general tenor of this discussion, namely that one can be a methodological naturalist without being forced into metaphysical naturalism. The so-called new atheists certainly feel that the one leads smoothly to the other. This is as it may be and one can only invite them to make their case and show this discussion wrongheaded if not outrightly mistaken.

Chapter 8, Reading 3

Naturalistic Explanations[1]

RELIGION OBVIOUSLY HAS HAD and in major respects still does have a huge hold on the human imagination. How can this be? Some naturalistic case for belief must be offered, a case where the truth value of the beliefs can be overridden; otherwise, the case is incomplete. If you are lucky, perhaps, you might even have a naturalistic explanation that suggests the religious beliefs are erroneous. In and after the First World War, many people went to spiritualists to communicate with the recent dead. And they did! "It's a'right, Mum. I'm happy now. I'm just waiting for you and Dad to pass over." I suspect that an explanation of these messages in terms of psychological stress combined with fraud gives both a causal answer to the responses and instills considerable skepticism about their authenticity. For the atheist or the skeptic, this is the philosopher's stone—an explanation that simultaneously explains and debunks.

The reflective believer is hardly going to be indifferent to these issues. Granted that ultimately it is God's gift of faith (and perhaps reason) that leads one to belief, how has God made it possible that we believe in the face of the difficulties? What is it that makes belief compelling? This is not to say that nonbeliever and believer will necessarily approach this question in exactly the same ways. One presumes that the religious person may be somewhat more circumscribed in his or her range of possible naturalistic causes. For the nonbeliever, almost anything is

1. Ruse 2015d. Reprinted with permission of Oxford University Press.

allowable. The only criterion is—Does it work? The believer presumably would be uncomfortable with an account of religion's origin that necessitated huge amounts of human suffering—although in the light of the Old Testament, one wonders just how big something has to be before it qualifies as huge. Obviously also, the believer would be uncomfortable with an explanation that carried within it a debunking of the main claims of religion. Supposing we accept that spiritualism can be shown false in all respects, a believer would be wary of a similar explanation of more conventional religion. Part of the problem here is about how similar "similar" has to be before you reject the claims. . . .

Why Religion?

Once you start to lay out things in the raw, with the ceremonies and the eating and drinking and birthing and marrying and dying, you can see why so much has been written about the supposed naturalistic underpinnings of religion. A rich culture like this just cries out for interpretation and understanding. Indeed, from around 1850, religion was often both cause and effect of the growth of the social sciences. The great French sociologist Emile Durkheim (1857–1917), writing a hundred years ago, was more penetrating on the natural causes of religion than any thinker before or since. He saw the extent to which religion gives meaning to life, makes the hardships of existence not just possible but in important respects explicable. With religion—and this surely screams out from my brief description of late medieval Christianity—we have a culture binding people and helping people and giving hope to all. "A religion is a unified system of beliefs and practices relative to sacred things, i.e., things set apart and forbidden—beliefs and practices which unite in one single moral community called a Church, all those who adhere to them" (Durkheim 2004, 1.1). Durkheim wrote of the collective consciousness that binds society together in a functioning whole. This fits medieval Christianity exactly. Think of how we have just seen rich and poor connect through shared roles, expectations, and obligations in pre-Reformation Britain. Religion for these people was totally and utterly a unifying force and experience. Birth was a time of group celebration as the baby was shown around and through the church welcomed into the community. Marriage again, very much a group celebration (as it is still today), was ordered and blessed by religion. Death, as we saw so clearly and emphatically in the last section, was above all embedded in culture and tradition. Because of the church, its rituals, its demands, its customs, this most terrifying of human events could be accepted and handled. The fears were shared and ameliorated.

Durkheim was not alone in writing about religion. Marx told us that religion is the "opiate of the people," Frazer examined shared myths across cultures in an effort to understand our religion, and Freud declared that God is a father figure, an illusion invented to help us curtail our animal passions. Much that has been written is highly controversial and today finds little favor. Perhaps the most famous, or notorious, of all the naturalistic claims about religion was that of the sociologist Max Weber (1864–1920), who argued that the rise of Protestantism and the rise of modern capitalism are inextricably linked—that the changes in society after the period we have just been looking at were all part of a single pattern, as people tried to show their devotion to God by accumulating capital rather than simply spreading the bounty through society. Even today, a hundred years after he advanced this thesis, scholars argue about its truth. Yet even when one questions some of the more striking claims about the natural causes of religion—Freud, for instance, suggesting that Moses was an Egyptian who brought monotheism to the Jews, for which he was killed, and his role and beliefs long suppressed—one encounters many penetrating insights. . . .

Nothing stands still. Recent work by American sociologists has questioned the kinds of explanations offered by Durkheim and his successors. It is one thing to explain a society where one and only one religion prevails, as in England and Wales before Henry VIII. It is another to explain societies with multiple religions, as is the case in the United States. Perhaps not surprisingly, economic models seeing rival sects as competing for customers in the marketplace have found much favor (Stark and Bainbridge 1996). Following such developments, however, would sidetrack us. Keeping our focus, let us turn to suggestions made by today's atheists (including the new atheists) who have argued that religion can be explained—generally explained away—naturalistically. For reasons that are perhaps as much historical as conceptual, we shall see that evolutionary biology plays a large role in such discussions.

Does Religion Start with Mistaken Perceptions?

The classic naturalistic account of religion's origins—one that appears again and again in one guise or another—is that of David Hume in his *Treatise of Human Nature*: "We find human faces in the moon, armies in the clouds; and by a natural propensity, if not corrected by experience and reflection, ascribe malice or good-will to everything, that hurts or pleases us" (Hume 1940, 78). In other words, religion begins in mistaken identification of the inanimate with the living. From there, presumably, it is all uphill (or downhill, as the fancy may take you) to a full-blown religious system. This kind of thinking was picked up a hundred years later by Charles Darwin in the *Descent of Man*. By this stage of his life (the *Descent* appeared in 1871 when Darwin was sixty-two), Darwin had slid into a fairly easy agnosticism, and it is surely significant that, although he devoted considerable time to the origins and nature of morality, he dealt with religion almost briskly. He wasn't scared to talk about it; it was just that as a late Victorian he was getting a bit bored in talking about it. Borrowing heavily from Hume (Darwin first read *Natural History* in the late 1830s), he thought it was all a matter of chance and confusion, thinking that the "tendency in savages to imagine that natural objects and agencies are animated by spiritual or living essences" was illustrated by the mistaken actions of his dog (a beast, Darwin tells us, who is "a full-grown and very sensible animal"). Snoozing on the lawn, the dog was upset by a parasol moving in the wind. Going on the attack "every time that the parasol slightly moved, the dog growled fiercely and barked. He must, I think, have reasoned to himself in a rapid and unconscious manner, that movement without any apparent cause indicated the presence of some strange living agent, and that no stranger had a right to be on his territory" (C. Darwin 1871, 1:67).

More recently, anthropologist Scott Atran has proposed a similar kind of by-product explanation of religion, thinking that it is all bound up with our mechanisms for detecting danger and showing fear. "Natural selection designs the agency-detection system to deal rapidly and economically with stimulus situations involving people and animals as wired to respond to fragmentary information under conditions of uncertainty, inciting perception of figures in the clouds, voices in the wind, lurking movements in the leaves, and emotions among interacting dots on a computer screen." He thinks that this kind of adaptation can all too easily go astray. "This hair-triggering of the agency-detection mechanism readily lends itself to supernatural interpretation of uncertain or anxiety-provoking events" (Atran 2004, 78).

Why Does Religion Persist?

Agree that religion may have started life with no direct biological function. This in itself does not make religion false. As it happens, although by the time Hume and Darwin were writing on the subject neither had any religious beliefs that were particularly pressing, both stressed that their arguments did not and could not disprove religion as such. Today we have Christian anthropologist-psychologist Justin Barrett, who combines a deep faith commitment with a scientific position much like that sketched in the last section. He argues that religion comes very much in a Humean fashion from the overactivity of what he calls "agency detection devices" (ADDs). "Our ADD suffers from some hyperactivity, making it prone to find agents around us, including supernatural ones, given fairly modest evidence of their presence. This tendency encourages the generation and spread of god concepts and other religious concepts" (Barrett 2004, 31). Faces in the moon, armies in the clouds. Eventually, all of this gets blown up into a fully functioning religion. And it could all be simply God's way of getting religion naturalistically. "Suppose science produces a convincing account for why I think my wife loves me—should I then stop believing that she does?"

One thing we can say with some confidence is that if religion did start as a by-product, this does not mean that it stayed that way. Indeed, if it had no function or even a slight negative function, it would be unlikely to persist for long. Very well known in evolutionary biology is the move to a positive function for which a feature was not originally designed—what have been called *exaptations* (Gould and Vrba 1982). Vertebrates did not first grow four legs because that is the most stable land configuration. They began life in the sea, and two limbs fore and two limbs aft was the best way of moving up and down in the water. Only later did the fourness of the limbs become adaptive on land. Likewise, whatever the function or not of religion as it began, it could be picked up quickly by natural selection and turned to other uses. And if Durkheim and the other social scientists who think his way are even half right, the use is obviously some kind of group cohesion, where people benefit from being part of a society and not individuals alone. To quote Franklin on the signing of the Declaration of Independence: "We must all hang together, or assuredly we shall all hang separately." Does this kind of thinking lend itself to an evolutionary interpretation? Many would think not. Surely the struggle for existence means that Hobbes was right; everyone is turned against everyone else. But from Darwin on, as was noted earlier, it has been realized that this is a simplistic understanding of the workings of natural selection. Often cooperation, what biologists call altruism, is a highly effective adaptive strategy, simply because the benefits of working together outweigh the costs (Ruse 2006).

Harvard-based Edward O. Wilson (b. 1929), evolutionary biologist and nonbeliever, has been prominent among those who emphasize that altruism is widespread through the animal world, that it is distinctively characteristic of humans, and that religious belief and practice is at the heart of all of this. He thinks (in major respects echoing Durkheim) religion is adaptive because of its power to confer group membership. "In the midst of the chaotic and potentially disorienting experiences each person undergoes daily, religion classifies him, provides him with unquestioned membership in a group claiming great powers, and by this means gives him a driving purpose in life compatible with his self interest" (E. Wilson 1978, 188). Wilson does admit that there may be something to cultural causes, but (and here he goes beyond Durkheim) essentially he thinks that it all comes back to biology. "Because religious practices are remote from the genes during the development of individual human beings, they may vary widely during cultural development. It is even possible for groups, such as the Shakers, to adopt conventions that reduce genetic fitness for as long as one or a few generations. But over many generations, the underlying genes will pay for their permissiveness by declining in the population as a whole" ((E. Wilson 1978, 178). Culture can play variations on the themes, but ultimately these themes are biological.

Without judging too severely the particular causal input of biology over culture, we can grant Wilson his empirical case. Had the medieval British had modern science and technology, perhaps they would not have needed their religion. But they didn't have it, and one does suspect strongly that if they simply lived existences bereft of meaning, just going about their duties day in and day out, they would have functioned a lot less efficiently than they did in real life. And this is quite apart from the meaning to their lives. (A matter we shall take up in more detail in the final chapter.) What is fascinating and suggestive is how much of the religion revolved around precisely the sorts of things that evolutionary biologists think important. Marriage is an important part of the Christian life, the commitment to the other and the expectation of family. (There were lots of saints who were specifically into the fertility business.) Notice that there was no theological nonsense about multiple marriages down here leading unacceptably to multiple spouses in the hereafter. A man with children whose wife has just died in childbirth needed a new mate, not the consolations of the promised hereafter, and the same goes for a woman whose husband has fallen from a tree and broken his neck. The Christmas story cherishes the arrival of a new member into the group: "Make we joy now in this feast. *In quo Christus natus est: E-ya!*" And don't forget the loving parents. For all that God pushed him aside when it came to impregnating Mary, Joseph has a major part in the story, caring for his wife and child, stressing the importance of the father in the family. Is it too outlandish to suggest that in an age when, because of short lifespans, social relationships and obligations often took over from biological relationships and obligations, Joseph was an exemplary role model?

Does Evolution Show Religion False?

Truly pertinent to our discussion is what this all means with respect to the truth content of religion. Having given an adaptive explanation of religion in evolutionary terms, does this now mean we can dismiss religion as false? Is religion, as was suggested might be the case for morality, no more than the collective illusion of the genes to keep us reproducing efficiently? Wilson has pursued this line of thinking, arguing that now traditional religion can be seen as dispensable. He thinks that humans are by nature religious and that we must have a religion. But with evolution having shown traditional religion false, we must turn now to new alternatives. "But make no mistake about the power of scientific materialism. It presents the human mind with an alternative mythology that until now has always, point for point in zones of conflict, defeated traditional religion." We must now have a kind of secular religion, based on evolution. "Its narrative form is the epic: the evolution of the universe from the big bang of fifteen billion years ago through the origin of the elements and celestial bodies to the beginnings of life on earth." He concludes: "Theology is not likely to survive as an independent intellectual discipline" (E. Wilson 1978, 192). . . .

So it wasn't just a matter of science pulverizing all before it. History shows that internal issues of theology and external issues of politics can count for much. Philosophically, one might also doubt Wilson's claim about the inevitable triumph of evolutionary materialism. Go back for a moment to morality, an absolutely fundamental part of medieval Christianity. You ought to cherish your children. You ought to be faithful to your spouse. You ought to care about those less fortunate than you. You ought to say prayers for the dead. An earlier chapter argued that from a naturalistic perspective, the objectivity of morality may be an illusion, and there is ontologically nothing but the genes doing their thing. But even if we assume that this argument is well taken and that it scotches the argument for God's existence from the fact of morality, none of this prevents the believer (who presumably believes on other grounds) from interpreting it all from a Christian perspective (Ruse 2010b). It is one thing to say that morality does not necessarily have to have God's backing; it is another thing to say that morality could not have God's backing. One simply argues that this is how God gets us to be moral, through the genes rather than (as may be the case with mathematics) through reading some script on the Platonic wall. In fact, traditional Catholic teaching about morality—natural law theory, going back to Aquinas, who in turn went back to Aristotle—is that morality doesn't exist out there independently, but is a matter of doing what is natural, where what is natural is dictated by the way things are. A mother suckling her hungry baby is a good thing because that is the way the human body is designed. Quarreling with your neighbors is a bad thing because it disturbs the functioning of society, and everyone (including the malcontent) suffers. If you want a theological reading of morality—and you may or may not—then this natural law position perfectly complements the Darwin explanation. (No big surprise really, given that Aristotle was a biologist before he turned philosopher.) Likewise in the case of religion. That it exists because it is adaptively useful suggests in some sense that it is natural, and while religion may not be true, there is nothing to stop the religious from arguing on other grounds that it is.

But What If Religion Is a Bad Thing?

Something like purgatory suggests that, even if one sees religion as generally biologically adaptive, matters could backfire and religion work against biological interest. In the extreme case, we saw how Wilson takes note of the Shakers, whose prohibition on sexual activity led to their extinction. One might tack all the way and argue that, far from having a positive cultural or biological value, religion is dangerous and exists because in some sense it manages to pervert the proper order of things. Perhaps our medieval peasants were taking too many holidays from their labors. They took forty to fifty feast days, a matter of some tension with employers, and something the Reformers clamped down on—they were particularly keen to keep things moving through the prime agricultural months of summer—causing significant if understandable resentment (Duffy 1992, 40). Student of culture Pascal Boyer thinks this is the norm, rather than the exception; religion in some way subverts features and habits that in their own rights are perfectly adaptive and functioning. "The building of religious concepts requires mental systems and capacities that are there anyway, religious concepts or not. Religious morality uses moral intuitions, religious notions of supernatural agents recruit our intuitions about agency in general, and so on. This is why I said that religious concepts are parasitic upon other mental capacities" (Boyer 2002, 311).

Writing in much the same vein, we have the new atheists. Dawkins, as a biologist, is particularly sensitive to this issue. His explanation goes as follows: "When a child is young, for good Darwinian reasons, it would be valuable if the child believed everything it's told. A child needs to learn a language, it needs to learn the social customs of its people, it needs to learn all sorts of rules—like don't put your finger in the fire, and don't pick up snakes, and don't eat red berries. There are lots of things that for good survival reasons a child needs to learn" (Dawkins 1997b, para. 3). So natural selection steps right up to the plate. "Be fantastically gullible; believe everything you're told by your elders and betters" (para. 4). Which, of course, is fine much of the time but open to invasion by parasites with their own interests in mind. It is the same sort of thing that happens with computers. Viruses invade with their own agendas, not necessarily in the interests of the hosts. Unfortunately, religion is right up there with the worst of the invaders. Humans are wide open to such silly ideas as: "'You must believe in the great juju in the sky,' or 'You must kneel down and face east and pray five times a day.' These codes are then passed down through generations. And there's no obvious reason why it should stop" (para. 6). Even worse is the fact that those viruses that are really good at their job are precisely those with the most awful and dangerous messages. "So, if the virus says, 'If you don't believe in this you will go to hell when you die,' that's a pretty potent threat, especially to a child" (para. 7).

Dawkins is offering a somewhat scaled-down version of what is perhaps today's most sophisticated theory of gene culture coevolution. Peter Richerson (b. 1943) and Robert Boyd (b. 1948) argue that biology informs and directs culture with two innate rules: follow the norms and practices of your society ("When in Rome, do what the Romans do") and, if in doubt, do what the leaders and the influential do: "Determining who is a success is much easier than to determine how to be a success. By imitating the successful, you have a chance of acquiring the behaviors that cause success, even if you do not know anything about which characteristics of the successful are responsible for their success" (Richerson and Boyd 2005, 124). All of this certainly seems to fit nicely with our picture of pre-Reformation England. Your mum and dad went to confession once a year; you should go to confession once a year. The squire or the wealthy wool merchant gives freely at feast time; you should treat your cowman or milkmaid with the same kind of thoughtfulness. Notice, however, that none of this implies that religion is necessarily false or dangerous. Of course, it could be. Do the bread and wine really turn into the body and blood of Christ? Let us grant at least that the truth status of this belief is up for grabs. And yet you believe in it because everyone else does. And sometimes beliefs could be really unsound. One doubts that forking over large sums to the church to avoid time in purgatory really served anyone but the beneficiaries. But all of this is surely to do with the truth content of the claims and not with natural reasons for holding the claims. The same sorts of reasons could be applied to beliefs our medieval people would have had that we surely think true, for instance, that parents have obligations to their children, and you ought to try to work things out with your neighbors. Dawkins has biased the discussion by slipping in the weasel word *gullible*. Yet, although we are all gullible sometimes, learning from others is not necessarily being gullible. As was the case for Edward O. Wilson's explanatory suggestions, whether religion is true or false, even a more developed theory of gene culture coevolution does not give a definitive answer.

What about Memetics?

Philosopher Daniel Dennett also pursues a negative line of thinking. He invokes the notion of a meme, an invention of Richard Dawkins (1976) intended to represent a heritable unit of culture analogous to the gene, a heritable unit of biology. Based on an understanding of religion that is (at the most charitable) sketchy, Dennett's conclusion is that religion is a meme of a rather unpleasant kind. With Boyer and Dawkins, he thinks it is a parasite, a virus on humans, just as much as the liver fluke is a parasite on sheep. "You watch an ant in a meadow, laboriously climbing up a blade of grass, higher and higher until it falls, then climbs again, and again, like Sisyphus rolling his rock, always striving to reach the top." Why does this happen? The ant gets nothing out of all of this activity. "Its brain has been commandeered by a tiny parasite, a lancet fluke (*Dicrocelium dendriticum*), that needs to get itself into the stomach of a sheep or cattle in order to complete its reproductive cycle. This little brain worm is driving the ant into position to benefit its progeny, not the ant's" (Dennett 2006, 3–4).

Religion likewise serves its own ends without regard for the well-being of the host in which it resides. It leads Muslims, Christians, and Jews to devote "their lives to spreading the Word, making huge sacrifices, suffering bravely, risking their lives for an idea" (Dennett 2006, 4). Huge sacrifices perhaps, but as earlier with the implication that the sacrifices are for naught. For all that it is false—perhaps because it is false—religion can and does do immense harm, and the naturalistic explanations show why. Unfortunately, even sympathizers wonder if meme talk is anything but common sense wrapped up in fancy language (Ruse 2009f). Memetics seems not to have led to new and surprising predictions, the sine qua non of good new theories. Moreover, Dennett is playing the same trick as Dawkins, sliding in negative implications under the cover of supposed objective discussion. Talking of parasites and viruses is using what we in the philosophical trade call "persuasive definitions." We have analogy and not the literal truth. In any case, who said that all viruses are bad? Viruses (bacteriophages) that attack harmful bacteria have their virtues. This is not to say that religions are like bacteriophages, but it is not to rule this out either. To be fair to Dennett (and to Dawkins), approaches like theirs are suggestive about how religions can get into flights of fancy and at times prove positively dangerous, to practitioners and to others. But because an idea is transmitted as a meme, it does not follow that all religion is false. It may be false. Let us grant that this is true of purgatory. It may be true. Love your children. Or it may almost certainly be false but have good consequences. Pray to Saint Apollonia because you have toothache. The placebo effect (of which, of course, you are unaware) may make you feel better.

What Should We Conclude?

Given the ease with which religion can be made to fit the various naturalistic explanatory theories, one is reminded of a warning from Karl Popper (1963). A theory that can explain everything explains nothing. In like manner, one hears still the voice of Gould from beyond the grave. With reason, most people will be a little wary of these naturalistic explanations, the negative ones as much as the positive ones. Like the religions they purport to explain, there are so many that one suspects that many, if not most, must be off the mark in some respect. One cries out for quantitative predictions and attempts to test them. Perhaps the right conclusion is the following. There is surely enough now to the naturalistic explanations to religion to think that there must be some adequate explanations, whether we are close to finding them or still far away. The nonbeliever need have no fears here. There is no reason to think that one must invoke a supernatural explanation. However, there is breathing room for the believer. Perhaps religion is a by-product, perhaps not. Perhaps religion is functional, perhaps not. Or more likely it is a bit of everything. But that is the nature of the world in which we live. The bottom line is that there is nothing in naturalistic explanations of religion per se that suggests all religion is false or that it must necessarily be false, and there is nothing inherent in the more plausible explanations that we now have to change this conclusion. In short, naturalistic explanations give comfort to the nonbeliever, but whether they refute the believer is another question.

Chapter 8, Reading 4

Sociobiology[1]

FOR EDWARD O. WILSON, as for Herbert Spencer and Julian Huxley before him, Darwinism is a substitute for Christianity: a secular religion for a new age. By now you should realize that you do not have to read Darwinism in this way—most professional evolutionists today cringe rather at this kind of activity—and that, even if you do, you are probably singing the same good old songs that have sustained Christians down through the ages. Indeed, you probably first learned the songs at Sunday school in your childhood! No one could doubt the authenticity of Wilson's deeply religious nature or the power of his burning moral vision, but his arguments purportedly showing Darwinism and Christianity to be mutually exclusive are simply not well taken. Indeed, if you insist on making a religion of your science, then your best strategy might be to join forces with Christianity rather than trying to set up your own church.

Is this then the end of matters? If you would prefer not to make a religion of your science, has Darwinism nothing more to say? Can modern evolutionary theory tell us nothing about morality, at either the substantive or metaethical level? A totally negative answer to these questions would be surprising, if only because the past thirty years have seen major advances in the Darwinian understanding of the evolution of social behavior. That area where morality most comes into play, the interactions between individuals in a cooperative or social manner, has

---

1. Ruse 2000e. Reprinted with permission of Cambridge University Press.

been the subject of intense scrutiny by Darwinians, who think that they have completely transformed our thinking on the question. It is the development of the science and the implications to be drawn that are the subjects of this chapter.

The Evolution of Social Behavior

Charles Darwin himself always recognized that behavior is important. It is no good being strong and handsome if you do not have the desire and ability. He recognized also that while all behaviors are significant, some are more difficult to explain than others. If a sheep flees a wolf, this is behavior by both predator and prey, and nothing very surprising. If an organism does something for some other organism, then this does call for special attention. If indeed life is a bloody struggle for survival and reproduction, why would an organism behave "altruistically" rather than "selfishly," caring only for itself? Epitomizing the puzzle, why do the social insects evolve as they do, with sterile workers devoting their whole lives to the good of the nests within which they live? Why not look after number one exclusively?

There is a simple answer to the evolution of altruism in general and of the social insects in particular. It is just a question of selection favoring the group—the ant nest, for instance—rather than the individual. A simple answer, but inadequate. Darwin realized that any adaptations favoring the group at the expense of the individual will prove highly unstable. They will always be at risk of crumbling under an individual-favoring alternative. Over the long run an adaptation might revert to individual benefit via the group; unfortunately, in the short term the individual who takes advantage of the efforts of others while not returning in kind will be at the greatest advantage (Ruse 1980, 1989).

It was the 1960s before evolutionists found ways to tackle Darwin's problems about sociality. A major breakthrough came thanks to the then graduate student William Hamilton, who pointed out that the best-known social insects, the hymenoptera (the ants, the bees, and the wasps), have a haploid-diploid reproductive system (1964). Males have only mothers and only a half set of chromosomes, whereas females have both mothers and fathers and a complete set of chromosomes. Hamilton was able to show how workers nevertheless serve their own biological interests: it is more profitable from an evolutionary perspective to raise fertile sisters than fertile daughters. He was able to show also that his model, which is now covered by the generic term *kin selection*, can be extended to other organisms. And indeed, his work is still today considered the paradigm of an explanation of biological *altruism*: features and behavior involving aid to others but which serve an individual's own reproductive ends.

Following Hamilton, other models were proposed for biological altruism: notably one called *reciprocal altruism*, which takes place between nonrelatives and which depends on a kind of mutual back-scratching (Trivers 1971). More generally, using ideas and concepts of game theory, workers in the field were able to set these and other models in an overall Darwinian context, showing when certain reproductive "strategies" were likely to succeed and to lead to the evolution of physical and behavioral traits which enable organisms to survive and reproduce among their fellows, and when they were not. Particularly significant was the idea of a *reproductively stable strategy*: a course taken by an organism because none other will benefit the organism more in the social situation within which it finds itself (Maynard Smith 1982). Biological altruism was now seen as part of a more comprehensive Darwinian picture.

Together with the theoretical ideas, the workers of the 1960s and early 1970s turned increasingly to detailed and long-term empirical studies, both in the wild and in experimental situations, showing how the new models function, where adjustments are needed, and how new directions are to be sought (Ruse 1999.) For all that he himself was marching to the beat of a (somewhat) different drummer than most of his fellows, a major figure was Edward O. Wilson, even then with fair claim to be the world's leading expert on the social insects. A man for whom interconnections and synthesis are the very lifeblood of intellectual advance, he took readily to the task of creating a coordinated integrated subject or discipline. Giving this vibrant new field—the study of social behavior from a Darwinian perspective—its official name, Wilson authored the magisterial *Sociobiology: The New Synthesis* (1975). Going right through the animal kingdom, this work surveyed the theoretical models and ideas and showed how they were finding confirmation in the real world. It is fair to say that, in the two decades subsequent to *Sociobiology: The New Synthesis*—and to *The Selfish Gene* (1976), a sparkling popularization by Richard Dawkins—sociobiology has come into its own as a full member of the Darwinian areas of scientific inquiry. New models, new hypotheses, new techniques, new findings, new studies, all have helped sociobiology to take its place alongside such fields as paleontology, biogeography, and systematics.

Humankind

What has made sociobiology controversial has been its extension to our own species, to Homo sapiens. In this century, the study of humankind from a biological perspective has been muted and often under a cloud for several reasons: the territorial ambitions of the social scientists, for one, and the dreadful distortions of human genetics by the Nazis for another. But nothing has deterred the sociobiologists, who have rushed in to claim that kin selection, reciprocal altruism, and related models are the keys to understanding human behavior, particularly as it occurs in group or social situations. Marriage relationships, family structures, parent-children interactions, social customs, religious beliefs, power structures, and more have been subjected to sociobiological analysis. Controversial though it may be, let there be no mistake that human sociobiology—something today often hidden under innocuous-sounding names like "evolutionary psychology"—is part of the general Darwinian picture: selection working on features powered by the genes. . . .

The Evolution of Morality

For our purposes, without prejudice, let us now assume not only that sociobiology is a viable and trustworthy branch of science, but that this applies also to human sociobiology, however named. Pertinent to our inquiry is the fact that sociobiologists—myself prominently included!—have argued that we are in a position to make plausible suggestions about the evolution of human morality (Ruse 1986b). We start with the idea of altruism and with the division one must make between two senses in which this notion is used.

The key component in ethics—especially ethics at the substantive level—is the fact that we feel the obligation to promote (through our actions) the good, meaning that we feel the need to act kindly toward others simply because this is the right thing to do. This is altruism in the literal sense, and no one would deny, especially not the sociobiologist, that we have these moral sentiments and that they are genuine. Mother Teresa truly wanted to succor the poor of India because it was the right thing to do. The rest of us may not always be so good, but each and every one of us knows the tug of moral obligation. The other sense of altruism is what has been introduced as the biologists' sense, referring to the actions performed by an organism toward others because there is expectation (not necessarily conscious expectation) of return from an evolutionary perspective. This kind of altruism is metaphorical. It is not Mother Teresa–type altruism. It is ant altruism, and it is this which has been the focus of sociobiological attention. I will refer to it, in quotation marks, as "altruism." One should not confuse altruism (good done for its own sake) with "altruism" (help given for biological returns), but neither should one belittle either notion. A metaphorical understanding is not an incorrect or inadequate understanding. It is just not a literal understanding.

Now it is an empirical fact that humans have evolved in such a way as to be highly "altruistic," and moreover to be greatly dependent on such "altruism." These are not disconnected points, for there has obviously been evolutionary feedback. Humans are (compared to other mammals) not particularly strong or agile or fast or many other physical things. We need to cooperate to survive. Our Pleistocene ancestors could do little by way of hunting alone, unlike the lion or the cheetah. On the other hand, we are good at cooperating, and we have built-in biological devices against spoiling things through intragroup violence and strife. We do not have imposing weapons of destruction, like fangs or claws. And our hormonal balance keeps us all relatively calm. Hard though it may be to imagine, the murder rate among humans—even taking into account the mass killings of the last century—is less than that among many mammals.

The point is that humans are social and that we need social adaptations, like language and the ability to resist disease. Remember the sad tales of the isolated indigenous peoples who were killed by the viruses of those more used to living in large groups. Most particularly, we need to be "altruists," and we obviously are. But now the question comes of proximate causes. How in particular do humans put their "altruism" into effect? There are at least three possibilities, and in some respects humans have gone down all three routes.

Hardwired "Altruism"

The first possibility is that we might be hardwired, as one might say, by the genes to act in cooperative ways. We do what we do without any choice because this is the way of our biology. This is the cause of the "altruism" among the social insects. They are not thinking beings. They do not heed the call of the categorical imperative. They simply do what they do as automata, because their biology tells them so. And in some respects this accounts for human "altruism." Parent-child relationships are frequently of this kind. One responds instinctively to the needs of one's own offspring: especially in the case of mothers for whom (like other animals) a kind of imprinting takes place, bonding them to their children. Out of instinct, one loves one's own children more than the children of others.

Hardwiring is important, but there are very good reasons why such hardwiring—innately motivated feelings and actions, if you like—cannot account for the whole of human "altruism." Hardwiring is fine if nothing goes wrong, but if something does go wrong—the unexpected occurs—then you are in trouble. Take the ants. When foraging, they follow pheromone (chemical) trails. If it rains, then they can be lost forever. But it hardly matters to the queen (or to her sterile daughters, for that matter), for she produces literally millions of offspring. There are lots more to fill the gap. Humans have gone the other route of producing just a few offspring; but the cost is that we cannot afford to lose them carelessly. It would be a disaster if it were fatal every time a child got caught in a rain shower. There have to be more sophisticated mechanisms motivating and controlling humans, so that our actions, including our "altruistic" actions, can respond to change and challenges.

Super-Brain "Altruism"

This suggests a second proximate mechanism, the very opposite of hardwiring. Perhaps to produce "altruism" humans have evolved as super brains, calculating carefully the costs of any social interaction and acting positively only if it is in our self-interest. Selfish in the sense of looking to and only to our own needs and desires, although not necessarily selfish in the sense of grabbing more than our share. It is often thought that super brains would be forces for evil—Darth Vader types—but this is not necessarily so, especially if everyone else is a super brain. I want to take everything for myself, but I know that you want to take everything for yourself, so we have to compromise.

Again, humans have clearly gone down this route to some extent. When you purchase something from a store, you and the storekeeper are interacting socially: being "altruists" in the sense of doing things for the other for your own ends. You do not love L. L. Bean, nor does he love you, but you get on perfectly well together doing things for him (giving him money) as he does things for you (giving you goods). And if you do not like his goods, you can take your custom elsewhere, and if he does not want your money, he can stop offering the goods you want. But obviously we are not complete super brains; this is not true even of professional philosophers. Perhaps part of the reason for this is physiological or a lack of evolutionary time. Super brains might not be that easy to produce biologically. But there are other, good, selective reasons. Even super brains are going to need time to calculate their self-interest. However, in life, time is money. You often do not have the luxury of infinite time to make decisions. A tiger approaches. You and your fellow human are in danger. Should you warn him? What will be the benefits? Will he give you something? Should you demand first? Are his promises reliable? By the time you have made your decision, both of you are in the tiger's stomach!

Super brains are a bit like those early chess-playing machines. They thought of all the options, but were useless precisely because they thought of all the options! After a move or two, they were paralyzed because there was so much to consider. What was needed was an approach that incorporated quick-and-dirty solutions: strategies that would generally win (based on past experience) but which could certainly be beaten, since they were not perfect. And now, of course, with Deep Blue, the machines have been improved to such a point that in real life even the very best chess players can be beaten. Which analogy takes us to our third option for achieving biological "altruism."

From Altruism to "Altruism"

We want a pragmatic solution, one that will generate the right moves most of the time, even though there will be mistakes and breakdowns and some actions will misfire. In the case of the machines, we have (as it were) certain strategies hardwired in, so that the moves will be made in a certain way given the initial conditions. The sociobiological claim is that in the case of humans we are genetically predetermined to think in certain ways, so that in specified situations we will incline to act in certain ways. And the genetic predetermination manifests itself as a moral sense:

an awareness of certain rules or guides that are binding upon us—the prescriptions of normative ethics. . . .

In other words, the claim is that in order to make us good biological "altruists," natural selection has made us moral altruists. And note that the claim is that this is a genuine morality. By nature, we are going to be selfish or at least self-serving. If selection did not make us this way, we would die out immediately. The person who has no concern for food and drink, no interest in sex (or who willingly steps aside in favor of a rival), may be a saint, but he or she is going to be a Darwinian flop. Yet, because we are social animals, we need something to break through this barrier, to make us interact with our fellow humans, to make us biological "altruists." And this something is going to be the moral urge: the feeling that we "ought" to do certain things, even though our nature is against it. Not that any sociobiologist wants to claim that biology does it all. It is always a function of biology in the environment: in the case of humans, of biology in culture. So one expects to find that there will be cultural differences across societies in time and place. If nothing else, different technologies and different challenges call for different solutions. But underlying it all is a shared moral base: the morality needed to make social animals biological "altruists."

Biological Normative Ethics

What of Christianity in the light of all of this? We are assuming that what has just been presented is well taken. It is a Darwinian empirical argument and in this context is not to be questioned. Our first question or set of questions must be about normative ethics. What has the sociobiologist to say about the details, and how do these compare to Christianity? In a reasonably straightforward sense, the mesh is going to be good. The sociobiologist is committed absolutely and completely to the genuine nature of human altruism. The causal process might make use of "selfish genes," to use Richard Dawkins's felicitous metaphor, but this does not imply selfish people. Indeed, the very crux of the sociobiological case is that we need real altruism to make us break through our usual selfishness. We need something as powerful as this, or "altruism" will not be achieved. For this reason, the sociobiologist endorses completely the Humean distinction between "is" and "ought" and thinks the naturalistic fallacy a genuine fallacy. Morality is different. "I love my children." "I ought to love my children." These are two quite different claims....

Supreme Principles

Turn to the basic or supreme moral principles: the supreme principles of substantive ethics, that is. What kind of "altruism"-promoting altruism is one going to get from natural selection? Hardly any surprises, I suspect, although it is probably wrong to seek one and only one principle of morality binding on all people. Because one is taking a naturalistic approach, one expects a range of emotions and obligations, within certain limits. One is going to expect a kind of commonsense morality, with an underlying base of reciprocation: reciprocation because it is right, not because I have done something for you. "Be kind to people." "Help children and the less fortunate, and try to do so in proportion to need." "Give priority to mothers." "Don't rape and/or use gratuitous violence towards women." "Keep your word." "Do not take what is not yours." "Try to moderate habits like boastfulness which are going to irritate others." "Stick up for your country or your group." And so forth (Mackie 1977).

Morality's Range

Sociobiological ethics meshes with Christian ethics. But is this not all a little bit too smooth and optimistic? Does not Christianity try to push you out and beyond the commonsense maxims embraced by Darwinism? Jesus was not addressing your average well-fed member of the Rotary, who does his bit for handicapped children and who then goes home well satisfied to enjoy the fruits of his business. Or rather, he was addressing such a person and saying that this is not enough. "Turn the other cheek." "Give all you have to the poor and follow me." "Think not for what you will put on ..." and much more. A naturalistic account of morality like that of the sociobiologist may go so far, but ultimately it cannot go as far as the Christian demands in the name of the Lord.

This is a serious objection, and it should not be minimized. It is indeed true that the sociobiological substantive ethic is going to be limited. Reciprocation works only between those who can reciprocate. Self-benefit does not demand that every recipient give something immediately in return. Your contribution may be like an insurance policy payment, that is, something never returned because you never need to draw from the common pot. And it may be that you or a relative may never be able to contribute, because of illness or whatever. Hence, you are (if able) willing to give to those in your group who can never reciprocate. "There but for the grace of God go I." But self-benefit does mean that your social fellows have to be in the same pool as yourself. People beyond the pale, people with a different insurance company, cannot expect to draw from you, from your policy. In less metaphorical language: you expect to find morality falls away as one leaves first the family group, and then the immediate social group, and so on out, to the country as a whole. David Hume, an enthusiast for a naturalistic ethics, spotted this point 250 years ago. "A man naturally loves his children better than his nephews, his nephews better than his cousins, his cousins better than strangers, where every thing else is equal. Hence arise our common measures of duty, in preferring one to the other. Our sense of duty always follows the common and natural course of our passions" (Hume 1978, 3.2.1.18, 483–84).

In this day and age particularly, this does not mean that you can and should be indifferent to the starving poor of Africa. Apart from anything else, there may be expediency reasons for worrying about them. Social diseases, for instance, have a nasty way of becoming worldwide. More pertinently for morality, in the age of television and jet travel and email, one's group stretches out to the whole world, in some sense. But it does mean that you have a bigger obligation to your own children and those in your local group than you have to the children of Africa. It also means that your sense of obligation to animals will be truncated. They cannot reciprocate in the way of other humans, and so one rather downgrades the sense of obligation felt toward them. One does have an obligation, but a limited one. Cruelty is wrong. Meat eating is an option. . . .

For the Christian moralist, relativism is anathema. One can certainly accept that different societies may well have different customs, but there has to be an underlying universality to morality. We are all made in God's image, and there cannot be one rule for one set of people and another rule for another set. There are "judgements of moral conscience, which Sacred Scripture considers capable of being objectively true" (John Paul II 1998, 82). Suttee (the widow joining her husband on the funeral pyre) is wrong, whether it be in Britain or in India, in this century or the last. The sociobiological account of morality is in agreement about relativity here on earth with respect to our species. Morality has to be something shared or it will not function, and inasmuch as it is biologically based, since we are all the same species there probably is not much variation. But we do now seem to be faced with an intergalactic relativism.

Probably the Christian will think that if this is the greatest threat that Darwinism can pose to Christian ethics, there is not much need for worry. Let us wait until we meet rational aliens. Or the Christian might point out that there may still be a place for shared virtues: sticking to one's convictions, for instance, whatever these convictions may be. In this respect, God lays the same mandates on us all. All I say is that if one rejects Progressivism, then one has an added task in trying to harmonize Darwinism and Christianity. It is not necessarily impossible, but it is a task which will need to be performed. The Darwinian can be a Christian, but both sides have to think about their absolutely bottom-line commitments, and about where and how they might be prepared to compromise or show flexibility to achieve harmony.

Chapter 8, Reading 5

Social Darwinism[1]

ALWAYS INDEPENDENT AND ECCENTRIC—HERBERT Spencer never married and lived in boardinghouses chosen deliberately for their dullness, so that he would not be distracted from his great projects. Spencer had overwhelming confidence in his abilities and insights, reinforced by his inability to read books with which he disagreed. (They gave him headaches.) Some training in mathematics fitted him to be a surveyor, a popular job in the 1840s as the railways expanded, and through this occupation (which involved significant earthworks) he encountered Lyell's *Principles*, the reading of which turned him into an evolutionist. He switched to journalism and from then on lived by his pen, bolstered by countless admirers ever ready to cater to his needs and demands—which were many, in food, entertainment, and above all adoration. Psychology, biology, sociology, philosophy—nothing escaped Spencer's gaze or eluded his world picture, which he presented in many, many volumes through the second half of the nineteenth century, describing, explaining, and fleshing out his so-called synthetic philosophy.

In a series of articles through the 1850s, Spencer established his credentials as an evolutionist, and just after the *Origin* appeared he summed up his views in a work modestly described as *First Principles* (1862). Drawing on Kant's notion of the thing-in-itself, the noumenal world that supposedly lies behind all reality, Spencer (with a Germanic flourish of

1. Ruse 2005b. Reprinted with permission of Harvard University Press.

capital letters) spoke of the Absolute or the Unknowable. Both science and religion, properly understood, point to this, and as we appreciate it in our lives and understanding, it reveals itself as being in constant motion—not just motion but evolution (Spencer popularized this term), and evolution ever upward in a progressive mode.

Spencer had read Von Baer on embryology, as well as Adam Smith and others on division of labor, and, combining insights from all, he saw progress as being a move from the undifferentiated to the differentiated, or (in his words) from the homogeneous to the heterogeneous:

> Now we propose in the first place to show, that this law of organic progress is the law of all progress. Whether it be in the development of the Earth, in the development of Life upon its surface, in the development of Society, of Government, of Manufactures, of Commerce, of Language, Literature, Science, Art, this same evolution of the simple into the complex, through successive differentiations, holds throughout. From the earliest traceable cosmical changes down to the latest results of civilization, we shall find that the transformation of the homogeneous into the heterogeneous, is that in which Progress essentially consists. (Spencer 1857, 463)

Everything obeys this law. Humans are more heterogeneous (complex) than other animals, Europeans more heterogeneous than savages, and the English language is more heterogeneous than the tongues of other peoples. Spencer's law had empirical support, but it was more an a priori truth about the world, for it follows directly from the nature of causation. Causes are always exploding outward in the effects that they produce—one cause has multiple effects, and so homogeneity produces heterogeneity. This truth is also mixed up with laws of physics, for the second law of thermodynamics shows that everything tends toward equilibrium, but a moving (or, as it came to be known, a dynamic) equilibrium. A system such as a species is a kind of unstable equilibrium; something disturbs it (any outside cause is going to have uneven effects on a system), and then it moves upward as it strives once again to achieve equilibrium—which it might, only to set off yet again.

Darwin and Spencer were worlds apart. Spencer's picture of the world was inherently progressive in a way that Darwin's was not. This comes through most vividly in their respective attitudes toward natural selection—the mechanism that was so corrosive to notions of absolute progress. For Darwin, selection was always the primary mechanism. Although he admitted other factors, including Lamarckism, they were secondary. Spencer discovered natural selection independently, as it happened, but he always thought of it as unimportant—it got only perfunctory mention in *First Principles*. He put far more emphasis on Lamarck's inheritance of acquired characteristics. Progress counted, not selection. The chief function of Malthusian pressure was to put a premium on effort. Some will strive more effectively than others, and they will therefore develop adaptive characteristics that will then be passed directly to their offspring. Spencer argued that the higher up the evolutionary scale one goes, the greater the brains required for survival, and therefore the less vital bodily fluid left over for producing offspring. Eventually, a point is reached where the numbers balance out, and the Malthusian pressures and consequent struggles are over. The greasy pole has been climbed.

Laissez Faire?

Although his philosophy came to be known as social Darwinism, not Spencerism, Herbert Spencer more than anyone else came to epitomize the attempt to draw a moral code for proper living from his beliefs about evolution. The basic pattern of Spencer and all who followed or imitated or worked in parallel with him was simple. The key was the supposed progressiveness of evolution—simple to complex, homogeneous to heterogeneous, blob to Briton. In his view, humans had a good thing going with evolution, and their moral obligation was to help the process along, not to stand in its way. We humans were living proof that when evolution was allowed to do its work, the outcome was positive. So far, so good. Spencer's thinking seems the perfect exemplification of social Darwinism.

All those who advocated state-supported amelioration of poverty were compounding the nation's problem, he believed. "Blind to the fact that under the natural order of things, society is constantly excreting its unhealthy, imbecile, slow, vacillating, faithless members, these unthinking, though well-meaning, men advocate an interference which not only stops the purifying process but even increases the vitiation—absolutely encourages the multiplication of the reckless and incompetent by offering them an unfailing provision, and discourages the multiplication of the competent and provident by heightening the prospective difficulty of maintaining a family" (Spencer 1851, 323–24).

But the story is more complex than this passage would suggest. Not only was Spencer no selection enthusiast, but he wrote these sentences before he had written anything about evolution. In other words, he did not make a simple deduction from evolution to morality. With Spencer as with everyone else, science and ethics were a package deal, with the one reinforcing the other and conversely. Rather than a case of cause and effect, Spencer's science and ethics had a common cause, stemming from the same set of convictions. And progress was the backbone of it all. Evolution was derivative; evolutionism was basic. Although struggle was all important in the process of upward development, in the end it drops away, according to Spencer. Thus, a major component of his world vision was adamant opposition to militarism and especially the arms race between Britain and Germany. He also resisted all kinds of tariffs and other impediments to the free exchange of goods. To quote Spencer's most admiring American supporter, Edward L. Youmans: "The essence of evolution is transformation—the substitution of higher agencies for lower in the unfolding economy of the world. War is certainly one of the things that must certainly be left behind" (Youmans 1879, 817). Again, more than just evolution was at work here, for the Spencer family had Quaker connections and sympathies. Ethics and evolution for Spencer and his acolytes were like hand and glove.

Social Philosophy

Not everyone who used evolution for social or moral purposes was an exclusive or ardent Spencerian. Ernst Haeckel had his own evolutionarily inspired moral system, and Thomas Henry Huxley opposed Spencer toward the end. The important point is that after the *Origin*, evolution provided a foundation or support or cover for many different people who were inclined to think about social and moral issues. The result was evolutionism, rather than just the fact or theory of evolution (Ruse 2000c, Bannister 1979). . . .

Decline

As the nineteenth century drew to an end, people began to worry about whether progress was indeed inevitable. Perhaps it had already ended, and now the world was in decline. The biologist E. Ray Lankester was a little obsessive on the subject, forever warning in sepulchral terms about the decay of Britain—in science, in education, in defense, and more, especially compared with Germany. "The traditional history of mankind furnishes us with notable examples of degeneration. High states of civilization have decayed and given place to low and degenerate states." We should not get complacent. "Possibly we are all drifting, tending to the condition of intellectual Barnacles or Ascidians" (Lankester 1880, 60).

Secular Religion?

After the *Origin*, a functioning science of evolutionary morphology was in place. It was not very Darwinian, not very high class, and increasingly unattractive to the best-quality student, but it was professional nevertheless. What about evolutionism, with its progress, moral exhortations, world pictures, and so forth? Is it proper to speak of this as a religion, joining with Hurree Babu in seeing Herbert Spencer as its chief god?

Of one thing we can be sure. No one now and (more importantly) few back then thought of this kind of evolutionism as a forward-looking, mature science, or of ever having any chance of becoming one. John Tyndall—superintendent of the Royal Institution in London, a friend of Huxley, and a member, with Spencer and others, of the X Club, a group that met for dinner in the 1860s and 1870s and organized British science while eating their roast beef—was busy articulating the norms and practices of good science. Like Herschel and Whewell a generation earlier, he worked to block out a space in Victorian society for the professional scientist, trained and worthy of employ. Tyndall stressed above all the need to be objective and to keep personal opinions and motives and desires out of science. "It is against the objective rendering of the emotions—this thrusting into the region of fact and positive knowledge of conceptions essentially ideal and poetic—that science ... wages war" (Tyndall 1892, 2:393). And here precisely is where evolutionism, including its social Darwinian element, failed. Indeed, it did not even attempt to compete, because it had other goals: to promote a world picture, an ideology of progress.

An ideology, to be sure. But would the term "religion" also be appropriate? Considering the nature of the beast, it truly seems so. The concept of a religion is notoriously hard to define, but one thinks in terms of a world picture, providing origins, a place (probably a special place) for humans, a guide to action, a meaning to life. There are other prominent features of many religions, such as belief in a deity and a formalized and recognized priesthood, but these features are not absolutely essential to the definition. Buddhists (and many Unitarians) would probably flunk the God question, and Quakers (by explicit design) have no clergy. Rather than getting too flustered by counterexamples, let us allow the oxymoron "secular religion" and cast our question in these terms. And the answer does seem positive. Popular evolution—evolutionism—offered a world picture, a story of origins, and a special place for humans in the scheme of things. At the same time, it delivered moral exhortations, prescribing what we ought to do if we want things to continue well (or to be redeemed and a decline reversed). These things hardly came by chance or in isolation. In asking about origins, evolutionism was answering a question posed by Christianity (and Judaism before this), and in focusing on the status and obligations of humans, evolutionism was trying deliberately to do better than Christianity.

The followers of Auguste Comte founded a "religion of humanity" based on positivism, with all sorts of silly practices like the naming of secular saints. Evolutionists, to their credit, did not go this far. In fact, they took much care not to go this far, for formal groups of nonbelievers (secularists) tended to have the stench of working-class radicalism and nonrespectability that scientists and their supporters—striving hard to make themselves indispensable (and powerful) members of Victorian society—were at great pains to avoid. Huxley disapproved personally and politically of Charles Bradlaugh, who, although elected a member of Parliament, was unable for many years to assume his seat because of his refusal to take a Christian oath of allegiance. Herbert Spencer refused to cross the Atlantic on the same ship as George Jacob Holyoake, a notorious freethinker: "I think it very probable that the fact of our association would be used against me" (letter of Apr. 4, 1882, in Budd 1977, 92).

Self-serving though this kind of behavior may have been, it fulfilled its end. In his novel *Born in Exile*, George Gissing wrote sardonically of "that growing body of people who, for whatever reason, tend to agnosticism but desire to be convinced that agnosticism is respectable; they are eager for antidogmatic books to be written by men of mark. They couldn't endure to be classed with Bradlaugh, but they rank themselves confidently with Darwin and Huxley" (Budd 1977, 181). Yet, for all that they wished to remain respectable, the evolutionists still could not forgo the sweet taste of morality. And so they settled for a religion substitute, or religion lite. Not only did Huxley sit comfortably at the top of the hierarchy of a "priesthood of science"—the press referred to him as "Pope Huxley"—but his letter to Kingsley shows that, for him, hands-on contact with the world substituted for holy communion. Darwin, particularly after being buried in Westminster Abbey, came pretty close to divine status. (History does not record whether disciples were hanging around three days later.) Strengthening the case for the religious nature of post-Darwinian evolutionism was its link to millennial thinking, which began to flourish in a major way in popular culture in the nineteenth century (Oliver 1978, Harrison 1979).

Many Protestants took the French Revolution and the Napoleonic Wars to be signs of great significance, especially given their negative implications for the Catholic Church. Perhaps the antichrist (that is, the pope) was now truly threatened. Those drawn to postmillennial varieties of apocalyptic prophecy, in which Christ comes after the millennium, generally had optimistic views of human nature. The brighter future was something to be achieved by us, rather than something imposed from above. We have seen that this kind of thinking was akin to, if not outrightly identical to, Progressivism of one sort or another. As in the century before, Christian eschatology and progressivist philosophies were often closely connected. Remember the tripartite division of Joachim of Flora and its echo in the theological/metaphysical/scientific schema of Comte.

Because the second coming was now postponed to some time in the indefinite future, increasingly such pictures appealed to people with few or no religious commitments. In Britain, the man who so influenced Alfred Russel Wallace, Robert Owen—who rejected all formal religious systems—was greatly influenced by millennial ideas. He famously (or notoriously) tried to set up a planned utopia in the New World (Harrison 1969). At times, Owen took on the mantle of the Messiah himself, writing of being compelled "to proceed onward to complete a mission" whereby "the earth will gradually be made a fit abode for superior men and women, under a New Dispensation, which will make the earth a paradise and its inhabitants angels" (R. Owen 1858, xlii–xliii). This kind of sentiment continued unabated into the post-*Origin* part of the century, except that now such thinking was fused with evolutionism. Typically, Herbert Spencer denied that his thinking was influenced by predecessors—he pretended that he reviewed a book by Comte that he had not read because he found the contents so distasteful.

But even if Spencer was telling the truth, others were in the stream of history. Wallace was not the only one who breathed a kind of inherited, secular postmillennialism (for many, as we shall see, not so very secular), with human effort being urged to bring about a brighter future. A form of evolution-based secular millennialism, complete with religious language and metaphors, was frequently invoked. Although identifying Darwin's bulldog (as Huxley was known) with the pontiff was intended to capture the way in which Huxley led the movement of science, one suspects that one or two in their country parsonages did think of Huxley as the antichrist. Admittedly, all sorts of different and contradictory things were being claimed in the name of progressive evolution, but in this respect it was no different from regular religions. Some Christians defend war on the basis of the gospel, others argue for pacifism; some defend capitalism, others argue for socialism; and some cling to female subordination (as with an all-male priesthood), whereas others believe that Christianity mandates gender equality. The point is that—for the secular millennialist—one must do something to bring about change and improvement.

Looking at matters another way, in Britain particularly many people, including professional scientists, wanted nothing at all to do with formal religion, and they saw science and religion locked in a deadly battle. As Huxley put it back in 1859, "Theology and Parsondom . . . are in my mind the natural and irreconcilable enemies of Science. Few see it but I believe we are on the Eve of a new Reformation and if I have a wish to live thirty years, it is that I may see the foot of Science on the necks of her enemies" (T. H. Huxley, 1859, para. 11). The years after the *Origin* saw such bestsellers as *The History of the Warfare between Science and Theology in Christendom*. Although the trial of Galileo would always be the main course in such orgiastic feasts as these, Darwin provided a tasty dessert—at least Darwinism did. Huxley wrote that new science has to demolish theological obstructions, rather as the infant Hercules strangled snakes while in his crib. For people subscribing to that nonbelief newly coined as agnosticism, evolutionism—with its progress, origins, human status, and exhortations—was just what was wanted in the new age.

Given his views on the Unknowable, Herbert Spencer could hardly be described as a materialist, but many evolutionists could. They were all at one in wanting to take God out of the equation and to let the laws of nature govern everything. Knowing that he had nothing of value to say on the origin of life issue, and knowing also (he wrote just at the time when Pasteur was putting the final boot into spontaneous generation) that critics would pounce on ignorance and wild speculation, Darwin wisely had said nothing on the subject in the *Origin*. But this did not stop others from jumping into the fray. In a way, they had to, because materialism was part of their overall picture. They were not just offering a theory, they were offering a system—a system with a beginning and a hoped-for end, an eschatology, no less than the system they were trying to replace.

By the last decade of his life, Thomas Huxley—who had deliberately called a set of his essays "lay sermons"—was losing faith in progress. But his ardor to see evolutionism as the proper alternative to the Christian religion burned no less brightly (White 2003). Thirty years had now passed since he had spoken of bringing on a new reformation, and he was not now about to abandon the god of science. But at this late point, a truly curious episode in Huxley's controversial life occurred. He attacked the Salvation Army and its leader, "General" William Booth, with a ferocity reminiscent of his long-ago attack on Chambers's *Vestiges*. In lengthy letters to the *Times*, Huxley accused Booth and the army of just about every sin in the book—totalitarianism, religious bigotry, fiscal mismanagement, and more. "Undoubtedly, harlotry and intemperance are sore evils, and starvation is hard to bear or even to know of; but the prostitution of the mind, the soddening of the conscience, the dwarfing of manhood are worse calamities" (letter of Dec. 1, 1890, in T. Huxley 2004, 283).

Huxley wrote not one word about the truly saint-making work that Booth and his followers had been doing with the poor and despised for over twenty years. Not one word about the soup kitchens and the beds, the care of drunkards and the sick and the truly desperate in society. Why? A little book collecting the letters, published by Huxley with the provocative title *Social Diseases and Worse Remedies*, gave away the game. The introduction was a reprinted essay about the virtues of science, and science education, and of moving from bigotry and ignorance, especially Christian bigotry and ignorance. Huxley (rightly) saw the Salvation Army as competing for his space. It was not part of the establishment—anything but. Like Huxley and his fellow evolutionists, the army was trying to address the failings that had come from the inadequacy of the establishment. Yet the army was doing it in exactly the opposite way—by sermons and sympathy, meetings and songs and brass bands and free grub rather than by workingmen's lectures and education; by fervent reliance on a simple reading of the Bible rather than by observation, experiment, and reason; by enrolling as its officers (its clergy) people from the lower ranks and impressing on them the need for conformity and obedience and a simple heart, rather than years of laborious, intellectual training. And what was worse, as Huxley's appalled reaction showed, the Salvation Army was having success and gaining credibility—especially with sympathizers among liberal Christians who were friends of the scientists and with the lower masses to whom Huxley's scientific priesthood was likewise catering. To use a phrase invented by Thomas Henry Huxley's biologist grandson, Julian Huxley, the evolutionists were truly in the business of providing a "religion without revelation"—and like all fanatics, they were intolerant of rivals.

9

Darwinian Ethics and Morality

Chapter 9, Reading 1

Darwinian Ethics[1]

Substantive Ethics

WE ALL KNOW WHAT morality is about. It is about helping other folk. It is about giving to the poor and to the sick. It is about loving your neighbor as yourself. It is about being decent and kind and truthful and honest and reliable, and a host of other things. It is about being a good person rather than a bad one. But how do we tie together all of these different feelings and insights? Obviously, if you believe that honesty is the best policy, you should not cheat on your income tax returns. Yet does this mean that you should scrupulously tell the truth to a dying child? Does honesty have anything to do with not swearing in front of maiden aunts (or uncles)? Can we spell all of this out without making reference to God, in some way?

To pick up first on the last point, there is little doubt that many (most?) people would indeed spell out their moral beliefs within some sort of religious context. They would refer you to the Ten Commandments, or to the Sermon on the Mount, or (if they were against sex) to the Epistles of Saint Paul. However, this only takes you so far. Grant that one ought not to kill and that one ought not to commit adultery. Are these moral absolutes, or do they derive from more powerful premises? Jesus reduced the Ten Commandments down to two—but why stop there?

1. Ruse 1998a. Reprinted with permission of Prometheus.

And what of those of us who do not subscribe to the Judeo-Christian religion? Are we beyond the moral pale? Obviously not!

Advocates of utilitarianism argue that pleasure or, in more refined versions, happiness, is the supreme and indeed only good. One's actions should be directed toward the promotion of happiness, for oneself and others and, conversely, toward the diminution of unhappiness, for oneself and others. The classic statement comes in John Stuart Mill's celebrated essay on the topic:

> The creed which accepts as the foundation of morals. Utility, or the Greatest Happiness Principle, holds that actions are right in proportion as they tend to promote happiness, wrong as they tend to produce the reverse of happiness. By happiness is intended pleasure, and the absence of pain; by unhappiness, pain, and the privation of pleasure. (Mill 1910, 6)

Thus, morally, all you have to do is ask: "Does my thought or act make for an increase in happiness?" If it does, it is right. If it does not, it is wrong. Helping the sick increases happiness. Therefore, such actions are right. Robbing widows decreases general happiness. Therefore, such actions are wrong.

Today, there are enthusiasts for two major variants of utilitarianism (Lyons 1965). So-called *act utilitarians* argue that it is the happiness/unhappiness of each individual act that counts. *Rule utilitarians*, on the other hand, argue that it is the general policy that counts. Does the policy normally increase happiness/decrease unhappiness, or not? Thus, an act utilitarian might decide that a particular situation calls for a lie, whereas the rule utilitarian might feel that the general policy rules out any untruth. Even though a certain deception might indeed reduce unhappiness, the general end of maximizing happiness is better served by sticking to the rule of honesty and plain speaking (both positions are discussed in Mackie 1977).

The major substantive ethical alternative to utilitarianism was most brilliantly articulated in the eighteenth century by Immanuel Kant (1949, 1959). Kant argued that, in the moral world, as rational beings we are subject to what he called the "categorical imperative." This supreme norm was presented in a number of different ways. In one version, Kant picked up on the notion of universality, which (as we learned earlier) is central to the very meaning of a moral claim: "Act only according to that maxim by which you can at the same time will that it should become a universal law" (Kant 1959, 39). Thus, for instance, borrowing money with no intention of repaying is wrong, because if everybody did that society would collapse. There would be "contradictions." If morality means anything, it means being prepared to hold out a helping hand to others. Christians, utilitarians, Kantians, and everyone else come together on this. (Duties to oneself are a different, more complex matter; but they need not detain us here.)

The problem, raised by Huxley and a host of others, is that natural selection and its products are prima facie the very antithesis of help and cooperation. We start with a struggle for existence, and go on to find that winning alone counts from an evolutionary perspective. Because of this, virtually all of our features, physical and mental, are directed to personal success. Selfishness personified! No wonder that Huxley wrote: "Let us understand, once for all, that the ethical progress of society depends not on imitating the cosmic process, still less in running away from it, but in combating it" (T. H. Huxley and Huxley 1947, 82). There is nothing moral in the process of evolution, and there is no morality in its effects.

However, as Charles Darwin himself and many followers down to and (especially) including Edward O. Wilson have pointed out (C. Darwin 1871, E. Wilson 1975), to assume that matters rest here is naivety personified. Supporters of evolutionary ethics, like Sumner, and opponents like Huxley, have been sharing false empirical premises. Natural selection does indeed promote features that rebound to self-benefit, but to conclude that we all spend our days like characters in a spaghetti Western, forever grinding opponents into the dust, is to show a farcically incomplete grasp of the evolutionary process. You can frequently further your own ends much more successfully by subtle alternative strategies. In particular, you can often get a lot more for yourself by aiding and working with others. In other words, selection can be expected to promote what biologists call altruism. . . .

It is necessary to tread carefully here. A good number of evolutionists, thinking they were working in the true spirit of Darwinism, have argued that humans (and other animals) help each other as a natural consequence of that inevitable spirit of friendship which binds members of the same species. This friendship supposedly evolves because it is of benefit to the whole group. Zebras working together against predators thus keep up the species. This was the refrain of natural selection's codiscoverer Alfred Russel Wallace (1916), and (as noted earlier) was stressed in most detail by the late nineteenth-century Russian anarchist

Prince Peter Kropotkin (1902). However, as we know, this approach was explicitly rejected by Darwin, as well as by his modern-day supporters. Any "group selection" analysis of behavior, including human behavior, falls before strong counterevidence.

The Darwinian insists that we stay with the individual. All help given must rebound ultimately to the individual's benefit. Any benefits that others receive should be seen as incidental, and might well be selected against. Nevertheless, even within these constraints, help and cooperation can evolve. We all know of the principle of enlightened self-interest, where I help others because I thereby get help in return. . . .

Against this background, Darwinians who study social behavior (*sociobiologists*) strongly favor two primary mechanisms supposedly capable of producing help and cooperation between humans (Alexander 1979, Trivers 1971, Ruse 1979). The first is the relatively obvious process whereby humans develop innate tendencies to work together, because the cost of cooperation is (on average) significantly less than the hope of return. Suppose we all stand in risk of drowning. I help you from drowning because of my biological urges to do so. Although this puts me at a one in twenty risk of drowning myself, I in turn avoid the one in two risk of drowning were you never to respond to my sometime cry for help. I may not need such help now, but we were all young once, we will all grow old someday, we all fall sick on occasion.

We all have a share of bad luck during our lives, and the same goes for our children, the most immediate bearers of our genes. This mechanism for promoting cooperative interactions between humans is called a *reciprocal altruism* (Trivers 1971). It can occur between genetic strangers, although in real life these could and may well be good friends. In theory, it can even occur between humans and members of other species (the shepherd and his dog). The important distinguishing feature is that, although help is given, returns are in some way anticipated. Pushing the insurance model, one does not necessarily expect immediate repayment for every kind act. Rather, one throws one's help into the general pool, as it were, and expects to be able to draw on the pool as needed. (Why not cheat? Why not take without giving? Because, if everyone behaved this way, the system would collapse. Nevertheless, because evolution is always looking for ways to get ahead, you expect a certain amount of cheating. Also, you expect the evolution of techniques for spotting and preventing cheating.) Darwin himself proposed reciprocal altruism as a possible causal factor behind the help that we humans give to each other: "As the reasoning powers and foresight of the members became improved, each man would soon learn from experience that if he aided his fellowmen, he would commonly receive aid in return" (C. Darwin 1871, 1:163). Note, however, that Darwin did not (and, because of his ignorance, could not) relate his insight to our underlying genetic nature. The second supposed mechanism of (biological) altruism is much more recent, and indeed could not have been developed without a proper knowledge of the principles of heredity. The key to evolutionary success lies in improving your gene ratios. This means passing on your genes at a higher rate than do others. But note that, literally, you do not pass on your genes. If successful, you leave behind copies of your genes.

My children do not have my actual genes. They each have a half set, which are exact replicas of mine. But this possession of genes exactly like mine holds true of people other than my children—parents, siblings, nephews, nieces, grandchildren, and more. All of one's blood relatives have, to greater or lesser extent, copies of the same genes as oneself. Therefore, inasmuch as relatives reproduce, copies of one's own genes are being passed on. Reproduction by proxy, as it were. What this all means is that help given to relatives in itself rebounds to the favor of one's own reproductive interests, even though these relatives may themselves reciprocate with little or no help. As a consequence, what you expect through this process, known as *kin selection*, is the evolution of help-giving attributes, without necessarily having the parallel evolution of attributes, expecting or enforcing tangible returns. Any return comes indirectly, via the genes. . . .

Summing up, Wilson labels the results of the mechanisms of help as *hard-core* and *soft-core* altruism. The former is the result of kin selection. It occurs between relatives, and there is no expectation of direct return. The latter is the result of reciprocal altruism. It occurs between nonrelatives, and there is expectation of return, or at least of the potential for such return. . . .

Neither regarding human sentiments caused by kin selection, nor regarding human sentiments caused by reciprocal altruism, is the Darwinian claim that we humans know wherein lie our biological ends. The person who helped another, consciously intending to promote his own biological advantage, would not be moral. He would be crazy. The Darwinian's point is that our moral sense is a biological adaptation, just like hands and feet. We think in terms of right and wrong. It so happens that the overall effects are biological. . . .

This is the (empirical) Darwinian case for morality, and for its biological underpinnings. Epigenetic rules giving us a sense of obligation have been put in place by selection, because of their adaptive value. Of course, as with scientific knowledge, no one is claiming that every last moral twitch is tightly controlled by the genes. In science, the claim was that human reason has certain rough or broad constraints, as manifested through the epigenetic rules. The application of these leads to the finished product, which in many respects soars into the cultural realm, transcending its biological origin. In the case of ethics, the Darwinian urges a similar position. Human moral thought has constraints, as manifested through the epigenetic rules, and the application of these leads to moral codes, soaring from biology into culture. . . . Darwinism insists that features evolve gradually, and something as important as morality should have been present in our (very recent) shared ancestors. Furthermore, if morality is as important biologically to humans as is being claimed, it would be odd indeed had all traces now been eliminated from the social interactions of other high-level primates. Conversely, if human morality does not have a biological base, and is for instance a cultural invention of humans some few thousand years ago then there would be no reason to find it present in our ape relatives.

Recent, extended studies of the apes, particularly of chimpanzees, must shake all but the most dogmatic defender of the uniqueness of the human moral capacity (Hrdy 1981). I emphasize that these studies are recent. Anecdotal reports of apes showing friendliness and concern are obviously of limited worth. Apart from anything else, if an ape is reared close to humans, you expect it to learn humanlike behaviors, including those simulating morality (Goodall 1971). The question is whether the brutes show moral-type behavior innately, without human intervention, in situations that are clearly directly or indirectly of biological value. And to answer such a question, you need long-term studies of apes in natural (or virtually natural) situations, seeing whether or not such behavior appears. At last, such studies are being performed, and the answer is strong and clear. Apes interact in remarkably humanlike fashions, including fashions that, were we to believe them true of humans (rather than apes), we would unhesitatingly label "moral."

I draw your attention to perhaps the most remarkable of all the studies, that of the Utrecht primate ethology group (Waal, 1982). For over a decade, its members have been observing the dynamics of some fifty semi-wild chimpanzees at the nearby Arnhem Zoo, recording almost every move. Time and again, the primatologists have seen behavior that differs not at all from human moral behavior. This is particularly true of the older females. Although it is the mature males who are the physically dominant members of the group, the older females have great authority. For instance, the males seek their aid in forming alliances, younger females look to them for protection, youngsters are wary of them and set them up as models. In a human group, these females would have an important stabilizing effect, helping to mediate disputes and to avoid quarrels. The same is precisely true of the Arnhem chimpanzees. . . . Chimpanzee groups are forever on edge of exploding into conflict, primarily because of tensions between alpha males. Group members are, accordingly, forever smoothing over differences, reconciling rivals, and binding psychic wounds. . . . These are but a few straws in the wind; yet they must suffice for now.

Summing up: there is strong, and growing, evidence through the animal world that members of the same species interact socially to their mutual reproductive benefits. The nature of these interactions fits well with the claim that kin selection and reciprocal altruism are important causal mechanisms. Our closest relatives, the chimpanzees, have complex social lives, and behave in precisely the ways one would expect were morality a legacy of our simian past, and were that legacy also inherited by other primates. We humans, especially in our preindustrial state, show that biology is a crucial causal factor affecting our social nature, and the ways we behave are precisely those expected if selection acts to maximize the reproductive potential of the individual. . . . In broad outlines, Darwinism meshes happily with utilitarianism. Does it favor rule utilitarianism or act utilitarianism? One suspects perhaps the former, given that the human mind seems to work by rule, rather than by deciding each issue anew. . . .

My sense is that biological feelings—including moral feelings—toward one's own children must be stronger than feelings toward the children of others, or to adults for that matter either. To the Darwinian, morality is an important factor in getting you to do what is in your evolutionary interests. I find it implausible that selection would fail to back up your parental emotions with an increased sense of moral obligation. Help from others may or may not prove useful. Saving the skin of close relatives is absolutely crucial. Thus I suggest that the Darwinian looks for a greater feeling of obligation to promote the happiness of your own children than those of anyone else. This is not to say that you have no obligations to the children of others. But it is to say that not only do your children most probably come first in your affections, but you feel that they should come first. Consequently, the Darwinian does not expect you to feel undue guilt about spending money on your children, rather than sending all to Oxfam. (No one denies that your present distribution of funds may be morally unacceptable.) . . .

We come now to the fundamental questions about the ultimate ground of morality. The initial, and perhaps most pressing question of all, centers on the very nature or meaning of morality. Thanks most particularly to Hume, we realize that when we make a moral claim, we simply do not mean the same thing as when we make a nonmoral claim. The statement "I find the thought of sex with small children upsetting" is quite different from "Pedophilic behavior is morally wrong." The second has a sense of obligation, of force about what you ought or ought not do, which the first does not have. I could have a real yen for sex with children, and yet think it grossly immoral. . . .

Now, at once the critic will charge that the Darwinian approach glosses right over this point, and thus flounders on its own insensitivity. Let us grant the scientific case, that epigenetic rules influence our social intercourse, and that thanks to kin selection and reciprocal altruism human behavior rebounds to the reproductive benefit of the actor. But this says nothing about morality! At most, we have wants, wishes, desires, hates, and associated actions. Moreover, these are all directed to individual self-gain. That is to say, they are selfish (Singer 1981, Trigg 1982). This is the very antithesis of morality. Huxley was right. Right action opposes the Darwinian course of nature. It does not complaisantly acquiesce in it. . . .

It is precisely because of the limited potential of normal feelings alone that the Darwinian posits the need for morality—something which goes beyond such feelings to obligation. Biologically, we need to back up our normal feelings, and to help and cooperate with others. We need a spur to make us change a diaper in the middle of the night, or to coach for the spelling test. We need to push a make us aid our neighbor when his barn burns down. Our sense of morality, our sentiment of obligation, comes in here. We have evolved epigenetic rules that make us do things because they are right, and abstain from other things because they are wrong. These rules drive us into social action, above and beyond and perhaps despite our inclinations. . . .

We must ask whether, to the Darwinian, morality is—because of the science, must be taken as—something objective, in the sense of having an authority and existence of its own, independent of human beings? Or whether morality is—because of the science, must be taken as—subjective, being a function of human nature, and reducing ultimately to feelings and sentiments—feelings and sentiments of a type different from wishes and desires, but ultimately emotions of some kind? I claim that, having accepted the natural evolution of morality, the Darwinian is forced to take the second option. The naturalistic approach, locating morality in the dispositions produced by the epigenetic rules, makes our sense of obligation a direct function of human nature. We feel that we ought to help others and to cooperate with them, because of the way that we are. That is the complete answer to the origins and status of morality. There is no need to invoke (and much against invoking) some Platonic world of values. Morality has neither meaning nor justification, outside the human context. Morality is subjective.

In a sense, therefore, morality is a collective illusion foisted upon us by our genes. Note, however, that the illusion lies not in the morality itself, but in its sense of objectivity. I am certainly not saying that morality is unreal. Of course it is not! What is unreal is the apparent objective reference of morality. I hasten to add that I am not now suggesting that morality is in any way a sign of immaturity. Nor would I have those of us who see the illusory nature of morality's objectivity throw over moral thought, as suggested by Plato's Thrasymachus or Nietzsche's Superman. Morality is part of human nature, and . . . an effective adaptation. Why should we forego morality any more than we should put out our eyes? I would not say that we could not escape morality—presumably we could get into wholesale, anti-morality, genetic engineering—but I strongly suspect that a simple attempt to ignore it will fail. . . .

The analogy with morality fails. At the least, the objectivist must agree that his/her ultimate principles are (given Darwinism) redundant. You would believe what you do about right and wrong, irrespective of whether or not a "true" right and wrong existed! The Darwinian claims that his/her theory gives an entire analysis of our moral sentiments. Nothing more is needed. Given two worlds, identical except that one has an objective morality and the other does not, the humans therein would think and act in exactly the same ways. . . . I did not choose my moral code. For the Darwinian, the very essence of morality is that it is shared and not relative. It does not work as a biological adaptation unless we all join in. Unless there is this joint participation, the illusion of (objective) morality will not keep afloat. It is only in my biological interests to have moral sentiments if you likewise have such sentiments. . . .

Moreover, when a prominent Darwinian writes as follows, all fears seem realized.

> The genes hold culture on a leash. The leash is very long, but inevitably values will be constrained in accordance with their effects on the human gene pool. The brain is a product of evolution. Human behavior—like the deepest capacities for emotional response which drive and guide it—is the circuitous technique by which human genetic material has been and will be kept intact. Morality has no other demonstrable ultimate function. (E. Wilson 1978, 167)

Despite Wilson's words, a quick conclusion that the Darwinian approach makes genuine ethics impossible is unwarranted. . . . As a function of our biology, our moral ideas are thrust upon us, rather than being things needing or allowing decision at the individual level. This is the claim. Just as we have no choice about having four limbs, so we have no choice about the nature of our moral awareness. (I will ignore obvious questions about genetic manipulation.) But who, other than perhaps some of the existentialists, ever really pretended that we have choice in this respect anyway? Kant is surely right in arguing that the supreme principle of morality is categorical—it is laid upon us, without any "ifs" and "buts." We are not free to choose what right and wrong are to be. Where the freedom comes, if it is to come at all, is in working within the given bounds of right and wrong: "A free will and a will under moral laws are identical" (Kant 1959, 65). . . .

A human's choice and action is a function of that person, as he/she . . . interacts with the environment. This is not non-caused freedom. Given all of the information from outside and given the way that we work, then our thoughts and actions will follow necessarily. But it is a freedom denied the (obviously unfree) bound prisoner, and it is a freedom denied the (equally obviously unfree) rigidly programmed ant. Morality gives us standards that we feel the demand to follow; but there is nothing within or without us that alone determines that we must or must not follow these moral demands. We can respond to morality, and depending on circumstances we may or may not follow it. This is our freedom.

The analogy I like is that of missiles zeroing in on a target. Ants are like those missiles that have their expected target position built in.

Their social behavior is firmly genetically controlled. There are benefits in doing things this way. It is simple and cheap (in whatever costs you have). Usually, such missiles/ants work just fine. But, of course, things do break down, particularly as conditions tend to complexity. Humans are like those missiles that have internal homing devices. They can respond to changes in target positions because they can pick up information and modify their courses. Their social behavior is not firmly genetically controlled. However, the genes through morality influence behavior, as do programs put into the missile's homing device. Neither humans nor missiles have to figure out everything, right from scratch. Neither is programmed to work blindly, but each is programmed to respond to certain guides. Again, there are benefits in doing things this way. But it is more costly and more prone to internal problems.

One can push this analogy in ways that a biologist rather likes. Note, for instance, that any sensible defense minister will want missiles of both kinds. This corresponds to the Darwinian's conviction that there is no uniquely "best" way of doing things. Ants and humans are both biological success stories. The point I make here, however, is that humans (unlike ants) have a dimension of freedom, just as do missiles with homing devices (unlike fixed-direction missiles). . . . A Darwinian approach to morality does not call for a repudiation of standards and values cherished by decent people of all nations. As with epistemology, Darwinism tells us much about ethics. It does not call upon us to repeat the fallacies of the past, or to provide new ones for the future.

Chapter 9, Reading 2

Evolutionary Ethics: The Debate Continues[1]

I BELIEVE THAT ETHICS is an adaptation, put in place by our genes as selected in the struggle for life, to aid each and every one of us individually. Because it is a social adaptation, I believe that essentially we (societies, but at some ultimate level the whole human species) share the same ethics, and that charges of relativism are ill taken. I believe also that ethics is genuine in the sense that people really do do things because they think them right (and conversely), and connected with this I would argue that there is a real difference between the language of ethics and the language of other aspects of human life, specifically those about matters of fact. However, my claim is that ethics is without justification or foundation—in this sense, I am a noncognitivist—although I do think that an essential component of ethics as an adaptation is that we believe that ethics does have a real foundation (we "objectify"). You might reply that this is precisely what one would expect from a Humean such as myself. My enthusiasm for something akin to a moral sense is no more than gut feelings by another name. And in a way this is true, and this is why I am glad to have Ayala as a foil to sharpen and clarify my own position. But I would also, vigorously, defend my position as the right one. Consider two men, one of whom unthinkingly shares

1. Ruse 1995. Reprinted with permission of Routledge.

his crust of bread with the starving child and the other who agonizes over the situation before sharing. Are we to say that only the second is the truly good? There are those who would say, and those with whom I would agree, that the first man is far closer to God.

Actually, my suspicion—confirming my evolutionism and belief that morality is not in fact a creation of an all-wise God but the imperfectly adaptive product of natural selection—is that we might have a place here where basic intuitions conflict. Of course one gets moral credit for worrying and for triumphing over temptation. Of course one gets moral credit for doing good spontaneously. In real life though, we need both of these supports. Only occasionally do they come into conflict. Perhaps emotions keep you going most of the time, but every now and then you have to invoke reason. I would not think much of a bachelor who agonized over sharing his crust with a child, even if he finally did so. I would not think much of a man with a large family who readily gave away what he had to the first comer, without thinking and calculating the effects on and obligations to his own children. My point is that my position can appreciate this difference and that Ayala's position cannot. The most detailed (and negative) analysis of my position has come from the collaborating colleagues, philosopher William Rottschaeffer and biologist David Martinsen. In the course of their biological critique, they accuse me of overextending the true consequences of Darwinism, arguing that although biology yields the moral sentiments needed for immediate action, it does not produce the full-blown moral dispositions needed for full moral personhood.

> Let us distinguish between . . . a bare disposition and both a cognitive/emotional disposition and moral disposition. A cognitive/emotional disposition can manifest itself in subjective feelings and cognitive states and is reportable intersubjectively. A moral disposition has the features of the cognitive/ emotional disposition as well as the features ascribed to moral sentiments by Ruse. If parents possess a moral disposition, they report that they feel morally obliged to care for their children and that by that intend that these feelings have the characteristics of full-fledged prescriptivity and universality proper to moral norms and principles. What Ruse, therefore, needs to make his case for the primacy of nature in the origin of moral sentiments is evidence that moral sentiments are moral dispositions and that such sentiments are included states. But Ruse has failed to provide this evidence. (Rottschaeffer and Martinsen 1990, 154–55)

Hence we learn that culture, in the form of moral education, must come into play, and only then will a person have the second-order dispositions through which he or she as a thinking being can attempt to order and control his or her basic sentiments (gut feelings, in my language). To which criticism I want to say two things, although first let me express my entire agreement that moral reasoning will often (always?) involve the evaluation of first-order sentiments against second-order principles. This is the basis of my (Humean) freewill defense. Yet on the one hand, I am appalled that my writing is apparently so unclear that anyone could think that I deny the existence or importance of moral training. Humans are animals that develop through the interaction of their innate dispositions, as cultivated through their childhood training. Language is the paradigm, and I see no reason to make morality an exception. This is not to say that I think that any training whatsoever will lead to a moral being, or that I think that every moral lapse is due to inadequate training. (Rottschaeffer and Martinsen report on some empirical studies supposedly showing that, because of their training, some people have not internalized moral principles but act solely out of fear of punishment [1990]. To which I can only reply that these people have been turned into moral cripples.) Let me switch now to those (staying still at the empirical level) who make claims that certainly do not threaten my position, and might indeed be used to augment it. . . . Some people have been making general points of clarification, with which I can only nod agreement. One is the biologist David S. Wilson (1992), who warns that we should take care in our use of such terms as "selfish" and "altruist."

> The fact that evolutionarily successful behaviors are not necessarily selfish, and that proximate mechanisms are designed to elicit evolutionarily successful behaviors regardless of whether they are selfish or altruistic, destroys any hope for a simple relationship between definitions based on fitness effects and definitions based on motives. Not only can altruistic behaviors (in the evolutionary sense) be selfishly motivated (in the psychological sense), but the reverse is also true; individuals that care truly for others can be selfish in the evolutionary sense. . . . These observations are elementary but they are not sufficiently appreciated by evolutionists or philosophers interested in the concept of altruism. (D. Wilson 1992, 66)

The Naturalistic Fallacy and Beyond

I move now to philosophy, starting with those for whom a position like mine is altogether too much. The traditional criticism is that any attempt at an evolutionary ethics falls on the naturalistic fallacy, or on an illicit move from "is" to "ought." This charge is certainly not absent from the recent literature, although it is perhaps surprising that this is the main complaint of Ayala (1987), . . . given that he above all others has made so much of his enthusiasm for biological progress (1974, 1988).

> Because evolution has proceeded in a particular way, it does not follow that that course is morally right or desirable. The justification of ethical norms on biological evolution, or on any other natural process, can only be achieved by introducing value judgments, human choices that prefer one rather than other object or process. Biological nature is in itself morally neutral. (Ayala 1987, 245)

To which I can only reply that this may be a problem that troubles the positions of others (in fact, you know that I think it is), and it may be a problem that should trouble me (in fact, others will argue that it is), but it is certainly not a problem to which I am insensitive. Seeing a difference between "is" and "ought" is where I start, not where I end, nor what I ignore. So unless someone makes a reasoned case against me, I shall slough off the traditional criticism—not because I think it without force, but precisely because I think I am using that force to my own ends.

Defending Justification

I turn next to one whose position is stronger than mine, for he talks confidently of "justification" and of driving through (rather than round) the naturalistic fallacy. I confess that I do somewhat dread the task of giving a fair and acceptable-to-the-author account of the position, for apparently it is a universal fault of hitherto fair-minded commentators that their expositions fail to do the position justice—a fact that may, of course, be due to the position's subtlety and innovation rather than to the fact that the author changes his thinking, on the jogging shoe as it were. But given that the author has shown an Indiana Jones enthusiasm for defending impossible positions against all comers, extricating himself from certain philosophical death when any reasonable person would have started to work on his or her own obituary, I feel certain that any misreadings of mine will be justly censured, to the edification of all and the amusement of many.

Like me, Robert J. Richards has a two-pronged argument, directly empirical (that is, drawing on the work of empirical scientists) and subsequently philosophical (1986a, 1986b, 1987, 1989). It is probably fair to say that, at the empirical level, in line with a general liking for group selection–type arguments at his home base of the University of Chicago (Wade 1978), Richards is inclined to a more holistic account of human evolution than I am. In particular, in what he truly notes is probably a position more closely Darwinian (in the sense of what Darwin actually held, rather than what Darwin should have held) than mine, Richards sees human morality as having emerged from a kind of selection between bands of protohumans, generally although not necessarily closely related. But, although I would probably give a greater role to reciprocal altruism, I think it fair to say that viewed from a distance, at this level, even friends would say that our points of overlap are closer than our points of difference. (Actually, if we are going to slug it out on the minutiae of Darwin scholarship, I would suggest that in major respects, my position is more in the spirit of Darwin [Ruse 1980, 1986b]).

Summarizing his position, Richards writes:

> My strategy is to reveal that any ethical framework that might be urged upon us depends on a variety of empirical assumptions. I attempt to show, for instance, that philosophers who argue for the adoption of any normative framework—even that of modern logic—employ a common strategy, namely to justify the adoption by showing that the framework sanctions certain empirical descriptions that are deemed well confirmed. This leads me to reject the common belief that inferring values from facts is ipso facto fallacious. (Richards 1989, 337)

Two arguments for the necessity of morality are particularly important. The first, modeled on the metaethical theory of the neo-Kantian Alan Gewirth (1986), attempts to get "ought" statements from factual "is" statements, a move that is apparently allowed when one has something "necessitated or required by reasons stemming from some structured context" (Richards 1987, 287). From this, Richards argues:

> Evolution provides the structured context of moral action: it has constituted men not only to be moved to act for the community

> good, but also to approve, endorse, and encourage others to do so as well. This particular formation of human nature does not impose an individual need, not something that will be directly harmful if not satisfied; hence, the question of a logical transition from an individual (or generic) need to a right does not arise. Rather, the constructive forces of evolution impose a practical necessity on each man to promote the community good. We must, we are obliged to heed this imperative. We might attempt to ignore the demand of our nature by refusing to act altruistically, but this does not diminish its reality. The inability of men to harden their consciences completely to basic principles of morality means that sinners can be redeemed. Hence, just as the context of physical nature allows us to argue "Since lightening has struck, thunder ought to follow," so the structured context of human evolution allows us to argue "Since each man has evolved to advance the community good, each ought to act altruistically." (Richards 1986a, 288)

Frankly, this all looks a bit on a par with John Stuart Mill's notorious "proof" of the greatest happiness principle—we all like to be happy, so we ought to promote it for all—although I think that in a sense Richards would not take this analogy as a criticism. One of his more subsidiary arguments is that fundamentally he is not doing anything different from that which any metaethicist does, and he stresses again and again that ultimately the only proof that you have for morality is that which (as mentioned above) you offer for logic, namely that people accept it. In conjunction with this point, Richards is keen to point out that although his argument depends on a move from factual premises to moral conclusions, this is not done in ignorance but is licensed by an inference rule along the lines of "given a certain factual situation, then certain moral imperatives will obtain." He notes that if one is to avoid an infinite regress in one's argumentation, then (as Lewis Carroll showed) one has to stop the demand for justification at some point. And ultimately, this means ending "in what are regarded as acceptable beliefs or practices" (Richards 1987, 285). That is: "All meta-level discussions, all attempts to justify ethical frameworks depend on such inference rules, whose ultimate justification can only be their acceptance by rational and moral creatures" (Richards 1987, 285).

However, in the specific case we are dealing with, Richards does seem to feel that he escapes the usual objections (centering on the naturalistic fallacy) because of the peculiarity of the situation, namely that we are dealing with matters of human evolution, and that these have been matters leading to specific facts about morality, most importantly that we humans endorse them! "Moral 'ought'-propositions are not sanctioned by the mere fact of evolutionary formation of human nature, but by the fact of the peculiar formation of human nature we call 'moral,' which has been accomplished by evolution" (Richards 1989, 288–89). Backing this first argument, Richards offers a second major argument: the evidence shows that evolution has, as a matter of fact, constructed human beings to act for the community good; but to act for the community good is what we mean by being moral. Since, therefore, human beings are moral beings—an unavoidable condition produced by evolution—each ought act for the community good.

New Ethical Principles?

I come to the final part of my discussion, where I look at those who are fundamentally sympathetic to a position such as mine, but who feel that they can do the job better than I. There are two critiques or advances of or on my work that I shall consider; but to prepare the way, let me first take up a criticism, endorsed by Elliott Sober (1994), which they apparently think shows the implausibility of a position such as mine. The point of dispute revolves around two possible claims, with respect to ethics, that one might advance on behalf of biology:

> (1) Sociobiology can teach us facts about human beings that, in conjunction with moral principles that we already accept, can be used to derive normative principles that we have not yet appreciated.

> (2) Sociobiology can lead us to revise our system of ethical principles, not simply by leading us to accept new derivative statements—as in (1)—but by teaching us new fundamental normative principles.

Sober is happy to accept (1), but reject (2)—which I am supposed to accept—as too strong. But is this a consequence, implicit or explicit, of my position? I rather think not. Let us take an example of where sociobiology might lead us to revise our thinking about human moral behavior, namely that centering on the relationship between stepparents and stepchildren. In a society that is basically ignorant of the significance of the biological bond—I speak now with some experience and feeling when I thus characterize England in the 1950s—a huge amount of guilt can ensue when stepparents and children do not mimic exactly the close relationships that one expects and generally finds between natural parents and their children. However, now—thanks to a much deeper understanding of human relationships, an understanding to which sociobiology has contributed—we realize that social relationships can rarely if ever replace biological relationships, and that to effect such social relationships as well as we can we need understanding and sympathy rather than condemnation and guilt.

However, before I turn to my critics, let me admit that there are senses in which I do want to go beyond proposition (1), even if not as far as proposition (2). For a start, and not particularly contentiously, I can well imagine that a better understanding of biology might make us more sensitive to and appreciative of moral feelings that we have already, just as psychoanalysis was supposed to make us aware of general feelings that are already there. In my own case, to give a real example, my understanding of biology has helped me to realize how insincere I was when I used to mouth conventional platitudes about obligations to the Third World. It is not that I am now less moral in my attitudes to life's unfortunates—in fact, I am inclined to think that from the point of view of action, I am if anything more moral—but I no longer claim what I do not believe. For a second, perhaps more contentiously, I have always admitted that biology might make us realize that we have to treat certain moral principles more warily, perhaps even rejecting them. My point is that I am not sure that one does this in the name of normative principles, and certainly not "new fundamental normative principles."

My point is that a better understanding of biology might incline us to go against morality—especially if, as I do, you think of morality very much as something working at the immediate, personal level. We would go against morality for the sake of long-term goals, which I suspect will often center on personal (including descendant) survival. The sort of example I have in mind is the forcible prevention of people from having children, especially as many as I have had, for the sake of world population. If world population is not limited, I and my descendants will probably suffer. My position is unfair. The attitude of the Chinese authorities, preventing more than one child per couple, seems morally repugnant. I am not sure, however, that I, or they, are unreasonable. (Of course, human nature being what it is, we find excuses—in my case that five well-educated children will be a benefit to the world—outweighing the actual addition of numbers. But it is a gloss.)

Generally, I think that morality is in our own best interests. We humans are social animals. Why go through life trying to cheat on those whose company and friendship we need as much as we need food and drink? Or, rather, why go through life trying to cheat on people more than we do? And where there is a real conflict between morality and self-interest, I just do not think that we can simply suppress our feelings at the dictate of our reason.

The point about morality, as I have characterized it and tried to demonstrate in my supporting arguments, is that there is going to be flexibility and a range (perhaps unlimited) of possibilities. If what I have said about the possibility of alternative moralities is correct, then you are not going to get "good" corresponding to "red" and "bad" to "blue" and so forth, because depending on the way that evolution has gone, good and bad could be different according to the circumstances of the case—in a way that would not be so of red and blue. Put it this way. I can imagine it being true that one ought to hate and try to cheat neighbors. I cannot imagine that the sky be red and the grass blue. (And as I have said, if you can give me physicochemical reasons why I am wrong, I argue for the increased subjectivity of color rather than the objectivity of morality.) But if what I am saying is true, then it is simply not the case that there is in any sense an objective morality, waiting for selection to mold our natures around it—nor will culture be able to complete the task that biology has begun. Morality comes with us, and it comes with us by courtesy of our biology. The problem as I see it is that, for all their claim that it is I who have not taken Darwinism truly seriously, Rottschaeffer and Martinsen are the ones who have not absorbed the true message of sociobiology.

Evolution works not just between individuals and the outside world, that is the world of things, but it works equally between individuals and individuals, considered as individuals. Morality is a creation of the genes to help us get on with our fellows, not to help us get on with physical creation. As such, we should not expect to find, as indeed we do not find, that morality has any existence beyond the relationships between individuals. And as always in evolution, although we may skin the cat pretty well, there are probably many other ways in which the job might have been done.

Chapter 9, Reading 3
Darwinian Evolutionary Ethics[1]

What Is Evolutionary Ethics?

SOME IDEAS ARE NOT simply wrong; they are morally and aesthetically rather grubby. You know that people who push them almost certainly have issues of one sort or another—generally with authority or, more specifically, with certain racial groups or some such thing. For the first hundred years after Darwin published his *Origin of Species* in 1859, most Anglophone philosophers felt very much that way about evolutionary ethics, the attempt to explain and justify moral feelings and behaviors on the basis of our simian pasts (Ruse 2009b, 2009d). Thus in the first year of the journal *Mind*, we find the noted utilitarian philosopher Henry Sidgwick arguing that when it comes to ethics, evolution is just not that relevant (1876). His student G. E. Moore famously made evolutionary ethics the paradigmatic example of wrongheaded arguments, the worst of all possible ways of committing the so-called naturalistic fallacy (1993). And his student C. D. Broad was still at it thirty or forty years later (1944).

Prima facie this is all a little bit odd, because everyone recognizes that the world after Darwin is very different from the world before Darwin (Ruse 2017g). Even those who went on being religious saw that much, and those who had already jettisoned conventional religious belief saw only

---

[1]. Ruse 2017b. Reprinted with permission of Cambridge University Press.

confirmation that the path they had taken was correct. Surely, therefore, evolutionary theory in general and Darwin's theory of evolution through natural selection in particular were going to have some major implications, and nowhere more than on the ways in which we think and behave about right and wrong? And yet, not so. There are various reasons why evolutionary ethics found little favor among the philosophers (Cunningham 1996). Whether these reasons are good or bad, they are matters of history and dealt with in detail in other contributions to this volume, so I will not dwell on them here. The question here is with the revival of evolutionary ethics and the positive case that can be made.

I should say that I see the main issue is that of getting over the Humean is/ought distinction, of which I take the naturalistic fallacy to be a variant—the claim that you cannot derive matters of value from matters of fact—and I see two different approaches to the problem. One favored by Robert J. Richards that I think can be found in Herbert Spencer (1892) and then in a chain including Julian Huxley (1943)—the biologist grandson of Thomas Henry Huxley and older brother of the novelist Aldous Huxley—down to Edward O. Wilson (1975, 1978). Here, the is/ought distinction is simply denied or downgraded. It is argued that the world itself has value—not just the forests and the fish in the ocean, but the very mountains and lakes and seas—and so almost expectedly human values emerge through the evolutionary process. It is a view going back through the German Romantics to Plato, particularly the *Timaeus*, and it is a view coming forward to Rachel Carson and *Silent Spring* and more recently James Lovelock and the Gaia hypothesis (Ruse 2013). I suspect that nine-tenths of my fellow members of the Sierra Club subscribe to it. And although I do not myself accept it, no longer do I think it silly or necessarily philosophically crude. A case can be made, is made very ably by Richards, and I shall leave things at that. Or rather, I shall leave it to other contributors to this volume to make the contrary case and leave things at that. The other approach is to accept fully the is/ought distinction but (in the language of sports) to do an end run around it. If I say that this is the position found in Darwin, that is really only a half-truth because I don't think that Darwin is interested in the problems of philosophers, particularly those of justification, then it is at least a half-truth because the position does depend heavily on Darwin's thinking, especially that of the *Descent* (1871). Hints can be found in later writers, particularly those of the paleontologist George Gaylord Simpson (1964), and then later the philosopher (and my fellow student at the University of Rochester) Jeffrey

Murphy (1982). I think J. L. Mackie was going in the same direction (1979, 1985), but he died before he could articulate his thinking fully. I myself set out as a fervent disciple of Moore, a position I articulated in *Sociobiology: Sense or Nonsense* (1979). However, a detailed discussion review of that book by Mackie (1980)—gratifyingly favorable for a totally unknown philosopher that yet took me to task on biology and morality—set me thinking in new ways. Taking advantage of the fact that I was unknown and hence had no reputation to lose, I announced my new position in an article in the science-religion journal *Zygon* (1986a) and then in a book, *Taking Darwin Seriously: A Naturalistic Approach to Philosophy* (1986b). I should say that the responses were so gratifyingly awful that I collected them together in a collage, making it the backdrop for invitations to my fiftieth birthday party—"Come and celebrate fifty years of unbroken success"—and sent copies to all of my critics. I should say that most of them responded in the good-humored way that the offer was extended. Already people were moving from ignoring me to writing refutations in journals that would never accept anything by me. The case I make for evolutionary ethics is simple—too simple, say the critics.

From Darwin on it has been virtually a truism that evolution by natural selection promotes "altruism." By this it is understood that the key to evolutionary success is adaptation—features that help their possessors to survive and reproduce—that behavioral features are as important as physical features, and while at times strife and combat may be good adaptive strategies, often cooperating pays major dividends. Half a cake is less than the whole cake but better than no cake at all. It is worth noting that the 1960s saw a quantum leap in interest by evolutionary biologists (all Darwinians) in social behavior and a number of powerful models to explain "altruism" were devised (Ruse 1979). These included *reciprocal altruism*—you scratch my back and I will scratch yours—an idea with roots in the *Descent*, and *kin selection*—help to relatives rebounds vicariously with the success of your own shared genes—an idea not found in Darwin because it requires understanding of modern genetics.

Note that I put "altruism" in quotes because this is not necessarily literal altruism—Mother Teresa altruism, where people consciously try to do the right thing. It extends to all social behavior of a reciprocal kind and indeed the paradigm examples are the hymenoptera—the ants, the bees, and the wasps—and no one thinks these creatures to be reflective. However, the claim is made by evolutionary biologists—starting with Darwin in the *Descent*—that genuine altruism is something promoted by natural selection to make us humans good "altruists." "It must not be forgotten that, although a high standard of morality gives but a slight or no advantage to each individual man and his children over the other men of the same tribe, yet that an advancement in the standard of morality and an increase in the number of well-endowed men will certainly give an immense advantage to one tribe over another" (C. Darwin 1871, 1:166). Hence, "There can be no doubt that a tribe including many members who, from possessing in a high degree the spirit of patriotism, fidelity, obedience, courage, and sympathy, were always ready to give aid to each other and to sacrifice themselves for the common good, would be victorious over most other tribes; and this would be natural selection" (C. Darwin 1871, 1:166). And so it follows that "at all times throughout the world tribes have supplanted other tribes; and as morality is one element in their success, the standard of morality and the number of well-endowed men will thus everywhere tend to rise and increase." Although it does not affect the discussion here, Richards and Ruse differ over the interpretation of "tribe" (Richards and Ruse 2016). Ruse takes tribes to be groups of interrelated humans and, hence, Darwin is promoting a kind of proto-kin selection—only proto because he didn't have genetics—whereas Richards thinks that tribal members need not be related and Darwin is invoking selection at the level of the group (without specification of relatedness).

Darwinian Substantive Ethics

In the case of the Darwinian approach to evolutionary ethics, at the level of substantive ethics, I don't think there is much cause for concern or special thinking. It is often believed that traditional evolutionary ethics—so-called social Darwinism—promotes attitudes favorable to warfare and extreme laissez faire economics—"widows and children to the wall and let the robber barons take all." But although there are certainly instances of such writing, by and large the prescriptions are much more inclined to cooperation and helpfulness (Russett 1976; Richards 1987; Ruse 2009b, 2009d) . . . . The more Darwinian approach is basically going to be one of commonsense morality—help others, try to avoid cheating by yourself and others, and so forth. Widows and children deserve more attention and help than prosperous businessmen. Interesting, the most influential American ethicist of the second half of the twentieth century, John Rawls, picked up on evolutionary biology and suggested that his position is one that emerges from the evolutionary process (1971).

Darwinian Metaethics

What of justification? What of metaethics? Here's the rub. If the is/ought barrier is impenetrable, then you simply cannot get justification out of the evolutionary process or its products. And if you refuse to look elsewhere—Rawls for instance is a Kantian thinking that morality emerges as a necessary condition for rational interaction—then there is no justification. The point is that the Darwinian does not take this as a mark of failure, but the starting point. There is no objective justification of substantive ethics. One is an ethical skeptic, meaning not skeptical about substantive ethics but skeptical about justifications—skeptical to the point of nonbelief.

In a way, of course, this is not so very radical a conclusion. The emotivists, analyzing moral sentiments in terms of emotion—"I don't like rape. Boo-hoo, don't you like rape either"—were also moral skeptics in this sense. In a moment, we shall see a bit of a difference with emotivism, but leave this for a moment. First, let us turn to the obvious objection (Nozick 1981). "You have derived substantive ethics from the evolutionary process," says the critic. These are basically scientific arguments and let us for the sake of argument grant you these. You still have not shown us why ethics has no foundation. Take the comparable case of epistemology. You want to argue that the claim that rape is wrong has no objective foundation because our beliefs that rape is wrong emerge from the process of natural selection. Why should we not equally argue that the speeding train bearing down on us has no objective foundation because our beliefs in the train emerge from the process of natural selection? Those who believed in trains survived and reproduced and those that did not, did not.

There is a difference. All roads lead to trains. If evolution does not make you believe in trains, you are going to die before reproducing. You might not know of the trains in the way that we do. Perhaps we could use a kind of radar like the bats. But in the end, trains win out. For moral claims, like love your neighbor, it is not the same. There is no absolute direction to evolution through selection. The equivalence to trains is not the love commandment but some kind of cooperation. If you could get the cooperation in a different way, then so be it. But suppose that, instead of "love your neighbor," you had "hate your neighbor but recognize that your neighbor thinks it a sacred duty to hate you, so you had better get along." This might well work, and indeed I call it the Dulles system of morality, after Eisenhower's secretary of state (John Foster Dulles) who hated the communists but knew they felt the same way about him and so cooperated right through the 1950s at the height of the Cold War. Perhaps you don't have any morality at all. You do it all on game-theoretic reasoning. Although whether this is practically viable is another matter.

Morality may not always work but it gives you a quick and dirty solution to the cooperation problem and that is enough to go on. There is nothing sacred about the way we have evolved any more than there is anything sacred about having five digits rather than eight (as was the case with some early vertebrates). So there is nothing sacred about fixing on to "love your neighbor" rather than "hate your neighbor." Hence, it seems now that you are faced with two options. Either you go straight to arguing that substantive morality has no backing. It is, as I have called it, an illusion of the genes put in place by natural selection to make us good cooperators (Ruse and Wilson 1986). To make us "altruists," nature has made us altruists. Or you say that there exists an objective morality but as like as not we don't know about it! Perhaps morality demands that we hate our neighbors but we deluded fools think that we should love our neighbors. In my opinion, if it is possible for humans to live full and satisfying lives doing the very opposite of what is demanded by objective morality, this gets pretty close to being a reductio ad absurdum of objective morality.

What about the objection of mathematics? It too seems nonnatural like substantive morality, yet we not only believe in it but also think that it could be the true mathematics. It is not an illusion of the genes, for if we don't obey it, we will be in big trouble. In response, I would reject mathematical Platonism, thinking mathematics exists out there independent of our thought or very existence. I just don't see how you could link up human brains with such a world of pure rationality. So at most (and at least), mathematics has to be something like relations between objects or some such thing. In other words, like morality, mathematics can be used but it doesn't to refer to anything—meaning any (objective) thing. The difference with mathematics is that I am not sure that alternative mathematics (akin to Dulles morality) works as well as the mathematics we have now. If there are such alternatives, mathematics is no objection, and if there are not, then the parallel between mathematics and morality breaks down.

So stressing what I think the critics often miss, the non-directionality of Darwinian evolution, I argue that there are no reasons to think that we have homed in on an objective morality and good reasons to think that there is no such morality to home in on. Of course, I have taken a somewhat Platonic stance on the objectivity of morality, namely that it exists outside our perceptions—the falling tree in the forest when no one is around—in the sense of rational entities as suggested by Plato or the will of an objective God or some such thing. What if you argue that the objectivity is more Kantian, not existing independently but as the necessary conditions of cooperative behavior and living? I have less of a quarrel here although I still don't think it works. I would concede that perhaps all morality ultimately must share the same formal structure of cooperation. If there is no reciprocation, it will not work. But as Kant himself pointed out (1959), this is not enough for morality. We need something filling it out, something referring to human nature. And it is here that I argue that both Christianity and the Dulles form of morality would do the job, and there is no reason to think that the evolutionary process will necessarily lead us to one option rather than the other. So my conclusion holds.

Why, finally, is such a strong and sensible argument of mine so difficult for so many intelligent people to grasp? Simply because their biology is working flat out to make my argument seem unconvincing. Go back to emotivism and ask why so many people—and I was very much one here—found it not just false but somehow rather immoral. Rape really is wrong, and not just a matter of emotions and directives. If it isn't objectively wrong, then it wasn't wrong for those without those emotions. Russian soldiers raping German women when they invaded East Prussia. So what is going on here? Morality has no foundation and yet I criticize philosophies that argue that morality has no foundation. The crucial point is that, although morality has no foundation, our biology makes us think that it has. To use an ugly term of Mackie, we "objectify" morality. The meaning of morality incorporates objectivity. "Rape is wrong" means it really is truly, objectively wrong to rape. It is not a matter up for grabs—we may not always obey morality but that is another matter. Its meaning is that transgression is wrong.

There are obvious reasons why biology would make us think this way. If we did not, if we simply could puzzle out that morality has no foundation, we would start to ignore it and soon others would too and before long it would break right down. So selection has tweaked the meaning of morality. I refer to it as the Raskolnikov problem, for it shows that what I have told you will not at once make you free to rape and pillage. In *Crime and Punishment* the young student murders for gain. The detective knows that he has done it but waits until he confesses himself. The truth does not set you free, or at least it does in one sense (confessing) but not in another (recognizing that morality has no foundation). We are biologically disposed to think morality objective, and so, even if philosophically we can puzzle out otherwise, we cannot live by denying our human nature. As always, Hume is ahead of us here. You worry that morality has no ultimate basis?

> Most fortunately it happens, that since reason is incapable of dispelling these clouds, nature herself suffices to that purpose, and cures me of this philosophical melancholy and delirium, either by relaxing this bent of mind, or by some avocation, and lively impression of my senses, which obliterate all these chimeras. I dine, I play a game of backgammon, I converse, and am merry with my friends; and when after three or four hours' amusement, I would return to these speculations, they appear so cold, and strained, and ridiculous, that I cannot find in my heart to enter into them any farther. (Hume 1940, 175)

Psychology trumps philosophy!

I once read a book in which the author stated in the preface with the sincere hope that nothing he was about to say was original. I am not quite that modest, but I know what he meant. Ultimately, it is all footnotes to Plato. More immediately, I see Darwinian evolutionary ethics as very much in the British empiricist tradition. It is, as I like to say, David Hume brought up to date by Charles Darwin. I am proud to be their spokesman.

Chapter 9, Reading 4

Morality for the Mechanist[1]

Human Moral Nature

How does the machine metaphor deal with morality? Many think there is no way that evolutionary theory—Darwinian evolutionary theory, that is—can lead to morality. Social Darwinism is fatally flawed, and the implication is that the same applies to any other approach. Thomas Henry Huxley is an articulate exponent of this position, one formulated explicitly in opposition to his old friend Herbert Spencer. In his Romanes lecture of 1893, "Evolution and Ethics," Huxley writes of human evolution: "For his successful progress, throughout the savage state, man has been largely indebted to those qualities which he shares with the ape and the tiger; his exceptional physical organization; his cunning, his sociability, his curiosity, and his imitativeness; his ruthless and ferocious destructiveness when his anger is roused by opposition" (T. Huxley 2004, 51). However, in a way, this has all been self-defeating, or perhaps more accurately, it means that those things that helped us to our present passion are no longer needed, they are in fact hindrances. "Whatever differences of opinion may exist among experts, there is a

---

1. Ruse 2021c. Reprinted with permission of Cambridge University Press.

general consensus that the ape and tiger methods of the struggle for existence are not reconcilable with sound ethical principles."

Lions and tigers and bears, oh my! Stirring stuff and, at one level, one hesitates to disagree. The Thrasymachus view of life is not the correct one. "Might is right" is deeply immoral. Yet, let us not rush to assume that Huxley got Darwin entirely right. He hadn't done so for most of his life, from the *Origin* on, so why assume he did so now? The whole point about humans is that we have taken a very different route from that of apes and tigers—or, more accurately I should say, from the route that the apes and tigers are supposed to have taken, for I am not sure that they are always so very different from us. Natural selection does not always promote conflict and aggression. In the case of humans, we know that this is very much not so. Say it again: the defining part of our nature is that we are social. You don't need to be a Darwinian to know this. The great metaphysical poet John Donne put my thoughts in words to which I can never aspire.

> No man is an island, entire of itself; every man is a piece of the continent, a part of the main. If a clod be washed away by the sea, Europe is the less, as well as if a promontory were, as well as if a manor of thy friend's or of thine own were: any man's death diminishes me, because I am involved in mankind, and therefore never send to know for whom the bells tolls; it tolls for thee. (Donne [1624] 1959, meditation 17, 108–9)

Don't ask me. Look at our biology. We get along by getting along. We don't have weapons of attack. I get cheesed off at you. I am unlikely to tear your throat out. We have social instincts. I see a small child, alone and crying in the park. I don't ignore it. I wonder where its parents are. We have adaptations for social life. Hard enough teaching logic as it is. Imagine if three of the class were in heat. Of course, being nice sometimes is hard and annoying. I won't say we always do what we should do—how often has a tired professor skipped out through the back door when a needy, pre-exam student is at the front?—but we know what we should do. Our biology directs us that way. "We have cooperative brains, it seems, because cooperation provides material benefits, biological resources that enable our genes to make more copies of themselves. Out of evolutionary dirt grows the flower of human goodness" (Joshua Greene 2013, 65).

There is no magic to any of this. It very much as it was when we were talking about knowledge. Successful protohumans took seriously "two plus two equals four." Successful protohumans took seriously the dictate to think of others. Helping this along, selection conferred a sense of self-worth dealing with the needs of others. "When people who are fairly fortunate in their material circumstances don't find sufficient enjoyment to make life valuable to them, this is usually because they care for nobody but themselves" (Mill 2008, 9). All very much in the spirit, almost the literal word, of Charles Darwin in the *Descent of Man*. Diametrically opposed to the Huxley view of humankind, he stressed that sociality is the key to human success. Tribes of people who get along and help each other do better than tribes who don't.

> It must not be forgotten that although a high standard of morality gives but a slight or no advantage to each individual man and his children over the other men of the same tribe, yet that an advancement in the standard of morality and an increase in the number of well-endowed men will certainly give an immense advantage to one tribe over another. There can be no doubt that a tribe including many members who, from possessing in a high degree the spirit of patriotism, fidelity, obedience, courage, and sympathy, were always ready to give aid to each other and to sacrifice themselves for the common good, would be victorious over most other tribes; and this would be natural selection. (C. Darwin 1871, 1:166)

Darwin speculated on the causes behind this evolution of the moral sense. One suggestion is today known as *reciprocal altruism*. You scratch my back and I will scratch yours. He wrote: "As the reasoning powers and foresight of the members [of a tribe] became improved, each man would soon learn from experience that if he aided his fellow-men, he would commonly receive aid in return" (C. Darwin 1871, 1:163). Critics, like David Stove, who complain that Darwinism means constant struggle—"If his theory or explanation of evolution were true, there would be in every species a constantly recurring struggle for life: a competition to survive and reproduce which is so severe that few of the competitors in any generation can win" (Stove 2007, 45)—simply do not know what they are talking about. Of course, the real hope of Stove and his comrades is that you do not know what Darwin is talking about.

This does not mean that we are always nice to everyone all the time. Obviously not! My worries at the end of the last chapter. Hatred, prejudice, violence, crime, war. Those, like me, who have lived so long in such peaceable societies, wonder constantly at their luck. Why are these horrible phenomena so common? In broad outline, the evolutionary answer is obvious. Darwin knew. Individual selection! "It is no argument against savage man being a social animal, that the tribes inhabiting adjacent districts are almost always at war with each other; for the social instincts never extend to all the individuals of the same species" (C. Darwin 1871, 1:85). The question really is not so much *that* tribes (which Darwin took to be groups of interrelated individuals) would be wary of other tribes, but precisely *why* this would happen. Clearly, it is either because they see the others as a threat, or because the others have things that they want. Agriculture! Moreover, in the evolutionary world, one always suspects that sex is not far from the surface—even if we are not all Don Giovanni—and one presumes that the women of others would be an attraction, as conversely your women would be an attraction to others. One thinks of the abuse of German women as the Russians started to invade Germany, East Prussia particularly. This was abuse sanctioned—encouraged—by the higher echelons of the Russian military.

Against this, notice how we have as part of our nature ways of dealing with these issues. We restrain violence thanks to strategies like just war theory. Obliteration bombing, indiscriminate bombing of civilians—notwithstanding all sides did it in the Second World War—is wrong. Coventry should not have been destroyed. Dresden should not have been destroyed. Hiroshima and Nagasaki should not have been destroyed. Sociality is important and we have ways to promote and safeguard it, even if those ways are not always successful.

Normative Darwinism

> Moral systems are sets of interlocking values, virtues, norms, practices, identities, institutions, technologies, and evolved psychological mechanisms that work together to suppress or regulate self-interest and make cooperative societies possible. (Haidt 2012, 314)

Agreed. Start now with questions to do with substantive ethics. We are subject to the suzerainty of the machine metaphor, so Hume's is/ought distinction reigns supreme. The fact that we do these things does not in itself make them moral. However, what is important is that so many of the things we do and think are precisely those things that we would expect of moral people, and, when they are not what we would expect of moral people, we have the conceptual tools to realize why they are not what we would expect of moral people. Morally, we realize we should be kind to small children. Morally, we realize that we are in the wrong if we do not pay our share into the coffee fund. John Rawls in his celebrated *A Theory of Justice* makes this point.

> In arguing for the greater stability of the principles of justice I have assumed that certain psychological laws are true, or approximately so. I shall not pursue the question of stability beyond this point. We may note however that one might ask how it is that human beings have acquired a nature described by these psychological principles. The theory of evolution would suggest that it is the outcome of natural selection; the capacity for a sense of justice and the moral feelings is an adaptation of mankind to its place in nature. (Rawls 1971, 502–3)

Rawls, like Plato in the *Republic*, is offering a contract theory—what's a good setup for a group of folk living together? One of the problems of such a theory is that, historically, it never seems very plausible that a group of wise men (and perhaps women) sat around and drew up the rules for proper conduct. It makes more sense to leave it to natural selection.

Still, an urge to be nice to small children is not a moral claim—you *ought* to be nice to small children. Rawls appreciates this, noting that even if biology does underlie the contract theory, it doesn't follow that it is morally obligatory to follow it. "These remarks are not intended as justifying reasons for the contract view." At most we seem to have a Darwinized version of the emotive theory of ethics (Ayer 1936). I don't like being unkind to small children. Boo-hoo! I don't like it when you are unkind to small children. For my peace of mind, I urge you not to be unkind to small children. Let us agree that in Darwin's world, moral sentiments are emotions. Very much akin to the categories/habits in epistemology, they are put in place to help us get on in life. However, and this is very Humean, phenomenologically as it were, moving on beyond raw emotivism, they must present themselves as more than mere emotions, desires. If they are just that, then because of cheating they are going to break down. Why should I respect your marriage? Why should you likewise respect my marriage? Canadian Philosophy Departments in the early years of the pill.

We must have some restraint on the naked emotions, and it is here that morality comes in. To use an ugly word of the late J. L. Mackie, we "objectify" our emotions—they present themselves as objectively binding (Mackie 1977). You ought not violate the marriage bonds of yourself or of others. This doesn't mean you never will, rather that you ought not. And because of the overall force of this moral prohibition, a kind of societal stability prevails.

Putting things together, at the substantive or normative level, Darwinian evolutionary theory generates the norms of morality. The norms of morality that we accept as, let us say, common sense. You ought to care about children, you ought not cheat on your wife sort of thing. Am I, almost deliberately, ignoring what philosophers through the ages have had to say about normative ethics? What about the categorical imperative? What about the greatest happiness principle? Do these have no place in my world picture? In the most important way, my position is intended to encompass them all! Philosophers make a living out of finding awkward counterexamples, but all the traditional suggestions agree on most issues. A Kantian, with thoughts about ends in themselves, would be horrified if—Nazi style—you intended to do horrendous operations on small children to make important medical discoveries. A utilitarian, especially of the John Stuart Mill ilk, would be equally horrified. Even if you get important discoveries in this case, the unhappiness of the child is all important, as is the sense of certainty that children (including your own) are not going to be whisked away in the cause of medical science.

What of those awkward counterexamples? To take one of those ethical dilemmas on which so many cases for tenure are built, what should we do if we find ourselves faced with a runaway railway truck, with six people tied up just downline (Thomson 1985). Should we pull the switch and divert the truck to a sideline on which only one person is tied up? What if there is no switch but we are standing on a bridge next to a fat man—he must be big, so self-sacrifice is not an option—and we can push him over, thereby killing him but stopping the truck? Intuition usually tells us that it is okay to pull the switch but not to upend our overweight neighbor. Why? Formally, they are identical. The Darwinian has little time for any of this. The answer lies not in ourselves but in our genes. In our past, we have often had situations where we must make decisions about the welfare of our neighbors. Selection has programed us to care about them, if only because we are their neighbors and we need them to care about us. In our past, rarely if ever have we had the problem of pulling a switch. We can avoid emotions and make a rational decision. We are going to act differently in different circumstances. Don't look to evolution to offer a unique proper solution. I regard this as a strength of the Darwinian case, not a weakness.

I spoke above of the hostility we have to outsiders, to the other. When violence occurs, particularly in the context of war, like the complexities of thinking epistemologically, we are primed to think through the complexities of morality. Our moralists spring into action, showing why and how violence can at times be justified. Agreed, but one final question. We are not hostile to outsiders all the time. I don't hate or fear the folk from Timbuktu. If there is no reason to feel hostility to outsiders, does this then mean we are neutral toward them, or, as Jesus rather implied, should we extend our moral concerns out to everyone? Darwin thought this too. For all that he thought violence toward others rooted in our past, he thought we could overcome our legacy. Remember how he told us that gradually we reach out from our close group to society at large, and then "there is only an artificial barrier to prevent his sympathies extending to the men of all nations and races" (C. Darwin 1874, 122). Perhaps so. My sense, however, is that, once again, Hume was right about these sorts of things. Morality works on a gradient.

> A man naturally loves his children better than his nephews, his nephews better than his cousins, his cousins better than strangers, where everything else is equal. Hence arise our common measures of duty, in preferring the one to the other. Our sense of duty always follows the common and natural course of our passions (Hume 1978, 3.2.1.18, 483–84)

Although there are suggestions in the Bible, for instance the story of the good Samaritan, that we might have equal obligations to all, more generally it agrees with Hume. "Anyone who does not provide for their relatives, and especially for their own household, has denied the faith and is worse than an unbeliever" (1 Tim 5:8 NKJV). The Darwinian thinks likewise.

Darwinian Metaethics

In a world of the naturalistic fallacy, what then of metaethics? You cannot justify Darwinian normative ethics by the fact of Darwinian evolution. G. E. Moore appealed to what he called nonnatural properties. The rules of morality exist in a kind of ethereal world, as do the rules of mathematics. If this sounds Platonic, it was intended so. "I am pleased to believe that this is the most Platonic system of modern times" (Moore 1990, 50). Speaking to an issue I brushed past at the beginning of the last chapter, the Darwinian is not very comfortable with this. For a Darwinian ethicist like me, the nonprogressive nature of the evolutionary process comes to the fore. In the case of epistemology—knowledge about the world, for better or for worse, our habits come together to say that there is a train bearing down on us. If our habits didn't do this, it would seem (to us using those habits of reasoning we do have) we are not long for this world. In the ethical case, you could have different systems—systems with rival demands—and they could all work. Darwin himself noted this.

> I do not wish to maintain that any strictly social animal, if its intellectual faculties were to become as active and as highly developed as in man, would acquire exactly the same moral sense as ours. In the same manner as various animals have some sense of beauty, though they admire widely different objects, so they might have a sense of right and wrong, though led by it to follow widely different lines of conduct. If, for instance, to take an extreme case, men were reared under precisely the same conditions as hive-bees, there can hardly be a doubt that our unmarried females would, like the worker-bees, think it a sacred duty to kill their brothers, and mothers would strive to kill their fertile daughters; and no one would think of interfering. Nevertheless the bee, or any other social animal, would in our supposed case gain, as it appears to me, some feeling of right and wrong, or a conscience (C. Darwin 1871, 1:73).

We are humans, not hive bees, and, thank God or Darwin, we males do not have to worry as winter approaches. But the relativity point is made. Platonic ideals don't have a dog in this fight.

Could the relativity point be made about humans as humans? Could we have different systems here? Not within the species. We saw in an earlier chapter, there is a lot of uniformity about humans. For all that I sometimes have down days, truly we are not hymenoptera. I see absolutely no reason to suppose men and women have or obey different moral codes. But what about the human species taken as a whole? Could it have different systems? Kant seemed to think so in some sense (2007). At least, he thought it logically possible that we have a society without morality, where everything is done rationally—I will give to you but what will you give to me. He just didn't think it could work. And he has a point. Apart from anything else, time is money. We couldn't function if we had to calculate at every point. The child runs in front of the bus. Can I save it? What are the chances of my getting hurt? Does it really matter if the kid is killed? Am I not more valuable than the kid? What will the parents give me if I save their kid? By the time I have made the calculations, the kid will be like the wretched Rebecca who slammed doors until she upset a statue above her: "It knocked her flat! It laid her out! She looked like that" (Belloc 2021, stanza 3).

Note the objection is pragmatic. It is not that using just reason is wrong. It wouldn't work. Could we have an alternative functioning morality? Suppose we had the John Foster Dulles system of morality (Ruse 1986b). Dulles was Eisenhower's secretary of state during the Cold War. He hated the Russians. He believed he ought to hate the Russians. But he knew that they felt the same way about him. So, they got along. Judged by today's standards, not too badly either. Notice what this all means. There is no direction to evolution. This means that we could have the Dulles morality, thinking that hating others is morally obligatory. We go all the way—born, live, die—with this. We are in total ignorance of the true objective morality. God's will, Platonic Forms, or whatever: "But I tell you, love your enemies and pray for those who persecute you" (Matt 5:44 NASB). Somehow this ignorance strikes me as a reductio of objective morality. I can accept that one or two psychopaths might be unable to perceive objective morality, but all of us? Objective morality must in some sense reach out to normal human beings. Plato, of all people, would be horrified if it did not. Hence, the non-directedness of Darwinian evolution means that there is no objective morality. Or, as it is sometimes put, it "debunks" the case for moral realism, directing you to moral non-realism. In that sense, there is no justification—meaning outside justification—for substantive ethics.

One final question. Should I be telling you all of this? Now that I have given the game away, why not go out and rape and pillage to your heart's content? At least, do whatever you want while being careful not to get caught. There is no objective morality, so make hay while the sun shines. The trouble is—or rather the good thing is—humans don't work that way. We have evolved to be part of the system. We cannot deny human nature even if we want to. Hume again. He is talking of the skepticism induced by his study of causation. "I dine, I play a game of backgammon, I converse, and am merry with my friends. And when, after three or four hours' amusement, I would return to these speculations, they appear so cold, and strained, and ridiculous, that I cannot find in my heart to enter into them any farther" (Hume 1940, 175). Same with morality. What our reason tells us is going to be capped by our emotions. In *Crime and Punishment*, Dostoyevsky tells of the student Raskolnikov who thinks he can escape conventional morality and who commits a murder. The police chief knows that he did it; but, knowing that he will not be able to live with himself, waits until the pressure becomes too great and, of his own free will, the student confesses. Perhaps, for a day or two, after reading this book, you will go out and do what you please. God help the library books because apparently Darwin won't. You'll be sorry! Sleepless nights, wracked by conscience, at that yellow highlighting of the *Critique of Pure Reason*. How could you destroy the joy of others as they set out, all pure and innocent and eager, for the first time into the transcendental analytic, to find the metaphysical deduction has been marked up like a copyeditor's proof? And that awful question is a good note on which to end this chapter.

Chapter 9, Reading 5

Morality[1]

"What should I do?" Let us turn now to the second of the great questions of philosophy. Does evolutionary theory, does Darwinian evolutionary theory, throw any light on this topic? Many people today think that it does, but how far the light actually penetrates is still a matter of great controversy. As with the problem of knowledge, it makes most sense to go back and start with Charles Darwin himself.

Darwin on Morality

Thanks to an extended discussion in the *Descent of Man*, Darwin had far more to say on the topic of morality and behavior than he had had to say on knowledge and its foundations. Moreover, his thinking on the issues takes us right to the heart of one of the most contested issues in contemporary evolutionary theory. But start as before with the fact that, although knowledgeable, Darwin was not a philosopher with a philosopher's questions. He was a scientist with a scientist's questions, and philosophy would be tackled (if at all) only tangentially. We see this at once in Darwin's treatment of morality. Philosophers distinguish between two major issues that must be addressed when dealing with moral thought and behavior. What should I do? Why should I do what

---

1. Ruse 2012a. Reprinted with permission of Cambridge University Press.

I should do? These two branches of the subject are usually referred to as *normative (or substantive) ethics* and *metaethics*, the first to do with directions and the second with foundations. By way of example in Christian ethics one finds normative questions along the lines of "Love your neighbor as yourself," and generally a metaethical answer in terms of God's will, "You should do that which God wants." Of course giving answers is only the beginning of inquiry. Who is one's neighbor? Why should one obey God? . . . Assume these, trying to explain them in the light of evolutionary biology. The popular mid-Victorian ethical theory was utilitarianism in some form or another. The greatest happiness principle: "You ought to promote the greatest happiness, and the least unhappiness, for the greatest number of people." John Stuart Mill's famous essay on the topic appeared in 1863, although historically the theory dates at least to the eighteenth century and Darwin's grandfather Erasmus endorsed a version. The progress of evolution "is analogous to the improving excellence observable in every part of the creation . . . such as the progressive increase of the wisdom and happiness of its inhabitants" (Erasmus Darwin 1801, 1:509). Charles Darwin accepted this philosophy, why would he not, but gave it a bit of a biological twist. "The term, general good, may be defined as the term by which the greatest possible number of individuals can be reared in full vigor and health, with all their faculties perfect, under the conditions to which they are exposed" (C. Darwin 1871, 1:98). To be honest, however, Darwin was not really tremendously reflective on what all of this might mean in practical terms, although it is fairly easy to infer that his values were those of an upper-middle-class Englishman, of a liberal persuasion. He was violently against slavery, and unlike many of his countrymen supported the North in the American Civil War. (A lot of the British support for the South came from the close connection between the cotton fields of the breakaway states and the cotton mills of industrial Lancashire.) However, equally, he was in favor of capitalism and we know that he regarded with horror proposals that workingmen should be allowed to form unions. This violates sound economics as well as intrudes on owners' liberties. Although by 1871 (the year of the *Descent*) Darwin's religious beliefs had moved over to agnosticism, like others (including the archetypal critic of Christianity, Thomas Henry Huxley) he endorsed Bible-based religious instruction on the grounds that this was an essential part of moral training. What Darwin did see was that, at the biological level, morality demands an element of sociability. We have got to have a feeling that we can and want to get on with our fellows.... Morality, he sees, is not just a matter of unreflective action, but of conscious deliberation leading to action. We have first-order desires, and then second-order reflections on the desires and the actions they promote. This is or leads to a conscience. When he was developing his theory in the late 1830s, Darwin had read Hume on animal reasoning and always saw a continuity between the ape and the human. But he did want to emphasize that it is this ability to second-order reflect that distinguishes a moral animal like a human from a mere brute....

Levels of Selection

As intimated above, before moving on to the thinking of others, we must first pause and turn more directly to the workings of Darwinian evolutionary biology. If you think about it, on the face of things, morality is profoundly non-Darwinian and should never have appeared. To hell with being nice to others! There is a struggle for existence between organisms and victory is all important. Win or die! This was certainly the conclusion of others, not the least of whom (in historical importance) was Adolf Hitler....

Being nice to others pays dividends to oneself. The social animal gives out care and sympathy, but can expect it back in return. There is a struggle for existence, but success in the struggle can be more subtle than simply beating the other into a bloody pulp. It can involve cooperation, because half a cake is a lot better than none at all. But this now raises the question of how exactly natural selection brings all of this about, and here we do start to dig right down into some of the most basic and important questions in the whole of evolutionary theory. Introducing the term "altruism" in the sense used by modern biologists, meaning help given to others at your own immediate (reproductive) cost, and recognizing that this is a metaphor and not necessarily "altruism" in the literal sense of helping others because you know that you should, there are two ways in which selection might promote altruism. In the first case, it might come about because the "unit of selection" is the group to which the altruist belongs. In other words, although the altruist pays, the group gains because of the altruist, and selection favors this group over others. In this case, where we have "group selection" at work, the group is rather like the organic body, and the individual is like a part (a heart or a lung) where what the part does is lost within the functioning of the whole. In the second case, you insist that the "unit of selection" is the individual organism. In this case, although the altruist puts out, it must be with the expectation that there will be a payoff later. Overall, the gain is to the individual altruist. In other words, altruism is enlightened self-interest. Darwin was fascinated by this issue, virtually from the moment that he discovered natural selection (Ruse 1980). Most particularly, he was fascinated by the challenge thrown up by the social insects. Everybody knew—certainly every schoolboy who had had to translate Virgil's *Georgics* knew—about the honeybee and about the sacrifice that the workers make for the good of the hive. Many wrote on and around these topics in the eighteenth and early nineteenth centuries and, of course, Darwin came from rural England, where beekeeping, a hive or two at the bottom of the garden, was commonplace. The social insects put the problem of altruism up front because the workers, members of the group devoting all of their energies to others, are sterile. It is the queens and the males who do all of the reproducing. How can this be? Really there are two questions here (Dixon 2008). First, how can sterility get passed on? How can you inherit it? Second, what adaptive advantage can sterility have? Surely being a reproducer is far better? A Lamarckian has trouble with the first question, because sterility is hardly an acquired characteristic that is transmitted. . . . Sterility in the social insects is not something they are born with, but rather something acquired during growth. In other words, in our terms there seems to be a genetic predisposition to sterility given the right circumstances. How could this be inherited? Simply by being passed on by those family members who do reproduce. They have the predisposition and they are selected, and they are not sterile. As Darwin pointed out, the situation is exactly what you find in animal breeding. You have a fine carcass from an ox and so you go back to the parent stock (or any close relative stock) and breed again. The relatives pass on the features, and if indeed instead of well-marbled meat it is a propensity to work hard and not to reproduce, then so be it. Nature is satisfied. . . .

"Hence, as more individuals are produced than can possibly survive, there must in every case be a struggle for existence, either one individual with another of the same species, or with the individuals of distinct species, or with the physical conditions of life" (C. Darwin 1859, 63). Why did he feel so strongly on this matter? In part, I strongly suspect, it was a consequence of his overall politico-economic philosophy. He was a Malthusian, and right after the passage just quoted, went on to say that he was endorsing the philosophy of that particular thinker. Darwin really did think of life as being a battle between individuals. This contrasts nicely with the codiscoverer of natural selection, Alfred Russel Wallace, who was a lifelong socialist and who happily embraced group selection. In part, perhaps, Darwin thought as he did because he wanted to be consistent. After all, sexual selection is virtually by definition an individual selection mechanism—a battle between organisms in the group—and so it is natural to think of natural selection the same way. As we shall learn, Wallace was a lot less keen on sexual selection. . . . Darwin was wary of group selection because of the matter that sticks in the craw of evolutionists today. Group selection is too open to cheating. If you have a collection of group altruists, then a variant cheater who does nothing for others is going to be at a selective advantage, and its offspring will spread at the expense of the altruists. How then can individual selection promote altruism? Today's evolutionists have proposed two main mechanisms. *Reciprocal altruism*, where help is given in anticipation of help received (Trivers 1971). "You scratch my back and I'll scratch yours." And something called *kin selection*, where altruists may not reproduce but by helping close relatives to reproduce they thereby get to pass on more copies of their own units of heredity, their own genes (Hamilton 1964). The latter is particularly important when it comes to explaining why, in the social insects, workers devote all of their efforts to raising the fertile offspring of their mother. It is a question of benefits outweighing costs. If one is in a place well defended and with food supplies, an individual's best reproductive strategy ("fortress defense") might be to raise related organisms (albeit not as related as one's own offspring) rather than venture afield with the risk of injury and death (and hence no offspring). In the hymenoptera, another strategy ("life insurance") seems important. Here it is in an individual's reproductive interests to stay with the nest because the coverage of the young (even if not yours) is better (thanks to the nest mates) than if you go it alone (and perhaps die before raising your young).

As noted, Darwin and Wallace then went on to discuss these issues in the 1860s. Hence, by the time Darwin came to write the *Descent*, he was fully sensitized to the topic, as of course he needed to be because moral behavior does involve altruism in an extreme form. (Note that the point of Darwin's discussion was to show that biologically we are altruists, but that the way in which we humans achieve our biological ends is through literal, ethical altruism. Doing good serves our reproductive goals. Not always, of course, but on average.) Does this mean that Darwin stuck to an individual selection stance throughout, or did he weaken (say, rather, modify) and allow that, at the human level, group selection might take over? He could certainly have used the excuse that humans with their powers of language and of thought so change the formula that all previous bets are off. (This is the position of Richerson and Boyd 2005.) And he recognized that there are some very knotty issues that do seem to call for group selection. "It is extremely doubtful whether the offspring of the more sympathetic and benevolent parents, or of those who were the most faithful to their comrades, would be reared in greater numbers than the children of selfish and treacherous parents of the same tribe" (C. Darwin 1871, 1:163). In part, he got around the problem by appealing to (individual selection–promoting) reciprocal altruism.

> In the first place, as the reasoning powers and foresight of the members became improved, each man would soon learn that if he aided his fellow-men, he would commonly receive aid in return. From this low motive he might acquire the habit of aiding his fellows; and the habit of performing benevolent actions certainly strengthens the feeling of sympathy which gives the first impulse to benevolent actions. Habits, moreover, followed during many generations probably tend to be inherited. (C. Darwin 1871, 1:163–64)

(Note the Lamarckism, something that Darwin always accepted along with selection. But then he did seem to bite the bullet and go in a group selection way.)

> It must not be forgotten that although a high standard of morality gives but a slight or no advantage to each individual man and his children over the other men of the same tribe, yet an advancement in the standard of morality and an increase in the number of well-endowed men will certainly give an immense advantage to one tribe over another. There can be no doubt that a tribe including many members who, from possessing in a high degree the spirit of patriotism, fidelity, obedience, courage, and sympathy, were always ready to give aid to each other and to sacrifice themselves for the common good, would be victorious over most other tribes; and this would be natural selection. At all times throughout the world tribes have supplanted other tribes; and as morality is one element in their success, the standard of

morality and the number of well-endowed men will thus everywhere tend to rise and increase. (C. Darwin 1871, 1:166)

Social Darwinism

Let us move on now from Darwin. We saw in the discussion of epistemology that there were two main approaches. One worked more at the cultural level and the other was the more literal approach, working from the supposed nature of the human mind. We find something very similar in the case of ethics. The first, more traditionalist approach is often today known as *social Darwinism*—a term incidentally invented in the 1940s (in other words unknown to the people whom it purports to represent) and with a somewhat unfortunate reputation as referring to a system or approach that is (from a normative perspective) pretty vile and, one can say with some relief, no longer fashionable. As is so often the case, history is a little more complicated. Herbert Spencer is the first and still the main representative of this approach. Essentially (so the story goes) he saw human societies in the way he saw animal and plant societies, namely as engaged in a bloody struggle for existence. In the human realm, this translated out into a form of laissez-faire economics, where the state stands aside and naked competition determines the winners and losers. Drawing on this, Spencer then argued that it is all a very good thing and what one ought to do. In other words, his normative ethics came straight out of the Darwinian selection processes. That there is some truth in all of this cannot be denied.

> We must call those spurious philanthropists, who, to prevent present misery, would entail greater misery upon future generations. All defenders of a Poor Law must, however, be classed among such. That rigorous necessity which, when allowed to act on them, becomes so sharp a spur to the lazy and so strong a bridle to the random, these pauper's friends would repeal, because of the wailing it here and there produces. Blind to the fact that under the natural order of things, society is constantly excreting its unhealthy, imbecile, slow, vacillating, faithless members, these unthinking, though well-meaning, men advocate an interference which not only stops the purifying process but even increases the vitiation—absolutely encourages the multiplication of the reckless and incompetent by offering them an unfailing provision, and discourages the multiplication of

the competent and provident by heightening the prospective difficulty of maintaining a family. (Spencer 1851, 323–24)

However, before we rush to judgment, we should note that sentiments like these were held before Spencer became an evolutionist; in any case, as we know already, although Spencer was an independent discoverer of natural selection, for him the main mechanism of change was always Lamarckism. The connections between this kind of thinking and Darwinian evolutionary biology are loose, to say the least....

This is not to deny that, after he became an evolutionist, Spencer was not slow to link his social thought with his biological thought, and that others were ready to follow in his path. A good example, one that shows that there is a lot more to traditional evolutionary ethics than what is usually thought to fall under the heading of social Darwinism, is the Russian anarchist exile Prince Peter Kropotkin. An unabashed group selectionist, he saw all organisms including humans as having developed a propensity to what he called "mutual aid" and held that this is and should be the basis of morality.

> Two aspects of animal life impressed me most during the journeys which I made in my youth in Eastern Siberia and Northern Manchuria. One of them was the extreme severity of the struggle for existence which most species of animals have to carry on against an inclement Nature; the enormous destruction of life which periodically results from natural agencies; and the consequent paucity of life over the vast territory which fell under my observation. And the other was, that even in those few spots where animal life teemed in abundance, I failed to find—although I was eagerly looking for it—that bitter struggle for the means of existence, among animals belonging to the same species, which was considered by most Darwinists (though not always by Darwin himself) as the dominant characteristic of struggle for life, and the main factor of evolution. (Kropotkin 1902, vi–viii)

What Kropotkin did see was one animal helping another. On the one hand, there was the appallingly harsh environment.

> On the other hand, wherever I saw animal life in abundance, as, for instance, on the lakes where scores of species and millions of individuals came together to rear their progeny; in the colonies of rodents; in the migrations of birds which took place at that time on a truly American scale along the Usuri; and especially

in a migration of fallow-deer which I witnessed on the Amur, and during which scores of thousands of these intelligent animals came together from an immense territory, flying before the coming deep snow, in order to cross the Amur where it is narrowest—in all these scenes of animal life which passed before my eyes, I saw Mutual Aid and Mutual Support carried on to an extent which made me suspect in it a feature of the greatest importance for the maintenance of life, the preservation of each species, and its further evolution. (Kropotkin 1902, viii)

Progress Again

Now swing round and ask about foundations. Why did (or do) people like Spencer and Kropotkin and Jack London and Bernhardi and Julian Huxley and Wilson feel so strongly that they could and must promote an evolutionarily based normative ethics? The answer—and by this stage you will hardly be surprised—is that they did it in the name of progress. Every one of these people was an ardent biological progressionist, as were their fellow travelers (Ruse 2009b). We have seen this in detail in Spencer and Huxley and Wilson, and the same is true of the others. They saw the evolutionary process as one that had direction, leading up to humans. They also saw the evolutionary process as one gaining in value as it rose upward, and that humans represent the peak of excellence, biologically and in all other respects. Hence, they saw it as our moral duty to cherish and preserve humans, at minimum to keep us at the level we are now, and perhaps even to improve things for us in the future. . . . Is it reasonable, in the light of Darwinian theory, to think that an ethical sense or ability might have evolved?

By now I think no one would deny an affirmative answer. Altruism in the biological sense is very well documented, and humans as completely social animals are very much in need of such altruism. If they had not developed it, they would be in deep trouble. But why, still speaking biologically, have we gone the extra step into literal altruism? Why are we not just like the ants, which seem to be programmed to do what they do, without any need of moral training or anything else of that nature? In one sense, presumably, the answer is "because we can." But that can hardly be the end of the matter. Surely, without dillying and dallying about these things, a lot of the time we would be far better off if we got on with things without thinking? A more satisfactory answer comes in another sense, where we appreciate the fact that humans have taken a reproductive route very different from most other organisms, especially the ants. For better or for worse, we simply cannot have many offspring. The physical needs of our children preclude that. Unlike the ants, where the queen can have literally hundreds of thousands of offspring, we humans are limited to about ten, give or take—and recognizing that the figures for men might be very different from those for women. What this means is that, if circumstances change, either the environment or fellow species members, we need to respond in a way that is not demanded of the ants. A queen ant can afford to lose a few thousand offspring in a shower of rain, washing away the pheromones that lead the nest members home. We cannot afford to lose even one child if it rains when they are away from home. We need the flexibility to rethink and reassess if things go wrong. That is what innate dispositions are all about. And in no place is this ability more crucial than when dealing with fellow humans and deciding what we should do in social situations, and what they should do also, and what we think they should do, and so forth.

Morality on this biological scenario, therefore, is an aid to decision-making in social groups. Should I help? Should others help me? Should I expect help? And so forth. In a way, it is all a bit like language, another social facilitator. Moral ability is innate, but we have to learn how to use it, and once acquired it is difficult to change it. It is the same as learning English or French. Perhaps you can push the analogy a bit further, suggesting that there is some kind of deep universal moral grammar and that growing up in a particular society you learn to apply it in somewhat different ways. People in an American small town in the Midwest share the deep structure with folk growing up under the Taliban in Afghanistan—reciprocation, and so forth—but how we learn to apply morality, say with respect to the role of women, is clearly going to be very different. A major criticism that is often brought against any kind of biological approach to humankind is that this locks us into a deterministic view of human nature. It is argued that we deny free will and see humans as if they were marionettes, dangling on strings controlled by the DNA. However, note that although there are deterministic aspects to this picture, it is far from one of crude "genetic determinism." We are determined by biology to have the dispositions that we have and by our culture in the ways that these are expressed. . . .

What Should We Do?

Turn now to the philosophical questions. First, what about normative morality. Note that the Darwinian is going to be asking about what we think we should do as opposed to asking what we really should do. These are not obviously one and the same, so keep this point in mind. The answer is going to be couched in terms something along the lines of commonsense morality. That is the whole point of what we think we should do. In other words, morality is going to be something along the lines of the love commandment. . . .

Objectification

One final point. Suppose the moral non-realist is right. Suppose there is no objective reality. Why do most people, including most philosophers, find this so hard to accept? To the Darwinian, the answer is obvious. If we did not believe that morality was objective, that it refers to real facts, then we would soon quit obeying it (Mackie 1977, 1979). Why should I be good when there is no reason? At least, no reason in my self-interest as against simply looking after number one? So, to use an ugly word introduced by the late John Mackie, we "objectify" normative morality, thinking it does have a foundation, even though it does not. This is a case where biology is deceiving us for our own good, because Kant was right: if we stop being moral, society breaks down and then we all lose.

So now that I have told you, why don't you go away and be immoral, pretending of course that you are being moral? Because we are not psychopaths. Our psychology would make us very uncomfortable were we to do this. Dostoyevsky showed this in *Crime and Punishment*. The police chief knows that the student Raskolnikov has murdered the two old women, because he thought he could transcend his moral nature. But in the end, the student confesses because he cannot go against himself. As the Freudian argues that you are denying the truth of the Freudian analysis because of your own problems, so the Darwinian argues that you assert the objective truth of morality because of your own nature, one that Darwinism has brought about! You can't win? Oh yes, you can, but only if you stay within the system!

Bibliography

Alexander, Richard D. 1979. *Darwinism and Human Affairs*. Seattle: University of Washington Press.
Allen, Colin, and Marc Bekoff. 1995. "Biological Function, Adaptation, and Natural Design." *Philosophy of Science* 62: 609–22.
Anselm. 2008. *Anselm of Canterbury: The Major Works*. Translated by Brian Davies and G. R. Evans. Oxford World's Classic. Oxford: Oxford University Press.
Appel, Toby A. 1987. *The Cuvier-Geoffrey Debate: French Biology in the Decades before Darwin*. Monographs on the History and Philosophy of Biology. New York: Oxford University Press.
Aquinas, Thomas. 1920–22. *Summa Theologica*. Translated by Fathers of the English Dominican Province. 10 vols. 2nd ed. London: Burns, Oates and Washbourne.
———. 1952. *The Summa Theologica of Saint Thomas Aquinas*. Vol. 2. Chicago: Encyclopedia Britannica.
———. 1975. *God*. Vol. 1 of *Summa Contra Gentiles*. Edited and translated by Anton C. Pegis. Notre Dame, IN: University of Notre Dame Press.
Aristotle. 1984a. "De partibus animalium." In *The Complete Works of Aristotle: The Revised Oxford Translation*, edited by Jonathan Barnes, 1:1087–110. Bollingen Series 71. Princeton, NJ: Princeton University Press.
———. 1984b. "Parts of Animals." In *The Complete Works of Aristotle: The Revised Oxford Translation*, edited by Jonathan Barnes, translated by W. Ogle, 1:991–99. Bollingen Series 71. Princeton, NJ: Princeton University Press.
Atran, Scott. 2004. *In Gods We Trust: The Evolutionary Landscape of Religion*. Evolution and Cognition. New York: Oxford University Press.
Augustine. 1991. *Confessions*. Translated by Henry Chadwick. Oxford's World Classics. Oxford: Oxford University Press.
———. 1998. *The City of God against the Pagans*. Edited and translated by R. W. Dyson. Cambridge Texts in the History of Political Thought. Cambridge: Cambridge University Press.
Ayala, Francisco J. 1974. "The Concept of Biological Progress." In *Studies in the Philosophy of Biology: Reduction and Related Problems*, edited by Francisco Jose Ayala and Theodosius Dobzhansky, 339–54. London: Macmillan.

———. 1987. "The Biological Roots of Morality." *Biology and Philosophy* 2: 235–52.
———. 1988. "Can 'Progress' Be Defined as a Biological Concept?" In *Evolutionary Progress*, edited by Matthew H. Nitecki, 75–96. Chicago: University of Chicago Press.
Babbage, Charles. 1838. *The Ninth Bridgewater Treatise: A Fragment*. 2nd ed. London: Cass.
Bacon, Francis. 1999. "New Organon" (1620). In *Selected Philosophical Works*, edited by Rose-Mary Sargent, 86–206. Indianapolis: Hackett.
———. 2000. *The Advancement of Learning*. Edited by Michael Kiernan. Vol. 4 of *The Oxford Francis Bacon*. Oxford: Oxford University Press.
Bada, Jeffrey L., and Antonio Lazcano. 2009. "The Origin of Life." In *Evolution: The First Four Billion Years*, edited by Michael Ruse and Joseph Travis, 49–79. Cambridge, MA: Belknap.
Bakker, Robert T. 1986. *The Dinosaur Heresies: New Theories Unlocking the Mystery of the Dinosaurs and Their Extinction*. 2nd ed. Toronto: Citadel.
Bannister, Robert C. 1979. *Social Darwinism: Science and Myth in Anglo-American Social Thought*. American Civilization Series. Philadelphia: Temple University Press.
Barrett, Justin L. 2004. *Why Would Anyone Believe in God?* Cognitive Science of Religion. Lanham, MD: AltaMira.
Barrow, John D., and Frank J. Tipler. 1986. *The Anthropic Cosmological Principle*. Oxford: Oxford University Press.
Barth, Karl. 1933. *The Epistle to the Romans*. Oxford: Oxford University Press.
———. 1957. *The Doctrine of God*. Edited by G. W. Bromiley and T. F. Torrance. Vol. 2 of *Church Dogmatics*. Edinburgh: T&T Clark.
Bavinck, Herman. 1951. *The Doctrine of God*. Translated by William Hendriksen. Grand Rapids: Eerdmans.
Beckner, Morton. 1969. "Function and Teleology." *Journal of the History of Biology* 2: 151–64.
Behe, Michael J. 1996. *Darwin's Black Box: The Biochemical Challenge to Evolution*. New York: Free.
Belloc, Hilaire. 2021. "Rebecca, Who Slammed Doors for Fun and Perished Miserably." English Verse. https://englishverse.com/poems/rebecca.
Benton, M. J. 1987. "Progress and Competition in Macroevolution." *Biology Review* 62: 305–38.
———. 2009. "Paleontology." In *Evolution: The First Four Billion Years*, edited by Michael Ruse and Joseph Travis, 80–104. Cambridge, MA: Belknap.
Bergson, Henri. 1907. *L'évolution créatrice*. Paris: Alcan.
Blane, Gilbert. 1819. *Elements of Medical Logic*. London: Bulmer and Nicol.
Bonhoeffer, Dietrich. 1979. *Letters from Prison*. Edited by Eberhard Bethge. Enlarged ed. New York: Macmillan.
Bonner, John Tyler. 1988. *The Evolution of Complexity by Means of Natural Selection*. Princeton, NJ: Princeton University Press.
Bowler, Peter J. 1976. *Fossils and Progress*. New York: Science History.
———. 1988. *The Non-Darwinian Revolution: Reinterpreting a Historical Myth*. Baltimore, MD: Johns Hopkins University Press.
———. 1989. *The Mendelian Revolution: The Emergence of Hereditarian Concepts in Modern Science and Society*. London: Athlone.

———. 2013. *Darwin Deleted: Imagining a World without Darwin*. Chicago: University of Chicago Press.
Boyer, Pascal. 2002. *Religion Explained: The Evolutionary Origins of Religious Thought*. New York: Basic.
Boyle, Robert. 1966. "A Disquisition about the Final Causes of Natural Things." In *The Works of Robert Boyle*, edited by T. Birch, 5:392–444. Hildesheim: Olms.
———. 1996. *A Free Enquiry into the Vulgarly Received Notion of Nature*. Edited by Edward B. Davis and Michael Hunter. Cambridge Texts in the History of Philosophy. Cambridge: Cambridge University Press.
Broad, C. D. 1944. "Critical Notice of Julian Huxley's Evolutionary Ethics." *Mind* 53: 156–87.
Brown, Dan. 2004. *The Da Vinci Code*. New York: Doubleday.
Brown, P., et al. 2004. "A New Small-Bodied Hominin from the Late Pleistocene of Flores, Indonesia." *Nature* 431: 1055–61.
Browne, Janet. 1995. *Charles Darwin: Voyaging*. New York: Knopf.
———. 2002. *Charles Darwin: The Power of Place*. New York: Knopf.
Budd, Susan. 1977. *Varieties of Unbelief: Atheists and Agnostics in English Society 1850–1960*. London: Heinemann.
Burchfield, Joe D. 1975. *Lord Kelvin and the Age of the Earth*. New York: Science History.
Bury, J. B. 1920. *The Idea of Progress: An Inquiry into Its Origin and Growth*. London: MacMillan.
Cain, Arthur J. 1954. *Animal Species and Their Evolution*. London: Hutchinson.
Callebaut, Warner, ed. 1993. *Taking the Naturalistic Turn, or How Real Philosophy of Science Is Done*. Chicago: Chicago University Press.
Calvin, John. 1960. *A Compend of the "Institutes of the Christian Religion."* Edited by Hugh T. Kerr. Philadelphia: Westminster.
Cannon, W. F. 1961. "The Impact of Uniformitarianism: Two Letters from John Herschel to Charles Lyell, 1836–1837." *Proceedings of the American Philosophical Society* 105: 301–14.
Carroll, Lewis. 1950. "The Hunting of the Snark." Poetry Foundation. https://www.poetryfoundation.org/poems/43909/the-hunting-of-the-snark.
Carroll, Sean B. 2005. *Endless Forms Most Beautiful: The New Science of Evo Devo and the Making of the Animal Kingdom*. New York: Norton.
Chambers, Robert. 1844. *Vestiges of the Natural History of Creation*. London: Churchill.
Chardin, Pierre Teilhard de. 1955. *Le phénomème humaine*. Paris: Seuil.
Clifford, W. K. 1901. "Body and Mind." In *Lectures and Essays of the Late William Kingdom Clifford*, edited by Leslie Stephen and Frederick Pollock, 2:31–58. London: Macmillan.
Clutton-Brock, Tim H., et al. 1982. *Red Deer: Behavior and Ecology of Two Sexes*. Wildlife Behavior and Ecology. Chicago: University of Chicago Press.
Cobb, John B., and David Ray Griffin. 1976. *Process Theology: An Introductory Exposition*. Philadelphia: Westminster.
Coleridge, Samuel Taylor. 1895. *Letters*. Edited by Ernest Hartley Coleridge. 2 vols. London: Houghton, Mifflin.
Conway Morris, Simon. 2003. *Life's Solution: Inevitable Humans in a Lonely Universe*. Cambridge: Cambridge University Press.
Coyne, J. A., et al. 1997. "Perspective: A Critique of Sewall Wright's Shifting Balance Theory of Evolution." *Evolution* 51: 643–71.

Coyne, Jerry. 2002. "Intergalactic Jesus." *London Review of Books* 24: 23–24. https://www.lrb.co.uk/the-paper/v24/n09/jerry-coyne/intergalactic-jesus.

Coyne, Jerry A. 2012. "You Don't Have Free Will." *Chronicle of Higher Education* (Mar. 18). http://chronicle.com/article/Jerry-A-Coyne/131165.

Cracraft, J. 1990. "The Origin of Evolutionary Novelties: Pattern and Process at Different Hierarchical Levels." In *Evolutionary Innovations*, edited by Matthew H. Nitecki, 21–46. Chicago: University of Chicago Press.

Cunningham, Suzanne. 1996. *Philosophy and the Darwinian Legacy*. Rochester, NY: University of Rochester Press.

Daeschler, Edward B., et al. 2006. "A Devonian Tetrapod-Like Fish and the Evolution of the Tetrapod Body Plan." *Nature* 440: 757–63.

Darwin, Charles. 1838. *Notebook M*. Cambridge Digital Library. DAR 125. https://cudl.lib.cam.ac.uk/view/MS-DAR-00125/1.

———. 1839. *Journal of Researches into the Geology and Natural History of the Various Countries Visited by H.M.S. Beagle*. London: Colburn.

———. 1842. *The Structure and Distribution of Coral Reefs*. London: Smith Elder.

———. 1851a. *A Monograph of the Fossil Lepadidae; or, Pedunculated Cirripedes of Great Britain*. London: Palaeontographical Society.

———. 1851b. *A Monograph of the Sub-Class Cirripedia, with Figures of All the Species: The Lepadidae; or Pedunculated Cirripedes*. London: Ray Society.

———. 1854a. *A Monograph of the Fossil Balanidae and Verrucidae of Great Britain*. London: Palaeontographical Society.

———. 1854b. *A Monograph of the Sub-Class Cirripedia, with Figures of All the Species: The Balanidae (or Sessile Cirripedes); the Verrucidae, etc., etc., etc.* London: Ray Society. https://ia601301.us.archive.org/7/items/monographonsubcl02darw/monographonsubcl02darw.pdf.

———. 1859. *On the Origin of Species by Means of Natural Selection, or Preservation of Favored Races in the Struggle for Life*. London: Murray.

———. 1860a. "Letter to Asa Gray, 22 May [1860]." Darwin Correspondence Project. https://www.darwinproject.ac.uk/letter/DCP-LETT-2814.xml.

———. 1860b. "Letter to Asa Gray, 5 September [1857]." Darwin Correspondence Project. https://www.darwinproject.ac.uk/letter/DCP-LETT-2136.xml.

———. 1861. *On the Origin of Species by Means of Natural Selection, or Preservation of Favored Races in the Struggle for Life*. 3rd ed. London: Murray.

———. 1862. *On the Various Contrivances by Which British and Foreign Orchids Are Fertilized by Insects, and on the Good Effects of Intercrossing*. London: Murray.

———. 1868. *The Variation of Animals and Plants under Domestication*. 2 vols. London: Murray.

———. 1871. *The Descent of Man, and Selection in Relation to Sex*. 2 vols. London: Murray.

———. 1878. *The Effects of Cross and Self-Fertilization in the Vegetable Kingdom*. London: Murray.

———. 1880. *The Power of Movement in Plants*. London: Murray.

———. 1881. *The Formation of Vegetable Mold, through the Action of Worms, with Observations on their Habits*. London: Murray.

———. 1903. *More Letters of Charles Darwin*. Edited by Francis Darwin and A. C. Seward. London: Murray.

———. 1958. *The Autobiography of Charles Darwin, 1809–1882*. Edited by Nora Barlow. London: Collins.

———. 1959. *The Origin of Species by Charles Darwin: A Variorum Text*. Edited by Morse Peckham. Philadelphia: University of Pennsylvania Press.

———. 1980. "The Red Notebook of Charles Darwin." Edited by Sandra Herbert. *Bulletin of the British Museum (Natural History): Historical Series* 7: 1–164.

———. 1985–2023. *The Correspondence of Charles Darwin*. Edited by F. Burkhardt et al. 30 vols. Cambridge: Cambridge University Press.

———. 1987. *Charles Darwin's Notebooks, 1836–1844*. Edited by Paul H. Barrett et al. Ithaca, NY: Cornell University Press.

———. 1993. *The Autobiography of Charles Darwin: 1809–1882*. New York: Norton.

———. 2013. "Letter 49 to J. D. Hooker. July 13th, 1856." In *More Letters of Charles Darwin*, edited by Francis Darwin and A. C. Seward, 1:44. Scotts Valley, CA: CreateSpace.

Darwin, Charles, and Alfred Russel Wallace. 1958. *Evolution by Natural Selection*. Foreword by Gavin de Beer. Cambridge: Cambridge University Press.

Darwin, Emma. 1915. *Emma Darwin: A Century of Family Letters*. 2 vols. Edited by Henrietta Litchfield. London: Murray.

Darwin, Erasmus. 1801. *Zoonomia; or, The Laws of Organic Life*. 2 vols. 3rd ed. London: Johnson.

———. 1803. *The Temple of Nature*. 2 vols. London: Johnson.

Darwin, Francis, ed. 1887. *The Life and Letters of Charles Darwin, including an Autobiographical Chapter*. 2 vols. London: Murray.

Davies, Brian. 2004. *An Introduction to the Philosophy of Religion*. 3rd ed. Oxford: Oxford University Press.

———. 2010. "Simplicity." In *The Cambridge Companion to Christian Philosophical Theology*, edited by Charles Taliaferro and Chad Meister, 31–46. Cambridge Companions to Religion. Cambridge: Cambridge University Press.

———. 2013. "Aquinas and Atheism." In *The Oxford Handbook of Atheism*, edited by Stephen Bullivant and Michael Ruse, 119–36. Oxford Handbooks. Oxford: Oxford University Press.

Davies, Brian, and Eleonore Stump. 2012. *The Oxford Handbook of Aquinas*. Oxford Handbooks. Oxford: Oxford University Press.

Davies, N. B., et. al. 1989. "Cuckoos and Parasitic Ants: Interspecific Brood Parasites as an Evolutionary Arms Race." *Trends in Ecology and Evolution* 4: 274–78.

Dawkins, Richard. 1976. *The Selfish Gene*. Oxford Landmark Science. Oxford: Oxford University Press.

———. 1983. "Universal Darwinism." In *Molecules to Men*, edited by D. S. Bendall, 403–25. Cambridge: Cambridge University Press.

———. 1986. *The Blind Watchmaker*. New York: Norton.

———. 1989. "The Evolution of Evolvability." In *Artificial Life: An Overview*, edited by Christopher G. Langton, 201–20. Redwood City, CA: Addison-Wesley.

———. 1992 "Progress." In *Keywords in Evolutionary Biology*, edited by Evelyn Fox Keller and Elisabeth A. Lloyd, 263–72. Cambridge, MA: Harvard University Press.

———. 1995. *A River Out of Eden*. New York: Basic.

———. 1997a. "Human Chauvinism: Review of *Full House* by Stephen Jay Gould." *Evolution* 51: 1019–20.

———. 1997b. "Religion Is a Virus." *Mother Jones* (Nov./Dec). Interview by Michael Krasny. https://www.motherjones.com/politics/1997/11/religion-virus/.

———. 2003. *A Devil's Chaplain: Reflections on Hope, Lies, Science and Love.* Boston: Houghton Mifflin.

———. 2006. *The God Delusion.* New York: Houghton Mifflin.

Dawkins, Richard, and John Richard Krebs. 1979. "Arms Races between and within Species." *Proceedings of the Royal Society of London B* 205: 489–511.

De Beer, Gavin, ed. 1960. "Darwin's Notebooks on Transmutation of Species." *Bulletin of the British Museum (Natural History)* 2: 4–200; 3: 29–176; 4: 17–175; 5: 19–79.

Dembski, William A. 1998a. *The Design Inference: Eliminating Chance through Small Probabilities.* Cambridge Studies in Probability, Induction and Decision Theory. Cambridge: Cambridge University Press.

———. 1998b. *Mere Creation: Science, Faith and Intelligent Design.* Downers Grove, IL: Intervarsity Academic.

Dembski, William A., and Michael Ruse, eds. 2004. *Debating Design: From Darwin to DNA.* Cambridge: University of Cambridge Press.

Dennett, Daniel C. 1984. *Elbow Room: The Varieties of Free Will Worth Wanting.* Cambridge: MIT Press.

———. 1995. *Darwin's Dangerous Idea.* New York: Simon and Schuster.

———. 2006. *Breaking the Spell: Religion as a Natural Phenomenon.* New York: Viking.

Descartes, René. 1964. *Meditations: Philosophical Essays.* Translated by Laurence J. Lafleur. Indianapolis: Bobbs-Merrill.

———. 1985. *The Philosophical Writings.* Translated by John Cottingham et al. 2 vols. Cambridge: Cambridge University Press.

———. 2014. *Principles of Philosophy.* Translated by John Veitch. Scotts Valley, CA: CreateSpace.

Desmond, Adrian. 1997. *Huxley: From Devil's Disciple to Evolution's High Priest.* Helix. New York: Basic.

Dick, Steven J. 1996. *The Biological Universe: The Twentieth-Century Extraterrestrial Life Debate and the Limits of Science.* Cambridge: Cambridge University Press.

Dickens, Charles. 1952. *Our Mutual Friend.* London: Oxford University Press.

Diderot, Denis. 1943. *Diderot: Interpreter of Nature.* Edited by Jonathan Kemp. New York: International.

Dijksterhuis, E. J. 1961. *The Mechanization of the World Picture.* Translated by C. Dikshoorn. Oxford: Oxford University Press.

Dixon, Thomas. 2008. *The Invention of Altruism: Making Moral Meanings in Victorian Britain.* Oxford: Oxford University Press.

Dobzhansky, Theodosius. 1937. *Genetics and the Origin of Species.* New York: Columbia University Press.

Dobzhansky, Theodosius, et al. 1977. *Evolution.* San Francisco: Freeman.

Donne, John. 1959. *"Devotions upon Emergent Occasions," Together with "Death's Duel."* Ann Arbor: University of Michigan Press.

Duffy, Eamon. 1992. *The Stripping of the Altars: Traditional Religion in England, 1400–1580.* New Haven, CT: Yale University Press.

Durkheim, Émile. 2004. *The Elementary Forms of Religious Life.* Edited by Mark S. Cladis. Translated by Carol Cosman. Oxford World's Classics. Oxford: Oxford University Press.

Edwards, Jonathan, et al. 2005. *"Sinners in the Hands of an Angry God" and Other Puritan Sermons*. Dover Thrift. New York: Dover.
Eldredge, Niles, and Stephen J. Gould. 1972. "Punctuated Equilibria: An Alternative to Phyletic Gradualism." In *Models in Paleobiology*, edited by T. J. M. Schopf, 82–115. San Francisco: Freeman, Cooper.
Epperson v. Arkansas. 1968. 393 U.S. 97. Justia. https://supreme.justia.com/cases/federal/us/393/97/.
Erwin, Douglas H., et al. 1987. "A Comparative Study of Diversification Events: The Early Paleozoic versus the Mesozoic." *Evolution* 41: 1177–86.
Farrar, Frederic W. 2007. *Eric, or Little by Little*. Gutenberg. https://www.gutenberg.org/cache/epub/23126/pg23126-images.html.
Feenstra, R. J. 2010. "Trinity." In *Christian Philosophical Theology*, edited by Charles Taliaferro and Chad Meister, 3–14. Cambridge: Cambridge University Press.
Fischer, John Martin, et al. 2007. *Four Views on Free Will*. Malden, MA: Blackwell.
Fisher, R. A. 1930. *The Genetical Theory of Natural Selection*. Oxford: Oxford University Press.
Fodor, Jerry. 1996. "Peacocking." *London Review of Books* 18: 19–20.
Ford, E. B. 1964. *Ecological Genetics*. London: Methuen.
Frazer, James G. 1906–15. *The Golden Bough: A Study in Magic and Religion*. 12 vols. 3rd ed. London: Macmillan.
Galen. 1968. *On the Usefulness of the Parts of the Body*. Translated by Margaret Tallmadge May. 2 vols. Ithaca, NY: Cornell University Press.
Gewirth, A. 1986. "The Problem of Specificity in Ethics." *Biology and Philosophy* 1: 297–305.
Ghiselin, Michael T. 1983. "Lloyd Morgan's Canon in Evolutionary Context." *Behavioral and Brain Sciences* 6: 343–90.
Gilbert, S. F., et al. 1996. "Resynthesizing Evolutionary and Developmental Biology." *Developmental Biology* 173: 357–72.
Gish, Duane, et al. 1960. "The Complete Amino Acid Sequence of the Protein of Tobacco Mosaic Virus." *Proceedings of the National Academy of Science* 46: 1463–69.
Goodall, Jane. 1971. *In the Shadow of Man*. London: Collins.
Gould, Stephen Jay. 1977. *Ontogeny and Phylogeny*. Cambridge, MA: Belknap.
———. 1981. *The Mismeasure of Man*. New York: Norton.
———. 1985a. *The Flamingo's Smile: Reflections in Natural History*. New York: Norton.
———. 1985b. "SETI and the Wisdom of Casey Stengel." In *The Flamingo's Smile: Reflections in Natural History*, 403–15. New York: Norton.
———. 1988. "On Replacing the Idea of Progress with an Operational Notion of Directionality." In *Evolutionary Progress*, edited by Matthew H. Nitecki, 319–38. Chicago: University of Chicago Press.
———. 1989a. *Full House: The Spread of Excellence from Plato to Darwin*. New York: Paragon.
———. 1989b. *Wonderful Life: The Burgess Shale and the Nature of History*. New York: Norton.
———. 2002. *The Structure of Evolutionary Theory*. Cambridge, MA: Harvard University Press.

Gould, Stephen Jay, and Richard C. Lewontin. 1979. "The Spandrels of San Marco and the Panglossian Paradigm: A Critique of the Adaptationist Program." *Proceedings of the Royal Society of London B* 205: 581–98.
Gould, Stephen Jay, and Elizabeth S. Vrba. 1982. "Exaptation: A Missing Term in the Science of Form." *Paleobiology* 8: 4–15.
Gould, Stephen Jay, et al. 1987. "Asymmetry of Lineages and the Direction of Evolutionary Time." *Science* 236: 1437–41.
Graham, Gordon. 2010. "Atonement." In *The Cambridge Companion to Christian Philosophical Theology*, edited by Charles Taliaferro and Chad Meister, 124–35. Cambridge Companions to Religion. Cambridge: Cambridge University Press.
Grant, B. Rosemary, and Peter R. Grant. 1989. *Evolutionary Dynamics of a Natural Population: The Large Cactus Finch of the Galapagos*. Chicago: University of Chicago Press.
Grant, Peter R. 1986. *Ecology and Evolution of Darwin's Finches*. Princeton Science Library. Princeton Legacy Library. Princeton, NJ: Princeton University Press.
Grant, Peter R., and B. Rosemary Grant. 2007. *How and Why Species Multiply: The Radiation of Darwin's Finches*. Princeton Series in Evolutionary Biology. Princeton, NJ: Princeton University Press.
Gray, Asa. 1876. *Darwiniana*. New York: Appleton.
Grayling, A. C. 2013. "Critiques of Theistic Arguments." In *The Oxford Handbook of Atheism*, edited by Stephen Bullivant and Michael Ruse, 38–52. Oxford Handbooks. Oxford: Oxford University Press.
Greene, John C. 1977. "Darwin as a Social Evolutionist." *Journal of the History of Biology* 10: 1–27.
Greene, Joshua. 2013. *Moral Tribes: Emotion, Reason, and the Gap between Us and Them*. New York: Penguin.
Greg, W. R. 1868. "On the Failure of 'Natural Selection' in the Case of Man." *Fraser's Magazine*: 353–62.
Grinnell, G. 1974. "The Rise and Fall of Darwin's First Theory of Transmutation." *Journal of the History of Biology* 7: 262.
Gruber, H. E., and P. H. Barrett. 1974. *Darwin on Man*. New York: Dutton.
Gruber, J. W. 1965. "Brixham Cave and the Antiquity of Man." In *Context and Meaning in Cultural Anthropology*, edited by Melford E. Spiro, 373–402. New York: Free.
Gunton, Colin E., ed. 1997. *The Cambridge Companion to Christian Doctrine*. Cambridge Companions to Religion. Cambridge: Cambridge University Press.
Haeckel, Ernst. 1866. *Generelle Morphologie der Organismen*. Berlin: Reimer.
Haidt, Jonathan. 2012. *The Righteous Mind: Why Good People Are Divided by Politics and Religion*. New York: Vintage.
Haldane, J. B. S. 1927. *Possible Worlds and Other Essays*. London: Chatto and Windus.
———. 1932. *The Causes of Evolution*. New York: Longmans, Green.
Hall, A. Rupert. 1983. *The Revolution in Science, 1500–1750*. London: Longman.
Hamilton, W. D. 1964. "The Genetical Evolution of Social Behavior." *Journal of Theoretical Biology* 7: 1–52.
Harari, Yuval Noah. 2015. *Sapiens: A Brief History of Humankind*. New York: Harper.
Harrison, J. F. C. 1969. *Quest for the New Moral World: Robert Owen and the Owenites in Britain and America*. New York: Scribner's.
———. 1979. *The Second Coming: Popular Millenarianism, 1780–1850*. New Brunswick, NJ: Rutgers University Press.

Heidegger, Martin. 1974. *An Introduction to Metaphysics*. Translated by Ralph Manheim. New Haven, CT: Yale University Press.

Hempel, C. G. 1965. *Aspects of Scientific Explanation*. New York: Free.

———. 1966. *Philosophy of Natural Science*. Englewood Cliffs, NJ: Prentice Hall.

Herbert, Sandra. 1971. "Darwin, Malthus, and Selection." *Journal of the History of Biology* 4: 209–17.

———. 1974. "The Place of Man in the Development of Darwin's Theory of Transmutation. Part 1. to July 1837." *History of Biology* 7: 217–58.

———. 1977. "The Place of Man in the Development of Darwin's Theory of Transmutation, Part 2." *Journal of the History of Biology* 10: 243–73.

Herschel, J. F. W. 1830. *Preliminary Discourse on the Study of Natural Philosophy*. London: Longman, Rees, Orme, Brown, Green, and Longman.

Hick, John. 1978. *Evil and the God of Love*. New York: Harper and Row.

———. 1961. "Necessary Being." *Scottish Journal of Theology* 14: 353–69.

Hobbes, Thomas. 1651. "Of the Natural Condition of Mankind." In *Leviathan*, 52–61. Harmondsworth, UK: Penguin.

Hodge, Charles. 1874. *What Is Darwinism?* Princeton, NJ: Scribner, Armstrong, and Co.

Hopson, J. A. 1977. "Relative Brain Size and Behavior in Archosaurian Reptiles." *Annual Review of Ecology and Systematics* 8: 429–48.

Hrdy, Sarah Blaffer. 1981. *The Woman That Never Evolved*. Cambridge: Cambridge University Press.

Hull, David L., ed. 1973. *Darwin and His Critics: The Reception of Darwin's Theory of Evolution by the Scientific Community*. Cambridge, MA: Harvard University Press.

———. 1988. *Science as a Process: An Evolutionary Account of the Social and Conceptual Development of Science*. Chicago: University of Chicago Press.

Hull, David L., and Michael Ruse, eds. 1998. *The Philosophy of Biology*. Oxford Readings in Philosophy. Oxford: Oxford University Press.

Hume, David. 1940. *A Treatise of Human Nature*. Edited by L. A. Selby-Bigge. Oxford: Oxford University Press.

———. 1947. *Dialogues Concerning Natural Religion*. Edited by Norman Kemp Smith. Library of Liberal Arts. Indianapolis: Bobbs-Merrill.

———. 1963. *Hume on Religion*. Edited by Richard Wollheim. Fontana Library. London: Fontana.

———. 1978. *A Treatise of Human Nature*. Edited by Lewis Amherst Selby-Bigge, revised by P. H. Nidditch. 2nd ed. Oxford: Oxford University Press.

———. 1999. *An Enquiry Concerning Human Understanding*. Edited by Tom L. Beauchamp. Oxford Philosophical Texts. Oxford: Oxford University Press.

———. 2000. *A Treatise of Human Nature*. Edited by David Fate Norton and Mary J. Norton. Oxford Philosophical Texts. Oxford: Oxford University Press.

Huxley, Julian S. 1912. *The Individual in the Animal Kingdom*. Cambridge: Cambridge University Press.

———. 1927. *Religion without Revelation*. London: Benn.

———. 1942. *Evolution: The Modern Synthesis*. London: Allen and Unwin.

———. 1943. *TVA: Adventure in Planning*. London: Scientific Book Club.

———. 1959. "Introduction." In *The Phenomenon of Man*, by Pierre Teilhard de Chardin, 11–28. London: Collins.

Huxley, Leonard. 1900. *Life and Letters of Thomas Henry Huxley*. 2 vols. New York: Appleton.
Huxley, Thomas Henry. 1859. "Letter from Huxley to F. Dyster, January 30, 1859." Clark University, n.d. HP 15.106. https://mathcs.clarku.edu/huxley/letters/59.html.
———, ed. 1860a. *Darwiniana*. London: Macmillan.
———. 1860b. "The Origin of Species." *Westminster Review* (Apr.) 541–70.
———. 1863. *Evidence as to Man's Place in Nature*. Cambridge: Cambridge University Press.
———. 1867–68. "Remarks upon the Archaeopteryx Lithographica." *Proceedings of the Royal Society of London* 16: 243–48.
———. 2004. *Evolution and Ethics*. Fairfield, IA: 1st World.
Huxley, Thomas H., and Julian S. Huxley. 1947. *Evolution and Ethics*. London: Pilot.
Jablonski, D., and D. J. Bottjer. 1988. "The Ecology of Evolutionary Innovation: The Fossil Record." In *Evolutionary Progress*, edited by Matthew H. Nitecki, 253–88. Chicago: University of Chicago Press.
Jackson, J. B. C., and F. K. McKinney. 1990. "Ecological Processes and Progressive Macroevolution of Marine Clonal Benthos." In *Causes of Evolution: A Paleontological Perspective*, edited by Robert M. Ross and Warren D. Allmon, 173–209. Chicago: University of Chicago Press.
Jerison, Harry J. 1973. *Evolution of the Brain and Intelligence*. New York: Academic.
John Paul II. 1998. *Fides et Ratio: Encyclical Letter of John Paul II to the Catholic Bishops of the World*. Vatican City: Osservatore Romano.
Johnson, Philip E. 1995. *Reason in the Balance: The Case against Naturalism in Science, Law, and Education*. Downers Grove, IL: Intervarsity.
———. 1997. *An Easy-to-Understand Guide for Defeating Darwinism by Opening Minds*. Downers Grove, IL: InterVarsity.
Kant, Immanuel. 1949. *Critique of Practical Reason*. Translated by Thomas Kingsmill Abbott. London: Longmans, Green.
———. 1959. *Foundations of the Metaphysics of Morals*. Edited by Robert Paul Wolff. Translated by Lewis White Beck. Indianapolis: Bobbs-Merrill.
———. 2000. *Critique of the Power of Judgment*. Edited and translated by Paul Guyer. Translated by Eric Matthews. Cambridge Edition of the Works of Immanuel Kant. Cambridge: Cambridge University Press.
———. 2007. *The Critique of Teleological Judgement*. Translated by James C. Meredith. Rev. ed. Oxford World's Classics. Oxford: Oxford University Press.
Kettlewell, Bernard D. 1973. *The Evolution of Melanism: The Study of a Recurring Necessity*. Oxford: Clarendon.
Kierkegaard, Søren. 1944. *Concluding Unscientific Postscript*. Translated by David F. Swenson. Princeton, NJ: Princeton University Press.
Kimler, William, and Michael Ruse. 2013. "Mimicry and Camouflage." In *The Cambridge Encyclopedia of Darwin and Evolutionary Thought*, edited by Michael Ruse, 139–45. Cambridge: Cambridge University Press.
Kitchell, J. A., and N. MacLeod. 1988. "Macroevolutionary Interpretations of Symmetry and Synchroneity in the Fossil Record." *Science* 240: 1190–93.
Kitcher, Philip. 2007. *Living with Darwin: Evolution, Design, and the Future of Faith*. Philosophy in Action. New York: Oxford University Press.
Kottler, Malcolm Jay. 1974. "Alfred Russel Wallace, the Origin of Man, and Spiritualism." *Isis* 65: 145–92.

Kropotkin, Peter. 1902. *Mutual Aid: A Factor of Evolution*. London: Heinemann.
Kuhn, Thomas S. 1962. *The Structure of Scientific Revolutions*. Chicago: University of Chicago Press.
Lakatos, Imré, and Alan Musgrave, eds. 1970. *Proceedings of the International Colloquium in the Philosophy of Science, London, 1965*. Vol. 4 of *Criticism and the Growth of Knowledge*. Cambridge: Cambridge University Press.
Lamarck, Jean-Baptiste de. 1809. *Philosophie zoologique*. Paris: Dentu.
Lankester, E. Ray. 1880. *Degeneration: A Chapter in Darwinism*. London: Macmillan.
Lawrence, D. H. 1915. *The Rainbow*. London: Methuen & Co.
Lewontin, Richard C. 1974. *The Genetic Basis of Evolutionary Change*. New York: Columbia University Press.
Lewontin, Richard C., et al. 1984. *Not in Our Genes: Biology, Ideology and Human Nature*. New York: Pantheon.
Lieberman, Daniel E. 2013. *The Story of the Human Body: Evolution, Health, and Disease*. New York: Vintage.
Liem, K. F. 1988. "Plausibility and Testability: Assessing the Consequences of Evolutionary Innovation." In *Evolutionary Progress*, edited by Matthew H. Nitecki, 147–70. Chicago: University of Chicago Press.
Limoges, Camille. 1970. *La sélection naturelle: Étude sur la première constitution d'un concept (1837-1859)*. Galien: Histoire et philosophie de la biologie et de la médecine. Paris: Presses Universitaires de France.
Locke, John. 1959. *An Essay Concerning Human Understanding*. Edited by Alexander Campbell Fraser. 2 vols. Dover Books on Western Philosophy. New York: Dover.
Lovejoy, Arthur O. 1936. *The Great Chain of Being*. Cambridge, MA: Harvard University Press.
Lyell, Charles. 1830–33. *Principles of Geology: Being an Attempt to Explain the Former Changes of the Earth's Surface by Reference to Causes Now in Operation*. 3 vols. London: Murray.
Lyons, David. 1965. *Forms and Limits of Utilitarianism*. Oxford: Oxford University Press.
Mackie, J. L. 1977. *Ethics: Inventing Right and Wrong*. Middlesex, UK: Penguin.
———. 1979. *Hume's Moral Theory*. Edited by Ted Honderich. International Library of Philosophy. London: Routledge and Kegan Paul.
———. 1980. "Review of *Sociobiology: Sense or Nonsense*, by Michael Ruse." *Erkenntnis* 15: 189–94.
———. 1985. "The Law of the Jungle: Moral Alternatives and Principles of Evolution." In *Persons and Values: Selected Papers*, edited by Joan Mackie and Penelope Mackie, 2:120–31. Oxford: Oxford University Press.
Maimonides, Moses. 1936. *A Guide for the Perplexed*. Translated by M. Friedländer. London: Routledge.
Malthus, Thomas Robert. 1826. *An Essay on the Principle of Population*. 6th ed. London: Everyman.
Mandelbaum, Maurice. 1958. "Darwin's Religious Views." *Journal of the History of Ideas* 19: 363–78.
Marshall, L. G., et al. 1982. "Mammalian Evolution and the Great American Interchange." *Science* 215: 1351–57.
Marx, Karl, and Friedrich Engels. 1965. *Selected Correspondence*. 2nd ed. Moscow: Progress.

Mayr, Ernst. 1942. *Systematics and the Origin of Species*. New York: Columbia University Press.
———. 1969. "Commentary." *Journal of the History of Biology* 2: 207–21.
———. 1988. *Toward a New Philosophy of Biology: Observations of an Evolutionist*. Cambridge, MA: Harvard University Press.
Maynard Smith, John. 1964. "Group Selection and Kin Selection." *Nature* 201: 1145–47.
———. 1981. "Did Darwin Get It Right?" *London Review of Books* 3: 10–11.
———. 1988. "Evolutionary Progress and Levels of Selection." In *Evolutionary Progress*, edited by Matthew H. Nitecki, 219–30. Chicago: University of Chicago Press.
———. 1992. "Taking a Chance on Evolution." *New York Review of Books* 39: 234–36.
Maynard Smith, John, and Eörs Szathmáry. 1995. *The Major Transitions in Evolution*. New York: Oxford University Press.
McGrath, Alister E., ed. 1995. *The Christian Theology Reader*. Oxford: Blackwell.
———. 1997. *Christian Theology: An Introduction*. 2nd ed. Oxford: Blackwell.
———. 2009. *A Fine-Tuned Universe: The Quest for God in Science and Theology*. Gifford Lectures. Louisville: Westminster John Knox.
McMullin, Ernan. 1983. "Values in Science." In *Philosophy of Science Association 1982*, edited by Peter D. Asquith and Thomas Nickles, 3–28. East Lansing, MI: Philosophy of Science Association.
———. 1993. "Evolution and Special Creation." *Zygon: Journal of Religion and Science* 28: 299–335.
McShea, Daniel W. 1991. "Complexity and Evolution: What Everybody Knows." *Biology and Philosophy* 6: 303–24.
———. 1996. "Metazoan Complexity and Evolution: Is There a Trend?" *Evolution* 50: 477–92.
McShea, Daniel W., and Robert N. Brandon. 2010. *Biology's First Law: The Tendency for Diversity and Complexity to Increase in Evolutionary Systems*. Chicago: University of Chicago Press.
Medawar, Peter. 1967. "Review of *The Phenomenon of Man*." *Mind* 70: 99–106.
Meister, Chad. 2010. "The Problem of Evil." In *The Cambridge Companion to Christian Philosophical Theology*, edited by Charles Taliaferro and Chad Meister, 152–69. Cambridge Companions to Religion. Cambridge: Cambridge University Press.
Merchant, Carolyn. 1980. *The Death of Nature: Women, Ecology, and the Scientific Revolution: A Feminist Reappraisal of the Scientific Revolution*. Scranton, PA: HarperCollins.
Meteyard, Eliza. 1871. *A Group of Englishmen, 1795–1815: Being Records of the Younger Wedgwoods and Their Friends; Embracing the History of the Discovery of Photography and a Facsimile of the First Photograph*. London: Longmans, Green.
Mill, John Stuart. 1910. *Utilitarianism, Liberty and Representative Government*. London: Dent.
———. 2008. *Utilitarianism*. Early Modern Texts. Edited by Jonathan Bennett. http://www.earlymoderntexts.com/assets/pdfs/mill1863.pdf.
Moore, Aubrey. 1890. "The Christian Doctrine of God." In *Lux Mundi: A Series of Studies in the Religion of the Incarnation*, edited by Charles Gore, 57–109. London: Murray.
Moore, G. E. 1939. "Proof of an External World." *Proceedings of the British Academy* 25: 273–300.

———. 1990. *G. E. Moore: Selected Writings*. Edited by Thomas Baldwin. International Library of Philosophy. London: Routledge and Kegan Paul.

———. 1993. *Principia Ethica*. Edited by Thomas Baldwin. Rev. ed. Chicago: University of Chicago Press.

Morison, Ian. 2014. *A Journey through the Universe. Gresham Lectures on Astronomy*. Cambridge: Cambridge University Press.

Morris, Henry M. 1989. *The Long War against God: The History and Impact of the Creation/Evolution Conflict*. Grand Rapids: Baker.

Moser, P. K. 2010. "Sin and Salvation." In *The Cambridge Companion to Christian Philosophical Theology*, edited by Charles Taliaferro and Chad Meister, 136–51. Cambridge Companions to Religion. Cambridge: Cambridge University Press.

Murphy, Jeffrie G. 1982. *Evolution, Morality, and the Meaning of Life*. Philosophy and Society. Totowa, NJ: Rowman and Littlefield.

Myers, F. W. H. 1881. "George Eliot." *Century Magazine* 23: 62–63.

Nagel, Ernest. 1961. *The Structure of Science: Problems in the Logic of Scientific Explanation*. New York: Harcourt, Brace and World.

Newman, John Henry. 1870. *A Grammar of Assent*. New York: Catholic Publishing Society.

———. 1973. *The Vatican Council, January 1870 to December 1871*. Vol. 25 of *The Letters and Diaries of John Henry Newman*. Edited by Charles Stephen Dessain and Thomas Gornall. Oxford: Clarendon.

Nitecki, Matthew H. 1988. *Evolutionary Progress*. Chicago: University of Chicago Press.

———, ed. 1990. *Evolutionary Innovations*. Chicago: University of Chicago Press.

Numbers, Ronald L. 2006. *The Creationists: From Scientific Creationism to Intelligent Design*. 2nd ed. Cambridge, MA: Harvard University Press.

Oakley, K. P. 1964. "The Problem of Man's Antiquity." *Bulletin of the British Museum (Natural History): Geology Series* 9: 83–155.

O'Collins, Gerald G. 1993. *The Resurrection of Jesus Christ: Some Contemporary Issues*. Pere Marquette Lecture in Theology. Milwaukee: Marquette University Press.

Oldroyd, D. R. 1984. "How Did Darwin Arrive at His Theory? The Secondary Literature to 1982." *History of Science* 22: 325–74.

Oliver, W. H. 1978. *Prophets and Millennialists: The Uses of Biblical Prophecy in England from the 1790s to the 1840s*. Auckland: Auckland University Press.

Ospovat, Dov. 1981. *The Development of Darwin's Theory: Natural History, Natural Theology, and Natural Selection, 1838–1859*. Cambridge: Cambridge University Press.

Owen, John. 1863. "On the *Archaeopteryx* of von Meyer, with a Description of the Fossil Remains of a Long-Tailed Species from the Lithographic Stone of Solenhofen." *Philosophical Transactions* 153: 33–47.

Owen, Richard. 1849. *On the Nature of Limbs*. London: Voorst.

———. 1858. *A Supplementary Appendix to the First Volume of the Life of Robert Owen: Containing a Series of Reports, Addresses, Memorials, and Other Documents Referred to in That Volume, 1808–1820*. London: Longmans, Green.

———. 1860. "Darwin on the Origin of Species." *Edinburgh Review* 111: 487–532.

———. 1992. *The Hunterian Lectures in Comparative Anatomy, May and June 1837*. Edited by Phillip Reid Sloan. Chicago: University of Chicago Press.

Paley, William. 1802. *Natural Theology*. London: Rivington.

Pannenberg, Wolfhart. 1968. *Jesus: God and Man*. London: SCM.

———. 1993. *Towards a Theology of Nature*. Louisville: Westminster/John Knox.
Peirce, Charles S. 1955. *Philosophical Writings of Peirce*. Edited by Justus Buchler. New York: Dover.
Pike, Nelson, ed. 1964. *God and Evil: Readings on the Theological Problem of Evil*. Prentice-Hall Contemporary Perspectives in Philosophy Series. Englewood Cliffs, NJ: Prentice-Hall.
Plantinga, Alvin. 1980. *Does God Have a Nature?* Milwaukee: Marquette University Press.
———. 1981. "Is Belief in God Properly Basic?" *Nous* 15: 41–51.
———. 1997. "Methodological Naturalism." *Perspectives on Science and Christian Faith* 49: 143–54.
———. 1999. "Augustinian Christian Philosophy." In *The Augustinian Tradition*, edited by Gareth B. Matthews, 1–26. Berkeley: University of California Press.
———. 2009. "Science and Religion: Why Does the Debate Continue?" In *The Religion and Science Debate: Why Does It Continue?*, edited by Harold W. Attridge, 93–123. Terry Lectures Series. New Haven, CT: Yale University Press.
Plato. 1997. *Plato: Complete Works*. Edited by John M. Cooper and D. S. Hutchinson. Indianapolis: Hackett.
Plotinus. 1992. *The Enneads*. Translated by Stephen MacKenna. LP Classic Reprint Series. Burdett, NY: Larson.
Popper, Karl R. 1963. *Conjectures and Refutations*. London: Routledge and Kegan Paul.
———. 1972. *Objective Knowledge: An Evolutionary Approach*. Oxford: Oxford University Press.
Powell, Baden. 1855. *Essays on the Spirit of the Inductive Philosophy*. London: Longman, Brown, Green, and Longmans.
Poythress, Vern S. 2006. *Redeeming Science: A God-Centered Approach*. Wheaton, IL: Crossway.
Provine, William B. 1971. *The Origins of Theoretical Population Genetics*. Chicago: University of Chicago Press.
———. 1988. "Progress in Evolution and Meaning in Life." In *Evolutionary Progress*, edited by Matthew H. Nitecki, 165–80. Chicago: University of Chicago Press.
Raup, David M. 1988. "Testing the Fossil Record for Evolutionary Progress." In *Evolutionary Progress*, edited by Matthew H. Nitecki, 293–318. Chicago: University of Chicago Press.
Rawls, John. 1971. *A Theory of Justice*. Cambridge, MA: Harvard University Press.
Reich, David. 2018. *Who We Are and How We Got Here: Ancient DNA and the New Science of the Human Race*. New York: Pantheon.
Reiss, Michael J., and M. Ruse. 2023. *The New Biology: A Battle between Mechanism and Organicism*. Cambridge, MA: Harvard University Press.
Re Manning, Russell, et al., eds. 2013. *The Oxford Handbook of Natural Theology*. Oxford Handbooks. Oxford: Oxford University Press.
Richards, Robert J. 1986a. "A Defense of Evolutionary Ethics." *Biology and Philosophy* 1: 265–93.
———. 1986b. "Justification through Scientific Faith: A Rejoinder." *Biology and Philosophy* 1: 337–54.
———. 1987. *Darwin and the Emergence of Evolutionary Theories of Mind and Behavior*. Chicago: University of Chicago Press.

———. 1989. "Dutch Objections to Evolutionary Ethics." *Biology and Philosophy* 4: 288–343.

———. 1992. *The Meaning of Evolution: The Morphological Construction and Ideological Reconstruction of Darwin's Theory*. Science and Its Conceptual Foundations. Chicago: University of Chicago Press.

———. 2002. *The Romantic Conception of Life: Science and Philosophy in the Age of Goethe*. Science and Its Conceptual Foundations. Chicago: University of Chicago Press.

———. 2004. "Michael Ruse's Design for Living." *Journal of the History of Biology* 37: 25–38.

Richards, Robert J., and Michael Ruse. 2016. *Debating Darwin*. Chicago: University of Chicago Press.

Richerson, Peter J., and Robert Boyd. 2005. *Not by Genes Alone: How Culture Transformed Human Evolution*. Chicago: University of Chicago Press.

Roberts, Alexander, and James Donaldson, eds. 1989. *The Ante-Nicene Fathers*. 10 vols. Repr., Grand Rapids: Eerdmans.

Rogers, K. A. 2010. "Incarnation." In *The Cambridge Companion to Christian Philosophical Theology*, edited by Charles Taliaferro and Chad Meister, 95–107. Cambridge Companions to Religion. Cambridge: Cambridge University Press.

Rolston, Holmes, III. 1999. *Genes, Genesis and God: Values and Their Origins in Natural and Human History*. Cambridge: Cambridge University Press.

Rorty, Richard. 1998. *Truth and Progress*. Philosophical Papers 3. Cambridge: Cambridge University Press.

Rosenzweig, Michael L., and Robert D. McCord. 1991. "Incumbent Replacement: Evidence for Long-Term Evolutionary Progress." *Paleobiology* 17: 202–13.

Rottschaeffer, W. A., and D. Martinsen. 1990. "Really Taking Darwin Seriously: An Alternative to Michael Ruse's Darwinian Metaethics." *Biology and Philosophy* 5: 149–73.

Ruse, Michael. 1973. *The Philosophy of Biology*. London: Hutchinson.

———. 1975a. "Charles Darwin's Theory of Evolution: An Analysis." *Journal of the History of Biology* 8: 219–41.

———. 1975b. "Darwin's Debt to Philosophy: An Examination of the Influence of the Philosophical Ideas of John F. W. Herschel and William Whewell on the Development of Charles Darwin's Theory of Evolution." *Studies in History and Philosophy of Science* 6: 159–81.

———. 1979. *Sociobiology: Sense or Nonsense?* Dordrecht: Reidel.

———. 1980. "Charles Darwin and Group Selection." *Annals of Science* 37: 615–30.

———. 1982. *Darwinism Defended: A Guide to the Evolution Controversies*. Reading, MA: Wesley.

———. 1986a. "Evolutionary Ethics: A Phoenix Arisen." *Zygon* 21: 95–112.

———. 1986b. *Taking Darwin Seriously: A Naturalistic Approach to Philosophy*. Oxford: Blackwell.

———. 1988a. *But Is It Science? The Philosophical Question in the Creation/Evolution Controversy*. Buffalo, NY: Prometheus.

———. 1988b. *Homosexuality: A Philosophical Inquiry*. Oxford: Blackwell.

———. 1988c. "Molecules to Men: Evolutionary Biology and Thoughts of Progress." In *Evolutionary Progress*, edited by Matthew H. Nitecki, 97–126. Chicago: University of Chicago Press.

———. 1995. "Evolutionary Ethics: The Debate Continues." In *Evolutionary Naturalism: Selected Essays*, 255–90. London: Routledge.
———. 1996a. "Charles Darwin and Progress." In *Monad to Man: The Concept of Progress in Evolutionary Biology*, 136–78. Cambridge, MA: Harvard University Press.
———. 1996b. *Monad to Man: The Concept of Progress in Evolutionary Biology.* Cambridge, MA: Harvard University Press.
———. 1998a. "Darwinian Ethics." In *Taking Darwin Seriously: A Naturalistic Approach to Philosophy*, 207–72. Amherst, NJ: Prometheus.
———. 1998b. "Evolution and Progress." In *The Philosophy of Biology*, edited by David L. Hull and Michael Ruse, 610–24. Oxford Readings in Philosophy. Oxford: Oxford University Press.
———. 1999a. "Charles Darwin and the *Origin of Species*." *The Darwinian Revolution: Science Red in Tooth and Claw*, 160–201. 2nd ed. Chicago: University of Chicago Press.
———. 1999b. *The Darwinian Revolution: Science Red in Tooth and Claw.* 2nd ed. Chicago: University of Chicago Press.
———. 1999c. *Mystery of Mysteries: Is Evolution a Social Construction?* Cambridge, MA: Harvard University Press.
———. 2000a. *Can a Darwinian Be a Christian? The Relationship between Science and Religion.* Cambridge: Cambridge University Press.
———. 2000b. "Design." In *Can a Darwinian Be a Christian? The Relationship between Science and Religion*, 111–28. Cambridge: Cambridge University Press.
———. 2000c. *The Evolution Wars: A Guide to the Controversies.* Controversies in Science. Santa Barbara, CA: ABC CLIO.
———. 2000d. "Naturalism." In *Can a Darwinian Be a Christian? The Relationship between Science and Religion*, 94–110. Cambridge: Cambridge University Press.
———. 2000e. "Sociobiology." In *Can a Darwinian Be a Christian? The Relationship between Science and Religion*, 186–204. Cambridge: Cambridge University Press.
———. 2003a. "The Argument from Design: A Brief History." In *Debating Design: From Darwin to DNA*, edited by William A. Dembski and Michael Ruse, 13–31. Cambridge, MA: Harvard University Press.
———. 2003b. *Darwin and Design: Does Nature Have a Purpose?* Cambridge, MA: Harvard University Press.
———. 2003c. "Design as a Metaphor." In *Darwin and Design: Does Evolution Have a Purpose?*, 271–90. Cambridge, MA: Harvard University Press.
———. 2003d. "Two Thousand Years of Design." In *Darwin and Design: Does Evolution Have a Purpose?*, 9–30. Cambridge, MA: Harvard University Press.
———. 2005a. *The Evolution-Creation Struggle.* Cambridge, MA: Harvard University Press.
———. 2005b. "Social Darwinism." In *The Evolution-Creation Struggle*, 103–28. Cambridge, MA: Harvard University Press.
———. 2006. *Darwinism and Its Discontents.* Cambridge: Cambridge University Press.
———. 2008a. *Charles Darwin.* Blackwell Great Minds. Malden, MA: Blackwell.
———. 2008b. *Evolution and Religion: A Dialogue.* Lanham, MD: Rowman and Littlefield.
———. 2008c. "The Origins of Religion." In *Charles Darwin*, 265–86. Blackwell Great Minds. Malden, MA: Blackwell.

———. 2009a. "Darwinism Explains Religion (?)." In *Defining Darwin: Essays on the History and Philosophy of Evolutionary Biology*, 199–214. Amherst, NY: Prometheus.

———. 2009b. "Evolution and Ethics: The Sociobiological Approach." In *Philosophy after Darwin: Classic and Contemporary Readings*, edited by Michael Ruse, 489–511. Princeton, NJ: Princeton University Press.

———. 2009c. "The History of Evolutionary Thought." In *Evolution: The First Four Billion Years*, edited by Michael Ruse and Joseph Travis, 1–48. Cambridge, MA: Belknap.

———. 2009d. "Introduction." In *Evolution and Ethics: Thomas Henry Huxley*, vii–xxxvi. Princeton, NJ: Princeton University Press.

———. 2009e. "The Origin of the *Origin*." In *The Cambridge Companion to the "Origin of Species,"* edited by Michael Ruse and Robert J. Richards, 1–13. Cambridge Companions to Philosophy. Cambridge: Cambridge University Press.

———, ed. 2009f. *Philosophy after Darwin: Classic and Contemporary Readings*. Princeton, NJ: Princeton University Press.

———. 2010a. "God." In *Science and Spirituality: Making Room for Faith in the Age of Science*, 181–207. Cambridge: Cambridge University Press.

———. 2010b. *Science and Spirituality: Making Room for Faith in the Age of Science*. Cambridge: Cambridge University Press.

———. 2012a. "Morality." In *The Philosophy of Human Evolution*, 155–84. Cambridge Introductions to Philosophy and Biology. Cambridge: Cambridge University Press.

———. 2012b. *The Philosophy of Human Evolution*. Cambridge Introductions to Philosophy and Biology. Cambridge: Cambridge University Press.

———. 2012c. "Progress." In *The Philosophy of Human Evolution*, 99–127. Cambridge Introductions to Philosophy and Biology. Cambridge: Cambridge University Press.

———, ed. 2013a. *The Cambridge Encyclopedia of Darwin and Evolutionary Thought*. Cambridge: Cambridge University Press.

———. 2013b. *The Gaia Hypothesis: Science on a Pagan Planet*. Chicago: University of Chicago Press.

———. 2013c. "Naturalism and the Scientific Method." In *The Oxford Handbook of Atheism*, edited by Stephen Bullivant and Michael Ruse, 383–97. Oxford Handbooks. Oxford: Oxford University Press.

———. 2015a. *Atheism: What Everyone Needs to Know*. Oxford: Oxford University Press.

———. 2015b. "Belief." In *Atheism: What Everyone Needs to Know*, 83–99. Oxford: Oxford University Press.

———. 2015c. "God and Humans." In *Atheism: What Everyone Needs to Know*, 67–82. Oxford: Oxford University Press.

———. 2015d. "Naturalistic Explanations." In *Atheism: What Everyone Needs to Know*, 188–210. Oxford: Oxford University Press.

———. 2016. "*L'état, c'est moi*: Fifty Years of History and Philosophy of Evolutionary Biology." *Theoretical Biology Forum* 109: 111–22.

———. 2017a. "Darwin and Design: Darwin Destroys Design." In *Science, Evolution, and Religion: A Debate about Atheism and Theism*, by Michael Peterson and Michael Ruse, 115–24. Oxford: Oxford University Press.

———. 2017b. "Darwinian Evolutionary Ethics." In *The Cambridge Handbook of Evolutionary Ethics*, edited by Michael Ruse and Robert J. Richards, 89–100. Cambridge: Cambridge University Press.

———. 2017c. "A Darwinian Pilgrim's Early Progress." *Journal of Cognitive Historiography* 4: 151–64.

———. 2017d. "A Darwinian Pilgrim's Middle Progress." *Journal of Cognitive Historiography* 4: 165–79.

———. 2017e. "A Darwinian Pilgrim's Late Progress." *Journal of Cognitive Historiography* 4: 180–98.

———. 2017f. "Darwinism and Belief." In *On Faith and Science*, by Edward J. Larson and Michael Ruse, 135–58. New Haven, CT: Yale University Press.

———. 2017g. *Darwinism as Religion: What Literature Tells Us about Evolution*. Oxford: Oxford University Press.

———. 2017h. "Evolutionary Directionality: No Direction to Evolution." In *Science, Evolution, and Religion: A Debate about Atheism and Theism*, by Michael Peterson and Michael Ruse, 125–35. Oxford: Oxford University Press.

———. 2017i. *On Purpose*. Princeton, NJ: Princeton University Press.

———. 2018a. "Darwinism." In *On Purpose*, 91–113. Princeton, NJ: Princeton University Press.

———. 2018b. *The Problem of War: Darwinism, Christianity, and Their Battle to Understand Human Conflict*. Oxford: Oxford University Press.

———. 2019a. "Darwinism as Religion." In *A Meaning to Life*, 97–132. Oxford: Oxford University Press.

———. 2019b. *A Meaning to Life*. Oxford: Oxford University Press.

———. 2019c. "The Unraveling of Belief." In *A Meaning to Life*, 11–53. Oxford: Oxford University Press.

———. 2019d. "Why Atheism?" In *Monotheism and Contemporary Atheism*. Cambridge Elements: Religion and Monotheism. Cambridge: Cambridge University Press.

———. 2021a. "The Arkansas Creationism Trial Forty Years On." In *Karl Popper's Science and Philosophy*, edited by Zuzana Parusniková and David Merritt, 257–76. Cham, Switz.: Springer.

———. 2021b. "Darwinian Evolution." In *A Philosopher Looks at Human Beings*, 49–67. A Philosopher Looks At. Cambridge: Cambridge University Press.

———. 2021c. "Morality for the Mechanist." In *A Philosopher Looks at Human Beings*, 150–62. A Philosopher Looks At. Cambridge: Cambridge University Press.

———. 2021d. *A Philosopher Looks at Human Beings*. A Philosopher Looks At. Cambridge: Cambridge University Press.

———. 2021e. "The Problem of Progress." In *A Philosopher Looks at Human Beings*, 122–50. A Philosopher Looks At. Cambridge: Cambridge University Press.

———. 2022. *Why We Hate: Understanding the Roots of Human Conflict*. Oxford: Oxford University Press.

———. Forthcoming. *Charles Darwin: No Rebel, Great Revolutionary*. Cambridge: Cambridge University Press.

Ruse, Michael, and Joseph Travis, eds. 2009. *Evolution: The First Four Billion Years*. Cambridge, MA: Belknap.

Ruse, Michael, and E. O. Wilson. 1986. "Moral Philosophy as Applied Science." *Philosophy* 61: 173–92.

Russell, Robert John. 2008. *Cosmology: From Alpha to Omega*. Theology and the Sciences. Minneapolis: Fortress.

Russett, Cynthia Eagle. 1976. *Darwin in America: The Intellectual Response, 1865–1912*. San Francisco: Freeman.

Schneider, John R. 2010. "Recent Genetic Science and Christian Theology on Human Origins: An 'Aesthetic Supralapsarianism.'" *Perspectives on Science and Christian Faith* 62: 196–212.

Sebright, J. 1809. *Art of Improving the Breeds of Domestic Animals in a Letter Addressed to the Right Hon. Sir Joseph Banks*. London: n.p.

Sedgewick, Adam. 1831. "Address to the Geological Society." *Proceedings of the Geological Society of London* 1: 281–316.

Sedley, David. 2008. *Creationism and Its Critics in Antiquity*. Sather Classical Lectures. Berkeley: University of California Press.

Sepkoski, John J., Jr. 1976. "Species Diversity in the Phanerozoic-Species-Area Effects." *Paleobiology* 2: 298–303.

Sheppard, Phil M. 1958. *Natural Selection and Heredity*. London: Hutchinson.

Sidgwick, Henry. 1876. "The Theory of Evolution in Its Application to Practice." *Mind* 1: 52–67.

Simpson, George Gaylord. 1944. *Tempo and Mode in Evolution*. New York: Columbia University Press.

———. 1964. *This View of Life*. New York: Harcourt, Brace, and World.

Singer, Peter. 1981. *The Expanding Circle: Ethics and Sociobiology*. New York: Farrar, Straus, and Giroux.

Smith, Roger. 1972. "Alfred Russel Wallace: Philosophy of Nature and Man." *British Journal for the History of Science* 6: 177–99.

Sober, Elliot. 1994. "Prospects for an Evolutionary Ethics." In *From a Biological Point of View*, 93–113. Cambridge Studies in Philosophy and Biology. Cambridge: Cambridge University Press.

———. 2000. *Philosophy of Biology*. 2nd ed. Boulder, CO: Westview.

Spencer, Herbert. 1851. *Social Statics; or, The Conditions Essential to Human Happiness Specified and the First of Them Developed*. London: Chapman.

———. 1852. "The Development Hypothesis." In *Essays: Scientific, Political and Speculative*, 377–83. London: Williams and Norgate.

———. 1857. "Progress: Its Law and Cause." *Westminster Review* 67: 445–65.

———. 1862. *First Principles*. London: Williams and Norgate.

———. 1892. *The Principles of Ethics*. London: Williams and Norgate.

Stark, Rodney, and William Bainbridge. 1996. *A Theory of Religion*. Newark: Rutgers University Press.

Stebbins, G. Ledyard. 1950. *Variation and Evolution in Plants*. Columbia Biological Series 16. New York: Columbia University Press.

Stove, David. 2007. *Darwinian Fairytales: Selfish Genes, Errors of Heredity, and Other Fables of Evolution*. New York: Encounter.

Sulloway, F. J. 1979. *Freud: Biologist of the Mind*. New York: Basic.

———. 1982a. "Darwin and His Finches: The Evolution of a Legend." *Journal of the History of Biology* 15: 1–53.

———. 1982b. "Darwin's Conversion: The *Beagle* Voyage and Its Aftermath." *Journal of the History of Biology* 15: 325–96.

———. 1985. "Darwin's Early Intellectual Development: An Overview of the *Beagle* Voyage (1831–1836)." In *The Darwinian Heritage*, edited by David Kohn, 121–54. Princeton Legacy Library. Princeton, NJ: Princeton University Press.
Swinburne, Richard G. 1970. *The Concept of Miracle*. New Studies in the Philosophy of Religion. London: Macmillan.
Taliaferro, Charles, and Chad Meister, eds. 2010. *The Cambridge Companion to Christian Philosophical Theology*. Cambridge Companions to Religion. Cambridge: Cambridge University Press.
Trigg, Roger. 1982. *The Shaping of Man: Philosophical Aspects of Sociobiology*. Oxford: Blackwell.
Trivers, R. L. 1971. "The Evolution of Reciprocal Altruism." *Quarterly Review of Biology* 46: 35–57.
Tyndall, John. 1892. *Fragments of Science*. 2 vols. London: Longmans, Green.
Valentine, James W., and Douglass H. Erwin. 1987. "Interpreting Great Developmental Experiments: The Fossil Record." In *Development as an Evolutionary Process: Proceedings of a Meeting Held at the Marine Biological Laboratory in Woods Hole, Massachusetts, August 23 and 24, 1985*, edited by Rudolf A. Raff and Elizabeth C. Raff, 71–107. MBL Lectures in Biology 8. New York: Liss.
Vanhoozer, Kevin J. 1997. "Human Being, Individual and Social." In *The Cambridge Companion to Christian Doctrine*, edited by Colin E. Gunton, 158–88. Cambridge Companions to Religion. Cambridge: Cambridge University Press.
Vermeij, Geerat J. 1987. *Evolution and Escalation: An Ecological History of Life*. Princeton, NJ: Princeton University Press.
Von Baer, K. E. 1828. "Fragments Relating to Philosophical Zoology." In *Scientific Memoirs*, edited and translated by Arthur Henfrey and Thomas Henry Huxley, 1:187–211. London: Taylor and Francis.
Vorzimmer, Peter. 1963. "Charles Darwin and Blending Inheritance." *Isis* 54: 371–90.
———. 1969. "Darwin, Malthus, and the Theory of Natural Selection." *Journal of the History of Ideas* 30: 527–42.
Waal, F. B. M. de. 1982. *Chimpanzee Politics: Power and Sex among Apes*. Baltimore: Johns Hopkins University Press.
Wade, M. J. 1978. "A Critical View of the Models of Group Selection." *Quarterly Review of Biology* 53: 101–14.
Wainwright, William J. 2010. "Omnipotence, Omniscience, and Omnipresence." In *The Cambridge Companion to Christian Philosophical Theology*, edited by Charles Taliaferro and Chad Meister, 46–65. Cambridge Companions to Religion. Cambridge: Cambridge University Press.
Wallace, Alfred Russel. 1864. "The Origin of Human Races and the Antiquity of Man Deduced from the Theory of Natural Selection." *Journal of the Anthropological Society of London* 2: clviii–clxxxvii.
———. 1870a. *Contributions to the Theory of Natural Selection: A Series of Essays*. London: Macmillan.
———. 1870b. "The Limits of Natural Selection as Applied to Man." In *Contributions to the Theory of Natural Selection: A Series of Essays*, 332–71. London: Macmillan.
———. 1916. *Alfred Russel Wallace: Letters and Reminiscences*. Edited by James Marchant. 2 vols. London: Cassell.
Ward, Keith. 1996. *God, Chance and Necessity*. Oxford: Oneworld.
Whewell, William. 1837. *The History of the Inductive Sciences*. 3 vols. London: Parker.

———. 1840. *The Philosophy of the Inductive Sciences*. London: Parker.
———. 1845. *Indications of the Creator*. London: Parker.
———. 2001. *Of the Plurality of Worlds: A Facsimile of the First Edition of 1853; Plus Previously Unpublished Material Excised by the Author Just before the Book Went to Press; and Whewell's Dialogue Rebutting His Critics, Reprinted from the Second Edition*. Edited by Michael Ruse. Chicago: University of Chicago Press.
Whitcomb, John C., and Henry M. Morris. 1961. *The Genesis Flood: The Biblical Record and Its Scientific Implications*. Philadelphia: Presbyterian and Reformed.
White, Paul. 2003. *Thomas Huxley: Making the "Man of Science."* Cambridge Science Biographies. Cambridge: University of Cambridge Press.
Whitehead, Alfred North. 1929. *Process and Reality: An Essay in Cosmology*. New York: Macmillan.
Williams, George C. 1966. *Adaptation and Natural Selection: A Critique of Some Current Evolutionary Thought*. Princeton, NJ: Princeton University Press.
———. 1989. "A Sociobiological Expansion of Evolution and Ethics." In *Evolution and Ethics: T. H. Huxley's "Evolution and Ethics" with New Essays on Its Victorian and Sociobiological Context*, by James J. Paradis and George Christopher Williams, 179–214. Princeton Legacy Library. Princeton, NJ: Princeton University Press.
Wilson, David Sloan. 1992. "On the Relationship between Evolutionary and Psychological Definitions of Altruism and Selfishness." *Biology and Philosophy* 7: 61–68.
———. 2002. *Darwin's Cathedral: Evolution, Religion, and the Nature of Society*. Chicago: University of Chicago Press.
Wilson, Edward O. 1975. *Sociobiology: The New Synthesis*. Cambridge, MA: Belknap.
———. 1978. *On Human Nature*. Cambridge, MA: Harvard University Press.
———. 1990. *Success and Dominance in Ecosystems*. Vol. 2 of *The Case of the Social Insects*. Excellence in Ecology. Oldenorf/Luhe, Germ.: Ecology Institute.
———. 1992. *The Diversity of Life*. Cambridge, MA: Harvard University Press.
———. 2000. *Sociobiology: The New Synthesis*. 25th anniv. ed. Cambridge, MA: Belknap.
———. 2006. *The Creation: A Meeting of Science and Religion*. New York: Norton.
Wright, Sewall. 1931. "Evolution in Mendelian Populations." *Genetics* 16: 97–159.
———. 1932. "The Roles of Mutation, Inbreeding, Crossbreeding and Selection in Evolution." *Proceedings of the Sixth International Congress of Genetics* 1: 356–66.
Youmans, W. J. 1879. "Science in Relation to War." *Popular Science* 14: 808–19.
Young, Robert M. 1985. *Darwin's Metaphor: Nature's Place in Victorian Culture*. Cambridge: Cambridge University Press.

www.ingramcontent.com/pod-product-compliance
Lightning Source LLC
Chambersburg PA
CBHW021231300426
44111CB00007B/502